CORRELATIONS, POLARIZATION, AND IONIZATION IN ATOMIC SYSTEMS

Proceedings of the International Symposium on (e,2e), Double Photoionization and Related Topics and the Eleventh International Symposium on Polarization and Correlation in Electronic and Atomic Collisions

Rolla, Missouri   25–28 July 2001

EDITORS
Don H. Madison
Michael Schulz
University of Missouri-Rolla

Melville, New York, 2002
AIP CONFERENCE PROCEEDINGS ■ VOLUME 604

Editors:

Don H. Madison
Michael Schulz

University of Missouri-Rolla
Physics Department
Rolla, MO 65409
USA

E-mail: madison@umr.edu
        schulz@umr.edu

L.C. Catalog Card No. 2001098447
ISBN 0-7354-0048-2
ISSN 0094-243X
Printed in the United States of America

CONTENTS

Preface

This volume combines the papers presented at two Atomic Physics meetings held jointly at the University of Missouri-Rolla, USA, from 25 – 28 July, 2001: the *International Symposium on (e,2e), Double Photoionization and Related Topics* and the *Eleventh International Symposium on Polarization and Correlation in Electronic and Atomic Collisions*. Both conferences were satellite meetings to the *XXII International Conference on Photonic, Electronic and Atomic Collisions* held in Santa Fe, New Mexico.

These symposia covered a broad range of currently hot topics in Atomic Physics. Rapid progress both in experimental and theoretical techniques leads to continuously growing insight into topics like many-body and electron-electron correlation effects in single and multiple ionization processes, or alignment and polarization effects in excitation and charge transfer processes. Furthermore, established techniques originally developed to study atomic collisions have been refined to a point where they are now being applied to chemical, solid state and even biological systems, thus providing links to other scientific areas. More than 100 scientists from around the world came together to discuss the most recent progress on these topics. In each symposium 30 invited talks were presented and further work was displayed in a joint poster session for both conferences.

The Editors gratefully acknowledge support from the National Science Foundation and the University of Missouri-Rolla. They also express their thanks to all participants for contributing to lively and fruitful discussions throughout the meetings. Last, but not least they are indebted to Pam Crabtree, Ellen Kindle, Ted Deskin, Brian Swift, Andy Pireaux, Matt Foster, Kittie Robertson and Sherry Adams for their valuable contributions to the organization of these symposia.

D.H. Madison
M. Schulz

Multiphoton Processes in a Two-Dimensional Model of Helium

Chris H. Greene, Mark Baertschy, Amber L. Young, and Brian E. Granger

Department of Physics and JILA, University of Colorado at Boulder, Boulder, CO 80309-0440

Abstract. The mechanism by which two electrons are ejected by a helium atom subjected to an intense laser field has been the subject of debate during the past decade. The consensus today in the community is that the second electron is ejected after the first one returns to the ionic core and rescatters. However, the specific details as to how this second electron is ejected remain uncertain. In this paper, we consider this process from two both quantal and semiclassical viewpoints, and propose a qualitative way of understanding how the second electron is ejected from a helium atom.

INTRODUCTION

A number of alternative models have been proposed for the process in which two electrons are ejected by a short, intense laser pulse. One early promising viewpoint developed by Corkum[1] involved a recollision between the He$^+$ ion and first electron that is ejected and subsequently accelerated outward before being driven back inward by the oscillating laser electric field. However, this was discredited by Walker et al.,[2] after their approximate implementation of this picture suggested that recollision underestimates the ionization rate of the second electron by at least one order of magnitude.

One alternative to the recollision picture was developed by Fittinghoff et al.[3], who proposed that some type of "shake-off mechanism" might be an operative mechanism responsible for double electron ejection. Their view was that the electric field changes rapidly at short distances, when the first electron tunnels out and ceases to be trapped. This was believed to cause a non-negligible fraction of the wavefunction to be projected onto the excited core states. Subsequently, other evidence – especially the fact that making the laser photons elliptically-polarized , which can "steer" the electrons and inhibit their recollision probability – has led most in the community to conclude that some type of a recollision plays a vital role in the double ionization of helium.[4]

Despite this consensus, a disparity remains among the alternative viewpoints. One influential paper developed the so-called "Crapola model".[5] That model treats the process as a two-step event, with each step involving dynamics of only one electron's wavepacket evolving under the time-dependent Schroedinger equation. Specifically, the first electron was viewed as tunneling out, creating a single electron wavepacket

CP604, *Correlations, Polarization, and Ionization in Atomic Systems*
edited by D. H. Madison and M. Schulz
© 2002 American Institute of Physics 0-7354-0048-2/02/$19.00

that fired outward before being turned around and recolliding with the He$^+$ ion. Then that wavepacket was viewed as providing a "source term", i.e., an instantaneous average potential field to which the bound electron of He$^+$ was subjected, in addition to the electron-nucleus and electron-field interaction potentials. Solution of the time-dependent Schrodinger equation for that inner electron moving in these combined potentials then yields an ionization rate that agrees reasonably well with experimental observations.

Despite the successes of the Crapola model, however, no consensus has yet emerged about the detailed mechanism for how the second electron is ejected from the He$^+$ ion when the recollision occurs. Unraveling this mechanism is the main goal of this study.

A ONE-DIMENSIONAL MODEL

In order to efficiently explore the parameter space, and unravel the qualitative mechanism for double ionization more completely, we develop a reduced-dimensionality model that has much of the correct physics of the three-dimensional problem. Specifically, we treat the two electrons in one dimension, with the electron-field interaction represented in the length gauge. In this model, the full potential energy of the system is:

$$V(x_1, x_2, t) = -\frac{2}{\sqrt{x_1^2 + \frac{1}{2}}} \cdot -\frac{2}{\sqrt{x_2^2 + \frac{1}{2}}} + \frac{1}{\sqrt{(x_1 - x_2)^2 + \beta}} + (x_1 + x_2)F(t). \quad (1)$$

Here the time-dependent electric field associated with the intense laser pulse is taken to be the following:

$$F(t) = F_0 \cos(\omega t)e^{-t^2/2\tau^2}. \quad (2)$$

This model Hamiltonian resembles the reduced-dimensionality models adopted by other researchers.[5-7] One minor difference from previous models is that we have chosen it to have the correct asymptotic form when either one or both electron positions is taken to infinity. Moreover, we have chosen it to generate the correct one-electron and two-electron binding energies when the time-independent Schrodinger equation is solved in the absence of any external field. The one-electron binding energy is found to agree with that of helium if we use $\beta \approx 0.3$.

CALCULATIONS AND THEIR INTERPRETATION

The intensity regime that we study here still requires the first electron to tunnel out through the barrier caused by the combination of the laser electric potential and the Coulomb potential. After the first electron of helium tunnels out, it is accelerated away, and is then driven back toward the residual He$^+$ ion by the laser electric field. We have arrived at the following interpretation of how the second electron is then ionized in the course of this recollision.

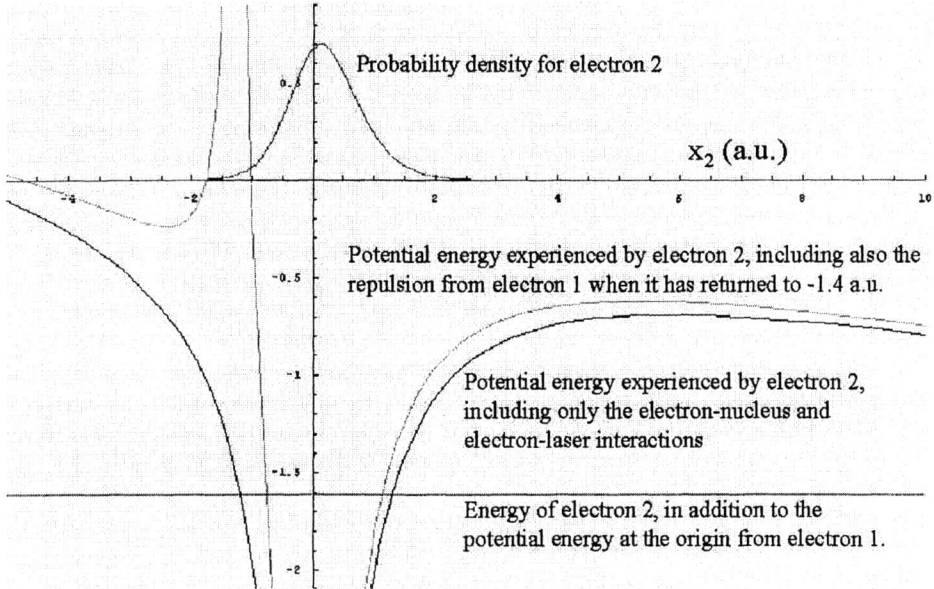

Probability density for electron 2

x_2 (a.u.)

Potential energy experienced by electron 2, including also the repulsion from electron 1 when it has returned to -1.4 a.u.

Potential energy experienced by electron 2, including only the electron-nucleus and electron-laser interactions

Energy of electron 2, in addition to the potential energy at the origin from electron 1.

FIGURE 1. Potential energy of the inner electron (#2) and its probability density as a function of distance in the one-dimensional model, at a particular time just before the outer electron (#1) is about to recollide with the core. The potential spike reflects that electron #1 is at a position of x_1=-1.4 a.u. at this time. The probability density of electron #2 is still hardly perturbed at this moment before the collision. In each of these Figs. 1-4, electron #1 is being treated classically after it tunnels out and is driven back into the He$^+$ ionic core, whereas electron #2 is treated quantum mechanically.

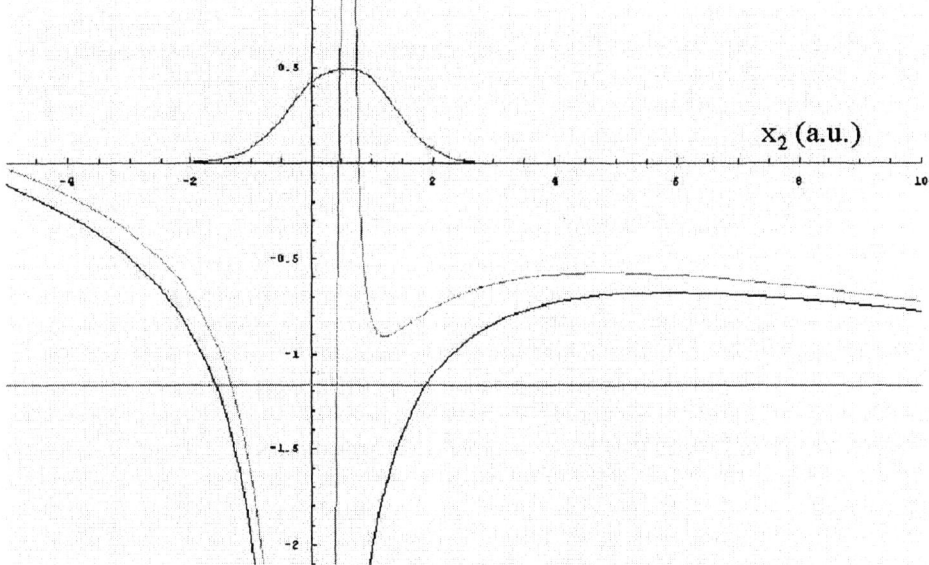

x_2 (a.u.)

FIGURE 2. This plot shows the same quantities plotted in Fig.1, except at a later time. Electron #1, moving from left to right, has passed the nucleus and is now at 0.6 a.u. Electron #2 has gained energy

3

from the potential due to electron #1, as is evidenced by the higher value of the horizontal line, as compared to Fig.1.

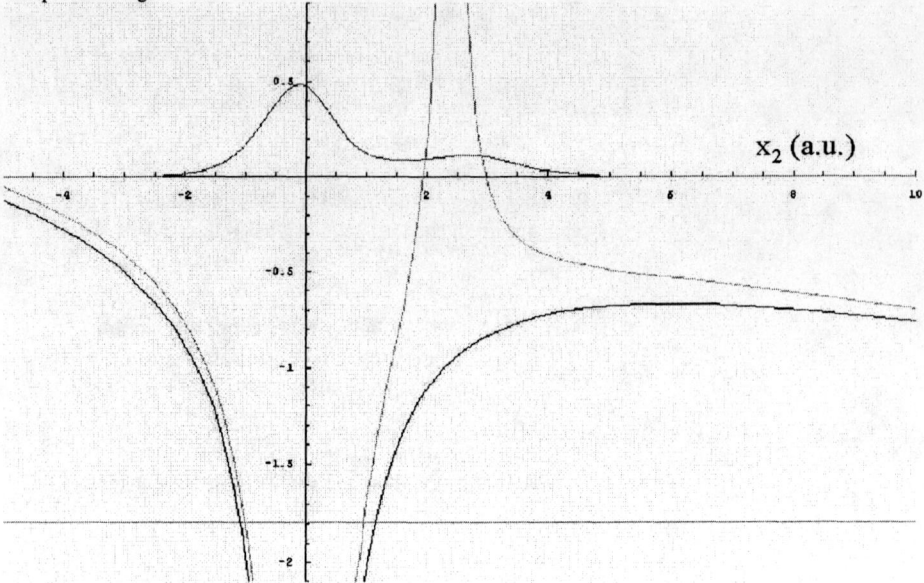

FIGURE 3. This plot shows the same quantities plotted in Fig.1, except at a still later time when electron #1 has now reached 0.6 a.u. A small wavefront of electron #2 is now apparent, as it begins to surf out, just to the right of the potential spike associated with the electron-electron repulsion.

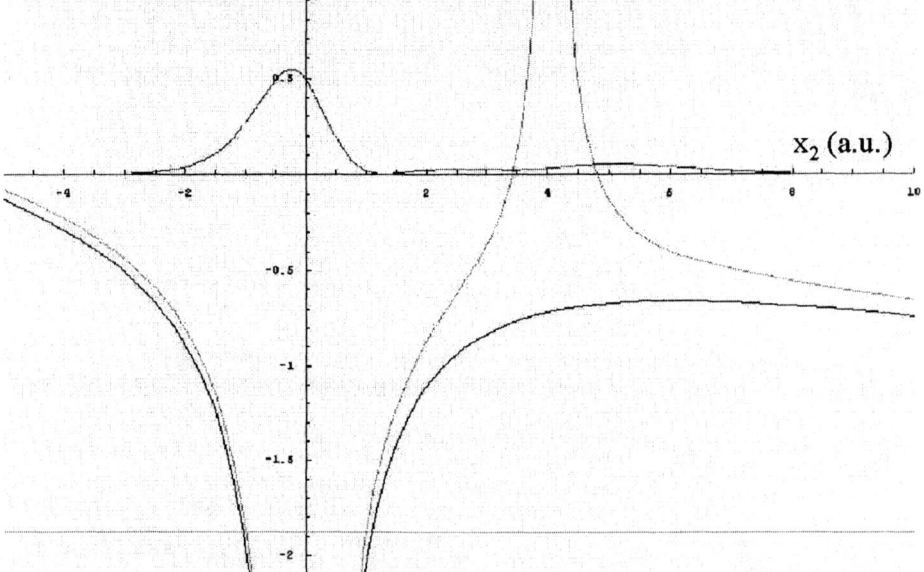

FIGURE 4. This plot shows the same quantities plotted in Fig.1, except at a still later time when electron #1 has now reached 4.1 a.u. The small wavefront of electron #2 is now clear, as it has surfed out of the Coulombic well and continues to be herded just ahead of the electron-electron repulsion. This helps to understand why the two electrons tend to emerge preferentially with equal momenta.

First of all, it is worth pointing out that the second electron is in the He^+ ground state and is far too strongly bound at these intensities to be able to tunnel out with any appreciable probability. One possibility we considered is that the potential energy of the returning first electron can play a large role when added to that experienced by the second (He^+) electron. Specifically, it might add enough of an electric field to permit the second electron to tunnel downstream, which is now in the direction opposite to the side where the first electron originally tunneled out. However, when this combined potential energy is plotted, it is clear that at the ground state He^+ energy of – 2 a.u., the barrier is not tipped over anywhere near enough for tunneling to play a strong role in ejecting the second electron. Accordingly, this first idea of a "Coulomb-repulsion-assisted tunneling" mechanism appears to be ruled out.

A second possibility that we have explored is that the residual electron of He^+ can perhaps be viewed as having an increasing energy when the first electron returns to the region close to the core. To understand this, recall that the tightly bound electron moves very rapidly compared to the outermost electron, at least until the outermost electron is within about 0.5 Bohr radii of the nucleus. This means that as the potential energy of repulsion grows during the return of the first electron, the "adiabatic energy" of the He^+ electron should be viewed as increasing commensurately. This picture is clearest when the inner electron energy is plotted as a function of time with the electron-electron repulsion included in that manner, along with its potential energy on the same graph.

The resulting picture shows that the energy of the ground state He^+ ion increases enough during the return of the first electron that it can simply "surf out" ahead of the first electron. One immediate implication of this observation is that the correlated momentum distribution should be peaked at equal momenta for the two electrons, which is in agreement with recent experimental measurements.[8] Further studies are underway to quantify this surfing aspect of two-electron ejection by an intense laser pulse.

ACKNOWLEDGMENTS

This work was supported by the Department of Energy, Office of Science.

REFERENCES

1. Corkum, P. B., *Phys. Rev. Lett.* **71**, 1994-1997 (1993).
2. Walker, B. *et al.*, *Phys. Rev. Lett.* **73**, 1227-1230 (1994).
2. Fittinghoff, D. N. et al., *Phys. Rev. Lett.* **69**, 2642-2645 (1992).
3. Yudin, G. L., and Ivanov, M. Y., *Phys. Rev. A* **63**, 033404-1 – 033404-14 (2001).
5. Watson, J. B., *et al.*, *Phys. Rev. Lett.* **78**, 1884-1887 (1997).
6. Sanpera, A., *et al. J. Phys. B* **31**, L841-L848 (1998).
7. Lappas, D. G., *et al. J. Phys. B* **29**, L619-627 (1996).
8. Weber, Th., *et al. Nature* **405**, 658-661 (2000).

Interplay of electron correlation and intense field dynamics in laser induced double ionization

A. Becker[*] and F.H.M. Faisal[†]

[*]Département de physique, de génie physique et de optique and Centre d'optique, photonique et laser, Université Laval, Québec (Québec), Canada G1K 7P4
[†]Fakultät für Physik, Universität Bielefeld, Postfach 100 131, D-33501 Bielefeld, Germany

Abstract. We discuss theoretical insights gained recently in understanding the experimental results of the observed momentum distributions of doubly charged recoil ions produced by femtosecond laser pulses in terms of a non-sequential ionization mechanism. The prominent characteristics are explained based on a combined influence of Coulomb correlation between the two electrons in the intermediate state and the field dressing in the final state.

INTRODUCTION

Probabilities of double ionization of atoms in intense femtosecond laser pulses have been found (e.g. [1, 2]) to exceed expectations based on a stepwise removal of one electron after another from the atom by many orders of magnitude. The large enhancements occur at laser intensities in the so-called sub-saturation regime, i.e. below the saturation intensity, at which the center of the interaction volume is depleted of neutral target atoms. It has been become clear from a number of theoretical investigations ([3] and References therein) that the large ionization probabilities are due to a non-sequential or direct double ionization process based on the interplay between intense field dynamics and electron correlation. Recent measurements of the momentum distribution of doubly charged He [4], Ne [5] and Ar ions [6] go beyond the formerly measured total ion yields and provide a test of the various theoretical models of non-sequential double ionization.

Theoretical analysis of the process has to account for the combined influence of the highly non-perturbative electron-field coupling, the long ranged Coulomb correlation between the electrons, as well as the quantum many-body nature of the atomic system. The attempt to numerically solve the corresponding Schrödinger equation involves the solution of a set of 2×3 dimensional partial differential equations on a large space-time grid. Much progress in this direction has been made recently (e.g. [7, 8] and References therein).

An alternative ab-initio approach towards this problem is a systematic approximation method. To this end, the "intense-field many-body S-matrix theory" (IMST) has been developed recently [9, 10, 11]. It is a reformulation of the usual S-matrix perturbation series, in which partitioning of the total Hamiltonian corresponding to arbitrary reference Hamiltonians for the initial, final *and* intermediate states of the system allow for including the influence of both Coulomb correlation and non-perturbative laser field interaction. The theory has been successfully applied to analyze total ion yields [3, 12]

CP604, *Correlations, Polarization, and Ionization in Atomic Systems*
edited by D. H. Madison and M. Schulz

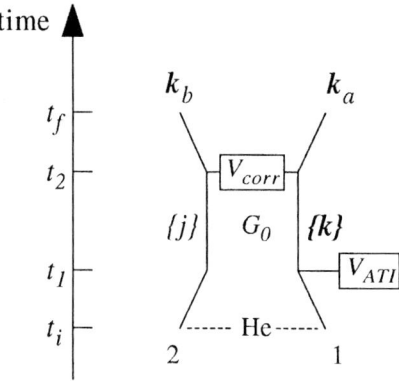

FIGURE 1. Feynman diagram for non-sequential double ionization of He in an infrared laser field.

and recoil ion momentum distributions [11, 13] in the non-sequential intensity regime of double and multiple ionization (as well as ion yields of single ionization of atoms [14] and of diatomic and polyatomic molecules [15]). Below, we discuss the physical mechanism of non-sequential double ionization emerging from the analysis with special emphasis on the results on the two-electron momentum distribution in He.

MECHANISM OF NON-SEQUENTIAL DOUBLE IONIZATION

Our previous investigations [3, 12] on the intensity dependence of the total double and multiple ionization yields using the IMST allowed us to identify a leading amplitude in the lowest order of the modified S-matrix series given by the Feynman diagram in Fig. 1. It contains [12b] in the quantum domain both the "antenna" model proposed by Kuchiev [16] and the classical "simpleman's rescattering" model proposed by Corkum [17].

The quantum-physical picture of the non-sequential double ionization process is visualized from the diagram by reading it from the bottom upward in the indicated direction of flow of time. At an initial time t_i, the two electrons (1 and 2) are in the ground state of the He atom. Then, at time t_1 (when the phase of the field is $\phi_1 = \omega t_1$), one of the two electrons (say 1) absorbs a large amount of field energy by interacting with the laser field (interaction operator V_{ATI}). The atom evolves in an intermediate state described by the two-electron Green's function G_0, which corresponds to the joint propagation of one electron in the virtual Volkov states of all momenta, $\{k\}$, and the other electron in the virtual states of the residual He$^+$ ion, $\{j\}$. At a subsequent time t_2 (phase of the field, $\phi_2 = \omega t_2$) the two electrons interact with each other via the electron correlation interaction (interaction operator V_{corr}), share the energy and may escape together with momenta $\mathbf{k}_a$ and $\mathbf{k}_b$ from the binding force of the nucleus to arrive at the detectors at time t_f. The final state could be affected by the field and is, therefore, described by the two-electron product Volkov states [18].

7

We note, that as long as the two electrons are not observed in the intermediate states in the experiment, the phases ϕ_1 and ϕ_2 remain undetermined and, hence, quantum-mechanically the corresponding double ionization amplitude has to be integrated for all values of the time interval $t_2 - t_1$. Thus, the diagram includes short-time "on the way out" scattering as well as long-time "rescattering" in a coherent way. The latter can give a significant contribution to the amplitude for $U_p \gg \omega$ and $\omega \ll E_B$, where $U_p = I/4\omega^2$ is the ponderomotive energy of a electron in a field of intensity I and frequency ω, E_B is the two-electron binding energy of the He atom. This gives the connection between the present quantum theory and the semiclassical "rescattering" model by Corkum [17]. We may parenthetically note that the above diagram can be understood as an intense field analog of the so-called two-step-one (TS1) process that is well known in weak-field double ionization of synchroton radiation (c.f., e.g. [19]).

TWO-ELECTRON SUM-MOMENTUM DISTRIBUTIONS

An exact evaluation of the non-sequential double ionization amplitude, corresponding to the diagram in Fig. 1, including all orders of the correlation and the field interaction operators is practically an impossible task. For the computations of the two-electron sum-momentum distribution [11, 13] we have, therefore, restricted ourselves to the lowest significant terms by replacing V_{corr} by $1/r_{12}$ and $V_{ATI}(t)$ by $-\hat{\mathbf{p}}_1 \cdot \mathbf{A}(t)/c$, where $\mathbf{A}(t)$ is the vector potential of the field. The expansion of the Volkov wavefunctions in terms of their Fourier components and evaluation of the time integration gives (in a.u., $e = m = \hbar = 1$, [11, 13]):

$$S_{fi}^{(2)}(\infty, -\infty)|_{NS} = -2\pi i \sum_{N=-\infty}^{\infty} \delta\left(\frac{k_a^2}{2} + \frac{k_b^2}{2} + E_B + 2U_p - N\omega\right) T^{(N)}(\mathbf{k}_a, \mathbf{k}_b), \quad (1)$$

where,

$$
\begin{aligned}
T^{(N)}(\mathbf{k}_a, \mathbf{k}_b) = & \sum_{n=-\infty}^{\infty} \sum_j \int \frac{1}{(2\pi)^3} d\mathbf{k} < \phi^0(\mathbf{k}_a, \mathbf{r}_1) \phi^0(\mathbf{k}_b, \mathbf{r}_2)| \frac{1}{r_{12}} |\phi_j^+(\mathbf{r}_2) \phi^0(\mathbf{k}, \mathbf{r}_1) > \\
& \times \frac{J_{N-n}\left(\alpha_0 \cdot (\mathbf{k}_a + \mathbf{k}_b - \mathbf{k}); \frac{U_p}{2\omega}\right) J_n\left(\alpha_0 \cdot \mathbf{k}; \frac{U_p}{2\omega}\right)}{\frac{k^2}{2} - E_j + E_B + U_p - n\omega + i0} \\
& \times (E_j - E_B - \frac{k^2}{2}) < \phi_j^+(\mathbf{r}_2) \phi^0(\mathbf{k}, \mathbf{r}_1)|\phi_{1S}(\mathbf{r}_1, \mathbf{r}_2) > .
\end{aligned}
\quad (2)
$$

$\phi_{1S}(\mathbf{r}_1, \mathbf{r}_2)$ is the ground state wavefunction of He atom, $\{\phi_j^+(\mathbf{r}_2)\}$ is the complete set of residual ionic states with energy E_j, $\phi^0(\mathbf{k}, \mathbf{r})$ is a plane wave state, and $J_n(a; b)$ is the generalized Bessel function of two arguments.

The six-fold space integrations are carried out analytically, the radial integration in k is performed by pole approximation and the integrals over the angles of $\mathbf{k}$ and the sum over n are performed numerically. Contribution of the lowest term of the sum over j (corresponding to the ground state of the ionic state) is retained only since it is found to

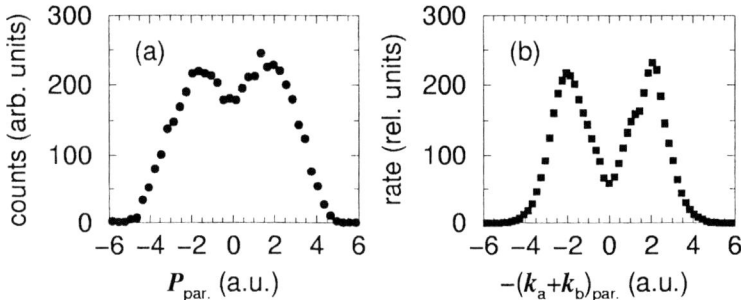

FIGURE 2. Recoil ion momentum distribution of He^{2+} parallel to the polarization direction, P_{par}, (experimental data, panel a, [4]) and the sum-momentum distribution of the two outgoing electrons in the opposite direction (present theory, panel b, [11, 13]).

dominate over the contribution from any excited state (c.f. [12a]). Finally, Monte-Carlo sampling method is used to evaluate the differential rate of double ionization:

$$\frac{d\Gamma^{(N)}}{d\mathbf{k}_a d\mathbf{k}_b} = 2\pi\delta\left(\frac{k_a^2}{2} + \frac{k_b^2}{2} + E_B + 2U_p - N\omega\right)|T^{(N)}(\mathbf{k}_a, \mathbf{k}_b)|^2. \tag{3}$$

Computations are carried out for He atom, and the distributions of the parallel and the perpendicular components of the sum of the two momenta, $(\mathbf{k}_a + \mathbf{k}_b)$, are determined, in each case by integrating over the remaining variables and summing the contributions from all significant Ns. We recall that under the condition of the experiment the He^{2+} recoil momentum $\mathbf{P} \approx -(\mathbf{k}_a + \mathbf{k}_b)$ [4].

In Fig. 2 we compare the experimental results (panel a, [4]) for the recoil momentum of He^{2+}, parallel to the polarization axis with the results of the present theory [11, 13]. The intensity of the field was 6.6×10^{14} W/cm^2 and $\lambda = 800$ nm. For the sake of comparision we have scaled the theoretical results to match with the experimental data, which are availble in arbitrary units only, at *one* point; this one-point fit determines the *relative* scale of the entire theoretical set in the Figure. All the essential features of the experimental data are reproduced by the theory. Namely, both the distribution show a broad double-hump distribution with maxima at the same positions. Also the width of both the distributions agree well with each other. However, there is a quantitative difference at the central minimum of the distribution, which might indicate significant contributions of higher orders in the present theory.

In Fig.3 we present the comparision of the experimental data (panel a) and the theoretical result for the component of the momentum distribution perpendicular to the polarization of the field. The characteristic features in the data are again reproduced by the theory, especially the width of the distribution in the perpendicular direction is found to be much narrower than that in the parallel direction.

We have further investigated the origin of the broad double-hump structure in the momentum distribution parallel to the polarization direction. To this end, we have deliberately neglected the Volkov field dressing of the two electrons in the final state

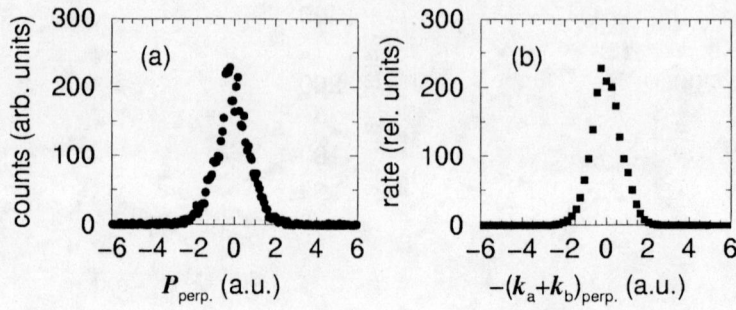

FIGURE 3. Same as in Fig. 2, but for the perpendicular component of the distributions (experimental data, panel a, [4]; theoretical results, panel b, [11, 13]).

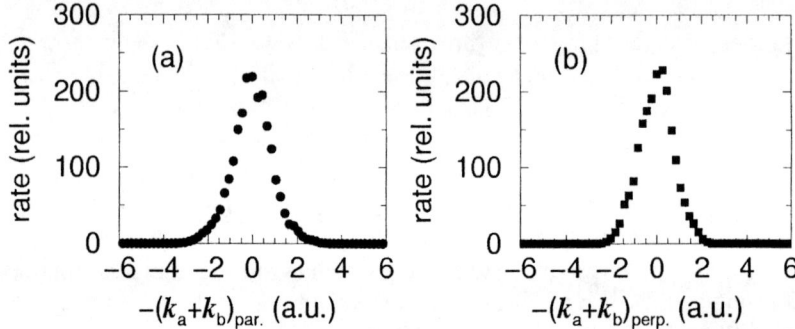

FIGURE 4. Theoretical results [11, 13] for the two-electron sum-momentum distributions obtained by neglecting the final state Volkov field dressing, a) parallel component, b) perpendicular component.

[11, 13]. The result so obtained are shown in Fig. 4, a) for the parallel component and b) for the perpendicular component. A comparision of the results of the numerical calculations for the *parallel* case *with* (Fig. 2b) and *without* (Fig. 4a) the final state field dressing shows that the double-hump structure collapses into a narrow single-hump in the *absence* of the Volkov dressing. This suggests that the final-state interaction of the two outgoing electrons with the field is primarily responsible for the broad distribution observed in the experiment. On the other hand for the perpendicular component both the distributions, with and without inclusion of the final-state Volkov dressing, are found to be very similiar. This might be understood from the fact that the force due to the electric field is negligible in the direction perpendicular to the polarization of the laser. We note, therefore, that in the latter case we have the interesting situation of observing the non-sequential double ionization process in the laboratory as if the laser field is switched off as soon as the two electrons are free from the atomic system.

CONCLUSIONS

We have briefly discussed the mechanism of non-sequential double ionization in an intense femtosecond infrared laser pulse as it emerges from the dominant Feynman diagram of a systematic analysis using the intense-field many-body S-matrix theory. Results of calculations for the recently observed momentum distributions of the doubly charged He ion are found to reproduce the main characteristics of the observations. It is shown that the broad double-hump structure of the distribution parallel to the polarization direction is due to an interplay of electron correlation and final state Volkov field dressing of both the electrons. The narrow distribution in the perpendicular direction is shown to be consistent with the absence of an effective final state interaction of the field with the outgoing electrons.

ACKNOWLEDGMENTS

A.B. acknowleges financial support by the Alexander-von-Humboldt Foundation (Bonn, Germany) via a Feodor-Lynen Fellowship. This work has been partially supported by the Deutsche Forschungsgemeinschaft (DFG, Bonn, Germany).

REFERENCES

1. Walker B., et al., *Phys. Rev. Lett.*, **73**, 1227-1230 (1994).
2. Larochelle S., Talebpour A., and Chin S.L., *J. Phys. B*, **31**, 1201-1214 (1998).
3. Becker A., and Faisal F.H.M., *J. Phys. B*, **32**, L335-L343 (1999).
4. Weber Th. et al., *Phys. Rev. Lett.*, **84**, 443-446 (2000).
5. Moshammer R. et al., *Phys. Rev. Lett.*, **84**, 447-450 (2000).
6. Weber Th. et al., *J. Phys. B*, **33**, L127-L133 (2000).
7. Muller H., *Optics Express*, **8**, 417-424 (2001).
8. Parker J., Moore L., and Taylor K., *Optics Express*, **8**, 436-440 (2001).
9. Faisal F.H.M., and Becker A., in *Selected Topics on Electron Physics*, edited by D.H. Campbell and H. Kleinpoppen, Plenum Press, New York, 1996, pp. 397-410.
10. Faisal F.H.M., Becker A., and Muth-Böhm J., *Laser Phys.*, **9**, 115-123 (1999).
11. Becker A., and Faisal F.H.M., *Optics Express*, **8**, 383-394 (2001).
12. Becker A., and Faisal F.H.M., (a) *J. Phys. B*, **29**, L197-L202 (1996); (b) *Phys. Rev. A*, **59**, R1742-R1745 (1999); (c) *Phys. Rev. A*, **59**, R3182-R3185 (1999); Faisal F.H.M., and Becker A., (d) *Comm. Mod. Phys. D*, **1**, 15-27 (1999); (e) Becker A. et al., *J. Phys. B*, **33**, L547-552 (2000); (f) Maeda H. et al., *Phys. Rev. A*, **62**, 035402 (2000).
13. Becker A., and Faisal F.H.M., *Phys. Rev. Lett.*, **84**, 3546-3549 (2000).
14. Becker A. et al., *Phys. Rev. A*, **64**, 023408 (2001).
15. Muth-Böhm J. et al., *Phys. Rev. Lett.*, **85**, 2280-2283 (2000); *Chem. Phys. Lett.*, **337**, 313-318 (2001).
16. Kuchiev M.Yu., *J. Phys. B*, **28**, 5093-5115 (1995).
17. Corkum P.B., *Phys. Rev. Lett.*, **71**, 1994-1997 (1993).
18. F.H.M. Faisal, *Phys. Lett. A*, **187**, 180-184 (1994); A. Becker and F.H.M. Faisal, *Phys. Rev. A*, **50**, 3256-3264 (1994).
19. McGuire J.H., *Electron Correlation Dynamics in Atomic Collisions*, Cambridge University Press, 1997

A complete experiment on photo-double ionization of helium

Bertold Krässig

Chemistry Division, Argonne National Laboratory, Argonne, IL 60439, USA

Abstract. A formalism is presented to experimentally determine the shapes, magnitudes, and relative complex phase of the correlation functions f_g and f_u which govern the photo-double ionization process in helium. The formalism is applied to a data set from a COLTRIMS measurement of helium double ionization at 99 eV photon energy. With the results of this procedure it is possible to predict all observables of helium photo-double ionization at this energy, with the exception of the direction of the circular dichroism and of the spins.

INTRODUCTION

Since the seminal work of Huetz et al. [1] and the break-through experiment of Schwarzkopf et al. [2], important advances have been made in the theoretical and experimental approaches to helium photo-double ionization (see the recent review [3]). The important quantity describing the absorption of a photon with the simultaneous emission of two electrons, in dipole approximation and disregarding spins, is the fivefold differential cross section (FDCS)

$$\frac{d^5\sigma}{dE_1 d\Omega_1 d\Omega_2} = C \left| \langle \mathbf{k}_1 \mathbf{k}_2 \,^1P^o; \mathrm{He}^{2+} | \mathcal{D} | \Phi_{\mathrm{He}} \rangle \right|^2, \tag{1}$$

with $|\Phi_{\mathrm{He}}\rangle$ and $\langle \mathbf{k}_1\mathbf{k}_2; \mathrm{He}^{2+}|$ being the initial bound and final continuum states of the system, $\mathcal{D}$ the dipole operator, and C a constant depending on the gauge of the transition operator. It is well established that the FDCS of eq. (1) for linearly polarized photons can be parametrized with correlation functions $f = f(E_1/E_{\mathrm{exc}}, \cos\theta_{12})$ which for a given photon energy depend only on the energy sharing and the mutual angle between the electrons, and simple angular terms [1]. Two alternative parametrizations are

$$\frac{d^5\sigma}{dE_1 d\Omega_1 d\Omega_2} = \left| f_1 \cos\theta_1 + f_2 \cos\theta_2 \right|^2, \tag{2}$$

and

$$\frac{d^5\sigma}{dE_1 d\Omega_1 d\Omega_2} = \left| f_g(\cos\theta_1 + \cos\theta_2) + f_u(\cos\theta_1 - \cos\theta_2) \right|^2, \tag{3}$$

with $f_g = (f_1 + f_2)/2$ and $f_u = (f_1 - f_2)/2$.

A substantial part of the work published on photo-double ionization in helium deals with the case of equal energy sharing between the two electrons. In this case $f_1 = f_2$

CP604, *Correlations, Polarization, and Ionization in Atomic Systems*
edited by D. H. Madison and M. Schulz
© 2002 American Institute of Physics 0-7354-0048-2/02/$19.00

and $f_u = 0$. The corresponding FDCS then depends on the absolute square of just the *gerade* correlation function, $|f_g|^2$. In the first experiment, at 20-eV excess energy and for equal energy sharing [2], it was confirmed that $|f_g|^2$ essentially has a Gaussian shape with respect to the angle θ_{12}. This shape has been predicted on the basis of the Wannier-Peterkop-Rau theory [4, 5, 6]. The characteristic width of this Gaussian shape, however, was found to be greatly overestimated by theory. The width θ_{fwhm} and the height c of a Gaussian ansatz for $|f_g|^2$ in eq. (3) are commonly used to quantify absolute and relative experimental data sets of the FDCS for equal energy sharing.

Not much is known about the *ungerade* correlation function f_u. Using a partial wave expansion [7, 8] for the correlation function $|f_u|^2$, truncated at $l_{max} = 3$, Soejima et al. reported a shape for the absolute square of the *ungerade* correlation which best reproduced their FDCS measurements with unequal energy sharing [9]. As an alternative to the (infinite sum) partial wave expansion, Cvejanović and Reddish proposed a parametrization with fewer parameters to describe unequal energy sharing FDCS data [10], based on Gaussians in θ_{12} for f_g and f_u with an added constant contribution.

In this report a formalism will be sketched out with which the complex correlation functions f can be obtained experimentally without making any assumptions on their shape.The formalism will be applied to an event-mode data set from a COLTRIMS experiment on helium photo-double ionization at 99-eV photon energy [11, 12]. The data were put on an absolute scale by normalizing to the integral cross sections for photo-double ionization of helium [13].

With the proposed formalism not only the absolute magnitudes of the correlation functions f are determined, but also the cosine of their relative complex phase. The correlation functions are closely related to the radial integral of the transition matrix element in eq. (1). These results therefore represent what has been termed a "complete experiment", because all observables (except for the spins) can be predicted solely on the basis of the the knowledge of the complex correlation functions f_g and f_u or f_1 and f_2. Knowing only the cosine, but not the sine of the complex phase, affects the predictive power of this approach only for experiments with circular polarization. For circular polarization the magnitude of the circular dichroism can be predicted on the basis of the present results, however, not its direction. All observables for polarizations other than circular depend solely on the cosine of the complex phase.

THE GIST OF THE FORMALISM

The formalism to extract the complex correlation function f from the COLTRIMS event mode data is based on the alternative parametrizations in eqs. (2) and (3). In the present form, it applies to linear polarization only. The approach is straightforward. According to the first parametrization in eq. (2), the subset of the total events with $\cos\theta_2 = 0$ is proportional to $|f_1|^2$. This subset is sorted into a two-dimensional histogram $H_{12}(E_1, \cos\theta_{12})$. The energy E_2 is determined by the energy conservation $E_2 = E_{exc} - E_1$. Similarly, a histogram $H_{21}(E_1, \cos\theta_{12})$ can be filled using the condition $\cos\theta_1 = 0$. With reference to the second parametrization in eq. (3), the total number of events can be sorted into two additional two-dimensional histograms $H_g(E_1, \cos\theta_{12})$ and $H_u(E_1, \cos\theta_{12})$ by imposing the conditions $(\cos\theta_1 - \cos\theta_2) = 0$ and $(\cos\theta_1 + \cos\theta_2) = 0$. The four histograms are proportional to the absolute squares of the four respective f functions. In practice, the number of events fulfilling the imposed conditions exactly is very small and less stringent conditions, e.g., $|\cos\theta_2| < \varepsilon$, etc., are applied. Resultingly, the contents of the corresponding histograms is not exactly proportional to a particular f function, but contains a small admixture of its sibling function. This impurity can be removed using an iterative approach.

The sorting of the total number of the double ionization COLTRIMS events is the equivalent to integrating the FDCS with appropriate integration limits. In order to determine the relationship between the $|f|^2$ functions and the histograms H this integration has to be carried out. In the form given in eqs. (2) and (3), the integration is not straightforward. One of the 5 angles $\theta_1, \phi_1, \theta_2, \phi_2, \theta_{12}$ would have to be expressed by the remaining four, complicating the integration. The integration can be carried out using a different set of angles, namely three Euler angles α, β, γ in addition to the inter-electron angle θ_{12} [1, 14]. Following integration over α and β, the resulting relations are of the form

$$\left.\frac{d^3\sigma}{dE_1 d\cos\theta_1 d\gamma}\right|_{\text{cond.}\gamma} = const|f|^2 a(\cos\theta_{12}). \tag{4}$$

The weight functions $a(\cos\theta_{12})$ depend on the particular case considered. The above sorting conditions in $\cos\theta_i$ translate into conditions for the Euler angle γ. The histograms H, after calibration to the integrated cross section and accounting for the finite bin widths, $\Delta E, \Delta\cos\theta_{12}, \Delta\gamma$, are to be identified directly with the differential cross sections in eq. (4). The cosine of the relative complex phase between f_g and f_u can be obtained from the four absolute squares $|f_g|^2, |f_u|^2, |f_1|^2, |f_2|^2$ using the relations given in the Introduction.

THE EXPERIMENTAL DATA SET

The experimental data set used to demonstrate the feasibility of the reported formalism stems from a COLTRIMS experiment on helium photo-double ionization performed at the Advanced Light Source (ALS) at Berkeley National Laboratory in 1996. Other aspects of the experimental results have already been published [11, 12]. For details on the experimental setup and the data reduction procedure, see Ref. [15]. The measurements were carried out at the undulator beam line BL7 at a photon energy of 99 eV, i.e. at an excess energy of 20 eV above the helium double ionization threshold. The degree of linear polarization at this beam line was measured to be 0.99(1)% [15]. A high degree of linear polarization is essential for the applicability of the formalism in the form described in the previous section.

RESULTS AND DISCUSSION

Using the formalism and the data set described in the previous two sections, the correlation functions f of photo-double ionization in helium with 99-eV photons have been

14

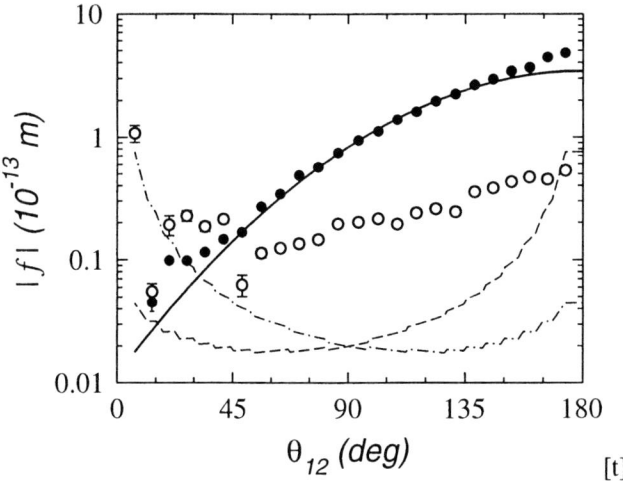

FIGURE 1. Absolute magnitudes of the correlation functions, integrated over E_1. Solid circles, $|f_g|$; solid line, Gaussian fit to $|f_g|$; open circles, $|f_u|$. The contributions of a single count per channel to the $|f_g|$ and $|f_u|$ functions are indicated by the broken and the dash-dotted curves, respectively.

determined. These functions are functions in two coordinates, the energy of the slower electron relative to the excess energy, $0 \leq E_1 \leq E_{exc}/2$, and the cosine of the relative angle between the two electrons, $-1 \leq \cos\theta_{12} \leq 1$. The functions f_g and f_u are symmetric with respect to $E_{exc}/2$; for the other two functions it holds $f_1(E_{faster}) = f_2(E_{slower})$.

As an example of these results, $|f_g|$ and $|f_u|$, integrated over E_1 from 0 to $E_{exc}/2$, are shown in Fig. 1 on a logarithmic scale as a function of the interelectron angle θ_{12}. In the 50°–150° region the function $|f_g|$ has a Gaussian shape as can be seen from the comparison with the parabolic curve included in Fig. 1 representing a Gaussian on a logarithmic scale. Near the the maximum at $\theta_{12} = 180°$ and at small angles $|f_g|$ diverges from that shape, similarly to the observation made in [2]. The function $|f_u|$ is also peaked at $\theta_{12} = 180°$, but it has a much smaller maximum value value than $|f_g|$ and it appears not to follow a parabolic, rather than a linear course in logarithmic representation. Towards small angles the experimental points of the function $|f_u|$ increase in value.

As was mentioned in the description of the formalism, the way to arrive at the functions $|f|^2$ is to divide the contents of the histograms H by the weight function $a(\cos\theta_{12})$. This division diverges at $\theta_{12} = 180°$ for $|f_g|^2$ and at $\theta_{12} = 0°$ for $|f_u|^2$. As a result, any uncertainties in the channels in the vicinity of these divergences are strongly amplified. The broken and dash-dotted curves in Fig. 1 show the equivalent of a single count in the original Histograms H_g and H_u as it contributes to the $|f_g|$ and $|f_u|$ results. From the dash-dotted curve in Fig. 1 it can be surmised that the rise of the function $|f_u|$ towards small angles θ_{12} is most likely due to spurious counts in those bins. Similarly, the enhancement of $|f_g|$ near 180° may also be caused by counts that have been erroneously sorted into those bins. The occurrence of these divergences is simply a manifestation for the fact that no double ionization events are created in these configurations.

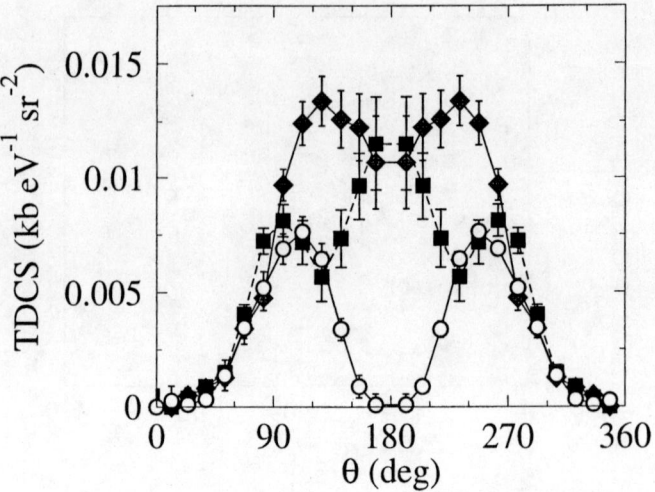

FIGURE 2. Absolute FDCS for equal and unequal energy sharing. Open circles, $E_1 = E_2 = 10$ eV; solid squares, $E_1 = 3$ eV, $E_2 = 17$ eV; open diamonds, $E_1 = 17$ eV, $E_2 = 3$ eV. Data points greater than $\theta = 180°$ are the mirror image of the points below $180°$.

The correlation functions f can now be used to generate absolute FDCSs that can be compared to published results. In Fig. 2 the FDCS for equal (open circles) and unequal (diamonds, squares) energy sharing are compared. In each case the direction of one electron is along the polarization direction, and the FDCS for finding the other electron at angle θ_2 (ϕ_2 fixed) is plotted. Fig. 2 is to be compared to the polar plots in figure 5 of Ref. [12]. Note that the result of Fig. 2 and the result of Ref. [12] are derived from the same set of raw data. In the present result a much larger subset of the raw data has been used to generate the FDCS, resulting in smaller error bars.

Similarly, the absolute FDCS of other experimental configurations can be generated. For example, the magnitude of the circular dichroism in photo-double ionization of helium at 99 eV [16] has been reproduced on the basis of the correlation functions f determined in this work.

CONCLUSION

A formalism has been presented to fully determine the complex correlation functions that govern the photo-double ionization process in helium. The only assumptions in the formalism are the validity of the dipole approximation and complete linear polarization. The formalism has been tested using a COLTRIMS data set taken at 99-eV photon energy. The magnitudes of the correlation functions and the cosine of their relative phase were obtained on an absolute scale. This finding establishes a *complete* experiment for photo-double ionization of helium as all other observables (with the exception of the spins and the direction of the circular dichroism) can be derived from these quantities.

This technique permits to study more closely the purely dynamical part of the photo-double ionization process. The correlation functions have mostly been modeled by Gaussians in the interelectron angle θ_{12}, which appears to be a reasonable approach for f_g, but may be insuffient for f_u. Modeling of the correlation functions in the variable $\cos\theta_{12}$, such as the partial wave expansions in [7, 8] appears to be more appropriate for this problem. A measurement of the correlation functions with better statistics than in the present case could reveal any modulations in the shape of the correlation functions caused by a specific makeup in the partial wave expansion or provide other hints leading to a full understanding of this process.

ACKNOWLEDGEMENTS

The author wishes to thank Dr. H. Bräuning for generating the required histograms from the COLTRIMS data. This work is supported by the Chemical Sciences, Geosciences, and Biosciences Division of the Office of basic Energy Sciences, Office of Science, U.S. Department of Energy under contract W-31-109-ENG-38.

REFERENCES

1. Huetz, A., Selles, P., Waymel, D., and Mazeau, J., *J. Phys. B: At. Mol. Opt. Phys.*, **24**, 1917–33 (1991).
2. Schwarzkopf, O., Krässig, B., Elmiger, J., and Schmidt, V., *Phys. Rev. Lett.*, **70**, 3008–11 (1993).
3. Briggs, J. S., and Schmidt, V., *J. Phys. B: At. Mol. Opt. Phys.*, **33**, R1–48 (2000).
4. Wannier, G. H., *Phys. Rev.*, **90**, 817–25 (1953).
5. Peterkop, R., *J. Phys. B: At. Mol. Phys.*, **4**, 513–21 (1971).
6. Rau, A. R. P., *Phys. Rev. A*, **4**, 207–20 (1971).
7. Manakov, N. L., Marmo, S. I., and Meremianin, A. V., *J. Phys. B: At. Mol. Opt. Phys.*, **29**, 2711–37 (1996).
8. Malegat, L., Selles, P., and Huetz, A., *J. Phys. B: At. Mol. Opt. Phys.*, **30**, 251–61 (1997).
9. Soejima, K., Danjo, A., Okuno, K., and Yagishita, A., *Phys. Rev. Lett.*, **83**, 1546–49 (1999).
10. Cvejanović, S., and Reddish, T. J., *J. Phys. B: At. Mol. Opt. Phys.*, **33**, 4691–709 (2000).
11. Bräuning, H., Dörner, R., Cocke, C. L., Prior, M. H., Krässig, B., Bräuning-Demian, A., Carnes, K., Dreuil, S., Mergel, V., Richard, P., Ullrich, J., and Schmidt-Böcking, H., *J. Phys. B: At. Mol. Opt. Phys.*, **30**, L649–55 (1997).
12. Bräuning, H., Dörner, R., Cocke, C. L., Prior, M. H., Krässig, B., Kheifets, A. S., Bray, I., Bräuning-Demian, A., Carnes, K., Dreuil, S., Mergel, V., Richard, P., Ullrich, J., and Schmidt-Böcking, H., *J. Phys. B: At. Mol. Opt. Phys.*, **31**, 5149–60 (1998).
13. Samson, J. A. R., Stolte, W. C., He, Z.-X., Cutler, J. N., Lu, Y., and Bartlett, R. J., *Phys. Rev. A*, **57**, 1906–11 (1998).
14. Nikitin, S. I., and Ostrovsky, V. N., *J. Phys. B: At. Mol. Phys.*, **18**, 4349–69 (2000), note the misprint in eq. (2.16); see Bogdanovich et al., *ibid.* **30**, 921–940 (1997).
15. Dörner, R., Bräuning, H., Feagin, J. M., Mergel, V., Jagutzki, O., Spielberger, L., Vogt, T., Khemliche, H., Prior, M. H., Ullrich, J., Cocke, C. L., and Schmidt-Böcking, H., *Phys. Rev. A*, **57**, 1074–90 (1998).
16. Achler, M., Mergel, V., Spielberger, L., Dörner, R., Azuma, Y., and Schmidt-Böcking, H., *J. Phys. B: At. Mol. Opt. Phys.*, **34**, 965–981 (2001).

17

Low Energy Inner Valence Ionization Of The Rare Gases

Matthew A. Haynes[1] and Birgit Lohmann[1]
and
D. A. Biava[2], R. P. McEachran[3], C. T. Whelan[4] and D. H. Madison[2]

[1]*Griffith University, Brisbane, Queensland, Australia*
[2]*University of Missouri-Rolla, Rolla, MO, USA*
[3]*Australian National University, Canberra, Australia*
[4]*Old Dominion University, Norfolk, VA, USA*

Abstract. Recent measurements are presented of the triple differential cross section for electron impact ionization of the inner valence shells of argon and krypton at low to intermediate energies, and both coplanar symmetric and asymmetric geometries. Comparison is made with some of the latest available theoretical calculations performed in the distorted wave Born approximation formalism.

INTRODUCTION

In recent years, measurements of the triple differential cross section (TDCS) for low energy electron impact ionization in coplanar symmetric or energy-sharing geometry have provided a considerable challenge to modern scattering theories. The majority of these measurements were performed on hydrogen and helium targets [1], although a few measurements exist for heavier atoms [2]. Very recently, the first low impact energy experiments on an argon target in the coplanar asymmetric geometry have been performed [3,4], for both 3s and 3p ionization, as well as new measurements [5] for 3s ionization in the coplanar symmetric geometry. The coplanar symmetric and asymmetric geometries are illustrated schematically in Figure 1. In Refs. [3] and [4], comparison of the coplanar asymmetric TDCS with two different distorted wave Born approximation (DWBA) calculations [6,7] showed very poor agreement, particularly for the case of 3s ionization. The level of disagreement was rather surprising, since for incident energies greater than about 100 eV, the DWBA had been shown to exhibit quite good agreement with experiment for ionization of the rare gases. No theoretical calculations were available for comparison at the time the data for 3s ionization in the coplanar symmetric geometry were published, but these kinematics are expected to offer an even more stringent test of the theory.

In this paper, we review the current status and level of agreement between the latest theoretical and experimental results for argon. The new DWBA calculations shown here explore the importance of exchange in determining the form of the cross section,

CP604, *Correlations, Polarization, and Ionization in Atomic Systems*
edited by D. H. Madison and M. Schulz
© 2002 American Institute of Physics 0-7354-0048-2/02/$19.00

18

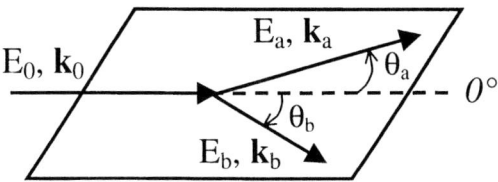

FIGURE 1. Diagram of the coplanar kinematics. The incident electron has energy and momentum E_0, k_0. The scattered electron is detected at a scattering angle θ_a with energy E_a, while the ejected electron is detected with energy E_b at angle θ_b. θ_a and θ_b are measured from $0°$, as shown. In the asymmetric measurements, θ_a is fixed while θ_b is varied. In the symmetric measurements both θ_a and θ_b are varied such that $\theta_a = \theta_b = \theta$.

and include results for coplanar symmetric argon 3s ionization. Details of the theory and the different approaches used to incorporate exchange may be found in Ref. [8]. We also present new measurements for the inner valence (4s) ionization of krypton, and contrast the form of the TDCS for this target with that obtained for 3s ionization in argon.

EXPERIMENTAL DETAILS

The electron coincidence spectrometer is shown schematically in overview in Figure 2. A detailed discussion of the apparatus may be found in Ref. [3]. Two identical hemispherical electron energy analysers, fitted with channel electron multipliers for electron detection, are mounted on concentric independent turntables. The analysers are mounted coplanar with the fixed electron gun; the incident electron beam crosses a target gas beam at right angles. Conventional coincidence timing electronics are employed. In the coplanar asymmetric measurements, the binary peak is measured by fixing the scattered electron energy analyzer at -15° with respect to the incident beam, and rotating the ejected electron energy analyzer through the desired angular range on the opposite side of the scattering plane. The recoil peak is measured by moving the scattered electron energy analyzer to +15° and again moving the ejected electron energy analyzer through the accessible angular range. The counting time at each ejected electron angle is normalized to a preset scattered electron count. As the measurements are not absolute, the binary-to-recoil ratio is determined in a separate measurement, as detailed in Ref. [3]. In the coplanar symmetric measurements, the two analyzers are rotated at equal angles on either side of the incident electron beam. Equal counting times are employed at each angle, and the incident current and gas pressure are carefully monitored to ensure no variation.

RESULTS AND DISCUSSION

Selected experimental and theoretical results for argon 3s and 3p ionization in the asymmetric geometry are presented in Figure 3. The incident energy is 113.5 eV and

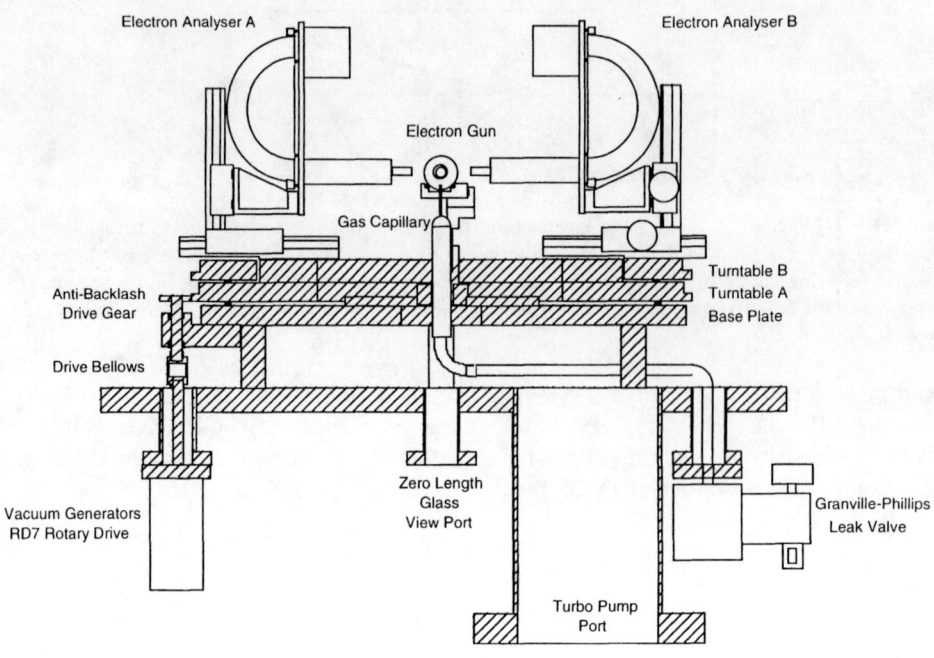

Electron Analyser A

Electron Analyser B

Electron Gun

Gas Capillary

Turntable B
Turntable A
Base Plate

Anti-Backlash
Drive Gear

Drive Bellows

Vacuum Generators
RD7 Rotary Drive

Zero Length
Glass
View Port

Granville-Phillips
Leak Valve

Turbo Pump
Port

FIGURE 2. Schematic diagram of the electron coincidence spectrometer.

the ejected electron energies are 2 and 5 eV. The scattered electron energies are adjusted to meet the energy balance, with binding energies of 29.3 eV (3s) and 15.8 eV (3p). The scattering angle is 15°. The solid curve is the DWBA with no exchange, while the dashed line is the DWBA with the Furness-McCarthy [9] approximation to the static exchange potential (DWBA-FM), with the triplet potential used for the incident and faster final state electrons and a combination of triplet and singlet potentials used for the slower ejected electron. The experimental results have been normalized to the peak value in the binary region in the latter calculation. It is apparent that the treatment of exchange is very important for 3s ionization, particularly at 2 eV, where only the DWBA-FM gives even qualitatively the right shape. Overall, the agreement between theory and experiment is considerably better for 3p ionization than for 3s ionization. In the former case, both theories give a good description of the shape of the binary peak, and of the recoil-to-binary peak ratio. The major discrepancy is in the theoretical versus experimental position of the recoil peak. This difference cannot be attributed to post collision interaction (not fully accounted for in either theory) which would tend to move both binary and recoil peaks towards 180°. In the case of 3s ionization, there are still significant differences between theory and experiment in both the binary and, more dramatically, the recoil region. As the maximum of the recoil peak in the measured data appears to lie at a smaller angle than is experimentally accessible, it is not possible to compare the experimental and

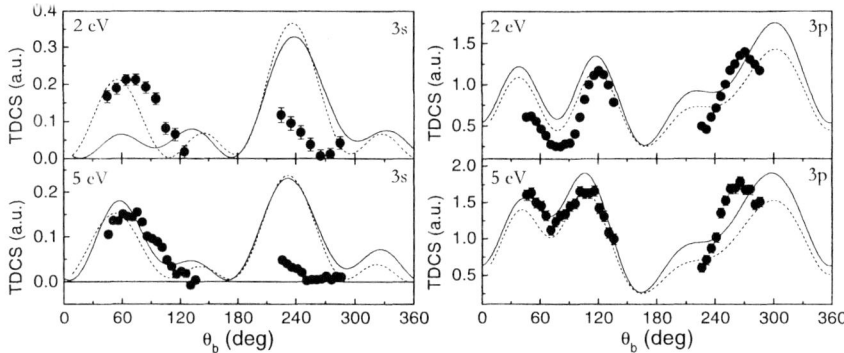

FIGURE 3. Triple differential cross section for argon 3s and 3p ionization in the coplanar asymmetric geometry. The incident energy is 113.5 eV, and the scattering angle is 15°. The ejected energy is shown on each graph, as is the orbital being ionized. The points are the experimental data. The solid line is a DWBA calculation without exchange, while the dashed line is a DWBA calculation with the Furness-McCarthy [9] exchange potential (see text). The experimental results have been normalized to the latter calculation in the binary region.

theoretical recoil-to-binary peak ratio, which normally provides a good test of the quality of the theory.

Figure 4 shows the results for argon 3s and krypton 4s ionization in the coplanar symmetric geometry. The solid line and dotted line are the DWBA and DWBA-FM calculations respectively. The dash-dot line is a DWBA calculation (labelled DWBA-1) we have performed using the McCarthy code [7]. The argon experimental results have been normalized to the DWBA-FM calculation at 40° for the case of 50, 20 and 10 eV outgoing energy. For the 4 eV case, the data have been normalized so as to give the same peak height as the DWBA-FM at backward angles. The krypton data have been normalized to the DWBA-1 at the forward peak. The DWBA-FM exhibits quite good agreement with the argon experimental data at 50 eV and 20 eV outgoing energy, even though the effect of PCI is underestimated at forward angles. At the lower outgoing energies, the DWBA (no exchange) appears to be in better shape agreement with the experimental data, although proper inclusion of PCI at these energies can be expected to change the form of the cross section substantially. The structure at forward angles in the TDCS has been attributed to a single electron-electron binary collision, with the structure at backward angles arising from a double binary collision in which the incident electron is first back-scattered from the nucleus and then undergoes a single binary collision with a bound electron [10]. The results show the evolution of the backward peak as the outgoing energy decreases, until at 4 eV the single binary peak has disappeared altogether and the double scattering peak has become the dominant structure. One interesting feature of the argon data at the higher energies is the forward position of the binary peak, which lies at a much smaller angle than 45°, where one would expect the peak to be located in an impulsive collision. In contrast, the krypton data at 85 eV, 50 eV and 20 eV outgoing energy

21

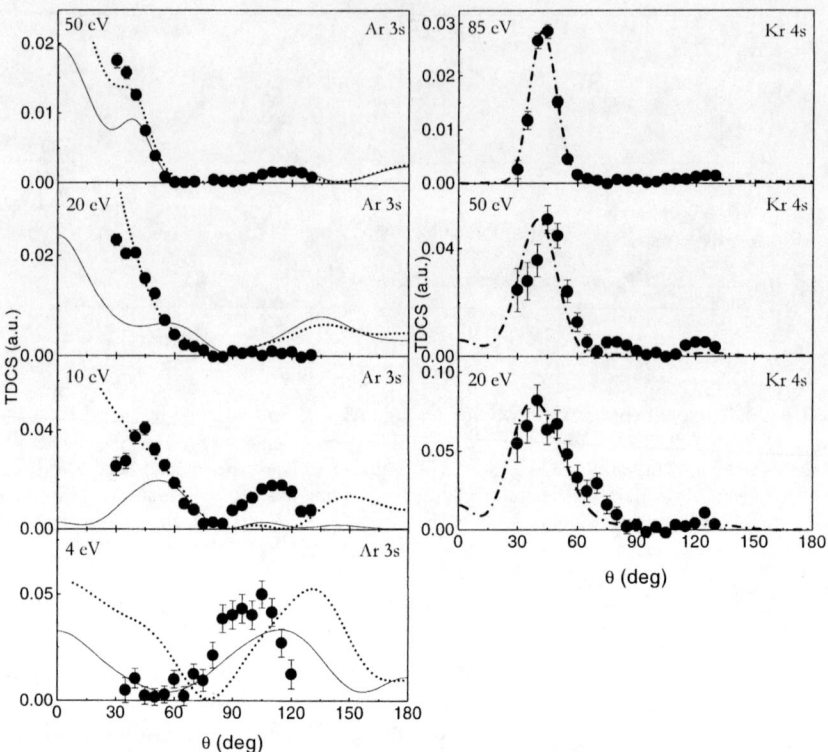

FIGURE 4. Triple differential cross section for argon 3s and krypton 4s ionization in the coplanar symmetric geometry. The outgoing electron energy is shown on each graph, as is the orbital being ionized. The points are the experimental data. The solid line is a DWBA calculation without exchange, the dotted line is the DWBA-FM calculation with the Furness-McCarthy [9] exchange potential and the dash-dot line is the DWBA-1 calculation using the code of McCarthy (see text). The experimental results have been normalized to the calculations as detailed in the text.

show a binary peak positioned near 45°. The DWBA-1 calculation is in very good agreement with the krypton experimental data, and has quite a different shape to the DWBA-FM calculation for argon at the same energies. The origin of the different behaviour of the TDCS for these two targets is not known.

ACKNOWLEDGMENTS

The authors would like to thank the Australian Research Council for providing funding for this research. MH was supported by an Australian Postgraduate Research Award.

REFERENCES

1. Röder J. et al *J. Phys. B* **29** 2103 (1996); Röder J. et al *J. Phys. B* **29** L67 (1996); Röder J. et al *Phys. Rev. A* **53** 225 (1996); Röder J. et al *J. Phys. B* **31** L525 (1998); Bray I. et al *Phys. Rev. A* **57** R3161 (1998); Rioual S. et al *J. Phys. B* **31** 3117 (1998).
2. Rösel T. et al *J. Phys. B* **24** 3059 (1991); Bell S. et al *Phys. Rev. A* **51** 2623 (1995); Rioual S. et al *J. Phys. B* **28** 5317 (1995); Rioual S. et al *J. Phys. B* **30** L475 (1997); Rouvellou B. et al *Phys. Rev. A* **57** 3621 (1998).
3. Haynes M. A. and Lohmann B. *J. Phys. B* **33** 4711 (2000).
4. Haynes M. A. and Lohmann B. *Physical Review A*, to be published.
5. Haynes M. A. and Lohmann B. *J. Phys. B* **34** L131 (2001).
6. Madison D. H. and Lang R. *J. Phys. B* **14** 4137 (1981).
7. McCarthy I. E. *Aust. J. Phys.* **48** 1 (1995).
8. Biava et al, submitted to *J. Phys. B*.
9. Furness J. B. and McCarthy I. E. *J. Phys. B* **6** 2280 (1973).
10. Whelan C. T. and Walters H. R. *J. Phys. B* **23** 2989 (1990).

Simple Ionization of Atomic and Diatomic Lithium by Electron Impact

B. Joulakian, F. Boudali and B. Najjari

*Laboratoire de Physique Moléculaire et des Collisions, Institut de Physique, 1 Bld Arago,
Technopôle 2000, 57078 Metz Cedex 3 France
Fax: 33 (0)387315801, Phone: 33 (0)387315858
e-mail: joulak@ipc.sciences.univ-metz.fr*

Abstract. The e,2e reaction of atomic and diatomic lithium is studied. In the atomic case, the problem is treated as a four electron system for which the conservation of the total spin produces a variety of exchange terms whose importance in the determination of the transition matrix element of the ionization process is studied systematically in a very large domain of incidence energy (50eV to 5keV) by applying in the small and intermediate energy domain an asymptotically exact distorted wave procedure. In the diatomic case, a model potential is employed for the K shell electrons and the problem is treated as a vertical transition from the lowest vibrational and rotational level of the fundamental electronic state $^1\Sigma_g$ of Li_2 to the fundamental $^2\Sigma_g$ state of $Li_2{}^+$. The comparison with experimental and recent theoretical results shows that the use of an all electron procedure improves the agreement with the experimental results.

INTRODUCTION

Complex systems, like many electron atoms, have been often studied as one or two active electron systems by neglecting the spectator electrons [1] or by considering, when possible, a model potential for the inner electrons. Even atomic lithium, which is one of the simplest many electron systems has been treated theoretically as a one (2s) electron target, where any exchange effect with the internal electrons is ignored [2,3].

In a non-relativistic treatment of the simple ionization of atoms or molecules by electron impact, the total spin of the system, constituted by the incident electron and the target, is supposed to be conserved. In the case of one or two electron targets, this permits to define direct and exchange terms, in which the roles of the different electrons in the transition matrix element change [4,5]. In this paper we define in a similar way the terms that one obtains in the treatment of lithium–like systems and look for the energy domains and the scattering directions for which the exchange with the inner electrons could be important. More, we want to show that, even in situations where exchange terms with the inner shell electrons are comparatively small, an all electron description of the target improves the agreement with existing experimental results. This will guide us for the choice of the energy domains in the calculations on the diatomic lithium and will be very useful for future calculations of the multiply differential cross sections of the e,3e and e,4e ionization of lithium.

CP604, *Correlations, Polarization, and Ionization in Atomic Systems*
edited by D. H. Madison and M. Schulz
© 2002 American Institute of Physics 0-7354-0048-2/02/$19.00

THEORY

The coupling of the spins of the incident and the lithium-like target in its fundamental state will result in four states, whose wave functions can be given by the following combinations of four electron Slater determinants

$$\psi_{i1}(S=0,M_s=0) = \frac{1}{\sqrt{4}}\left\{\left|\overline{\chi}_i\varphi_1\overline{\varphi}_2\varphi_3\right| - \left|\overline{\chi}_i\overline{\varphi}_1\varphi_2\varphi_3\right| - \left|\chi_i\varphi_1\overline{\varphi}_2\overline{\varphi}_3\right| + \left|\chi_i\overline{\varphi}_1\varphi_2\overline{\varphi}_3\right|\right\} \quad (1.a)$$

$$\psi_{i2}(S=1,M_s=1) = \frac{1}{\sqrt{2}}\left\{\left|\chi_i\varphi_1\overline{\varphi}_2\varphi_3\right| - \left|\chi_i\overline{\varphi}_1\varphi_2\varphi_3\right|\right\} \quad (1.b)$$

$$\psi_{i3}(S=1,M_s=0) = \frac{1}{\sqrt{4}}\left\{\left|\overline{\chi}_i\varphi_1\overline{\varphi}_2\varphi_3\right| - \left|\overline{\chi}_i\overline{\varphi}_1\varphi_2\varphi_3\right| + \left|\chi_i\varphi_1\overline{\varphi}_2\overline{\varphi}_3\right| - \left|\chi_i\overline{\varphi}_1\varphi_2\overline{\varphi}_3\right|\right\} \quad (1.c)$$

$$\psi_{i4}(S=1,M_s=-1) = \frac{1}{\sqrt{2}}\left\{\left|\overline{\chi}_i\varphi_1\overline{\varphi}_2\overline{\varphi}_3\right| - \left|\overline{\chi}_i\overline{\varphi}_1\varphi_2\overline{\varphi}_3\right|\right\} \quad (1.d)$$

satisfying the condition $\hat{S}^2\psi = S(S+1)\psi$ and $\hat{S}_z\psi = M_s\psi$ for the total spin operators. Here $\left|\chi_i\varphi_1\overline{\varphi}_2\varphi_3\right|$ represents a Slater determinant with χ_i describing the incident electron and φ_j the different atomic orbitals. The bars indicate the spin -1/2.

In the final state we have the same spin states, whose wave functions are constructed by the same combinations using the orbitals of the scattered electron, χ_s, that of the ejected electron, χ_e and the two bound electrons represented by the orbitals ϕ_1 and ϕ_2. The multiply differential cross section is given by :

$$\sigma^{(3)} = \frac{d^3\sigma}{d\Omega_s d\Omega_e d(k_s^2/2)} = (2\pi)^4 \frac{k_s k_e}{k_i} \frac{1}{4} \sum_{v=1}^{4} \left|T_v\right|^2 \quad (2)$$

where the summation runs over the four spin states. Here k_i, k_s and k_e represent the moduli of the wave vectors of the incident, scattered and ejected electrons respectively and Ω_s and Ω_e, the solid angles of the scattered and the ejected electrons respectively. T_v represents the transition matrix element, which we will write following [6] in the form:

$$T_v = \left\langle \psi_{fv}^- \left| V \right| \psi_{iv}^+ \right\rangle = \left\langle \psi_{fv}^- \left| U(r_0,r_{01},r_{02},r_{02}) - u(r_0) \right| \psi_{iv}^+ \right\rangle \quad (3)$$

Here $U(r_0,r_{01},r_{02},r_{02})$ represents the potential "seen" by the incident electron, where r_0 and r_{0j} are the respective distances of the incident electron from the nucleus and the three atomic electrons and $u(r_0)$ is the initial state distortion. Replacing the expression of the different wave functions in Equation (3) we can write:

$$\sigma^{(3)} = (2\pi)^4 \frac{k_s k_e}{4k_i} \left\{ 3 \left| f_1 + g_2 + h_1 - f_2 - g_1 - h_2 \right|^2 + \left| f_1 + g_1 + 2h_3 - f_2 - g_2 - h_1 - h_2 \right|^2 \right\} \quad (4)$$

where f1 and f2 represent the two types of direct terms, g1 and g2 the respective exchange terms and h1, h2 and h3 the three types of exchange terms with the residual bound electrons. The complete definitions and the details of the calculations of these different terms for different energy domains are given in [7].

In the diatomic case, we consider the ionization as a vertical transition from the lowest vibrational and rotational level of the fundamental electronic state $^1\Sigma_g$ of the Li_2 target to the fundamental electronic $^2\Sigma_g$ state of Li_2^+ and write the multiply differential cross section of the (e,2e) ionization of Li_2 in the form [8]

$$\sigma^{(3)} = \frac{d^3\sigma}{d\Omega_e \, d\Omega_s \, dE_s} = \frac{(2\pi)^4}{k_i} \frac{k_s k_e}{4\pi} \int d\Omega_\rho \left| T_{fi}(\theta_\rho, \varphi_\rho) \right|^2 \quad (5)$$

where θ_ρ, φ_ρ define the orientation of $\bar{p}$ the inter-nuclear axis in the laboratory frame. The electronic transition matrix element $T_{fi}(\theta_\rho, \varphi_\rho)$ will be given by:

$$T_{fi}(\theta_\rho, \varphi_\rho) = \left\langle \Psi_f \left| V \right| \Psi_i \right\rangle \quad (6)$$

with

$$V = \frac{1}{r_{01}} + \frac{1}{r_{02}} + V_m(r_{0a}) + V_m(r_{0b}) \quad (7)$$

Here r_{0j} with j=1,2,a,or b represents the distances between the projectile (0) and the valence electrons 1 and 2 and the two Li^+ cores a and b described by the following model potential [3]:

$$V_m(r) = -\frac{1}{r} - (Z_N - 1) \frac{e^{-\alpha r}}{r} - \beta e^{-\gamma r} \quad (8)$$

with $Z_N = 3$, $\alpha = 7.9$ $\beta = 10.3$ and $\gamma = 3.893$ for the Li^+ core. Ψ_i in Equation (6) represents the electronic wave functions in the initial state, which will be given for fast incident electrons (~keV) by an appropriate product of a plane wave and $\Psi_{Li_2}(\vec{r}_1, \vec{r}_2)$ the electronic wave function of the target. On the other hand Ψ_f, the final state wave function, will be given by the appropriate product of a plane wave for the fast scattered electron, $\chi_e(\vec{k}_e, \vec{r}_1)$ describing the slow ejected electron, and $\Psi_{Li_2^+}(\vec{r}_2)$ the electronic wave function of the residual ion. Following Schulz [5] we can write

$$T_{fi}(\theta_p, \varphi_p) = (2f - g - h) + (g - h) \qquad (9)$$

in which we will consider for high incident energy values only the direct term given by

$$f(\vec{k}_e, \vec{k}_s) = \frac{1}{(2\pi)^3} \left\langle \Psi_{Li_2^+}(\vec{r}_2) \chi_e(\vec{k}_e, \vec{r}_1) e^{i\vec{k}_s \cdot \vec{r}_0} \left| V \right| \Psi_{Li_2}(\vec{r}_1, \vec{r}_2) e^{i\vec{k}_i \cdot \vec{r}_0} \right\rangle \qquad (10)$$

The bound state wave functions are constructed by linear combinations of Slater type orbitals. Consequently, for the mono-electronic residual ion we have

$$\Psi_{Li_2^+} = c_1(e^{-\alpha_1 r_a} + e^{-\alpha_1 r_b}) + c_2(e^{-\alpha_2 r_a} + e^{-\alpha_2 r_b}) + c_3(r_a e^{-\alpha_2 r_a} + r_b e^{-\alpha_2 r_b}) \qquad (11)$$

with r_a and r_b representing the respective distances of a given electron from the two centers. For Li_2 we construct a variational solution (see F. E. Harris [9]), whose space part can be written in the form of a product of two mono-electronic functions having the same form as that of the preceding equation

$$\Psi_{Li_2}(1,2) = \psi(1)\psi(2) \qquad (12)$$

$\chi_e(\vec{k}_e, \vec{r})$ in Equation (10) corresponding to the ejected electron in the field of the residual diatomic ion will be given by a partial wave development in which the phase shift Δ_ℓ and the radial part $F_\ell(k, r_d)$ obtained numerically by solving the radial Schrödinger equation of the model potential given in Equation (8). Here $r_d = r_a$ or r_b depending on the center of the initial state wave function.

RESULTS

The systematic study of the influence of the different exchange terms with inner electrons like f2, g2, h1, h2 and h3 on the TDCS for a large incident energy domain varying from 50eV to 5keV shows that, the so called capture terms h1, h2 and h3 are always very small. On the other hand, f2 and g2 are negligible only for high incidence energy values and small scattering angles. This seems reasonable as the incident electron, which is weakly deviated, has a very big impact parameter, which favours interaction with the valence electron only. To show this effect quantitatively we present, on figures 1a and 1b, the variation of the triple differential cross section (TDCS) of lithium with the ejection angle θ_e obtained for a fixed scattering angle of $12°$ and an ejection energy value of 5eV by three different methods: 1) by the above described all electron calculation, 2) by a frozen core calculation, where the incident electron sees only the valence electron and a nucleus with charge Z=1, and 3) by a

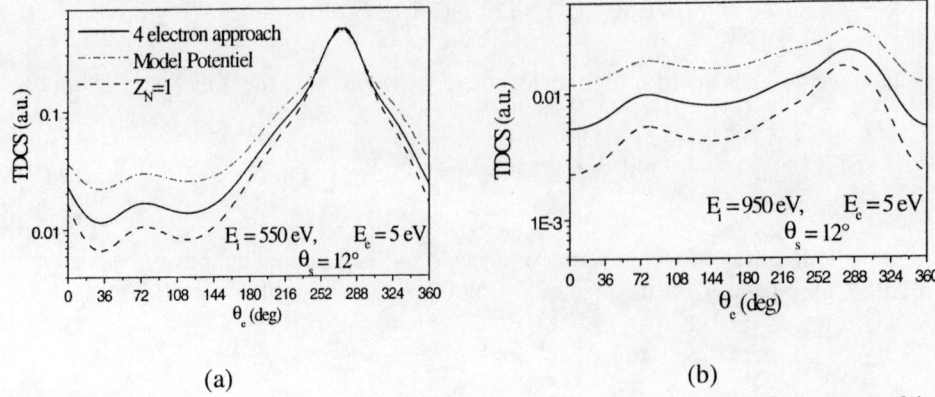

FIGURE 1. The variation of the triple differential cross section of the ionization of Li in terms of the ejection angle. In (a) $E_i=550eV$ and in (b) $E_i=950eV$

model potential approach described in [3]. On Fig. 1a the incident electron energy has the value of 550eV and the three methods give the same binary peak at the same ejection angle, which is oriented towards the momentum transfer $\vec{k} = \vec{k}_i - \vec{k}_s$. On Fig. 1b, where we have increased the incidence energy to 950eV, we observe differences between these three methods in magnitude and in ejection direction on the binary peak.

Another interesting situation is met, when one varies in an equal energy-sharing regime, the scattering direction, keeping the angle between the scattered and the ejection directions fixed. This gives in general very small values for the triple differential cross section (TDCS), but reveals interesting structures, which are attributed to the electronic structure of the target.

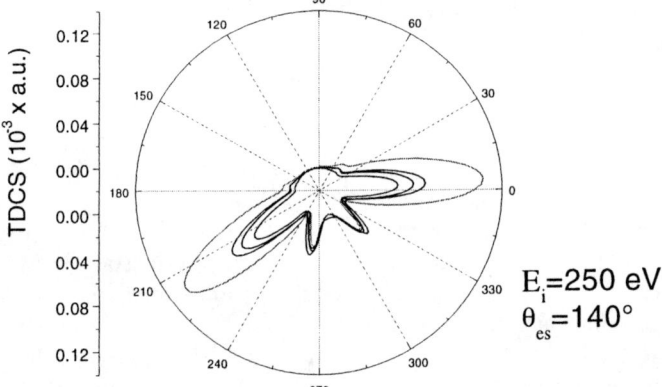

FIGURE 2. The variation of the TDCS of the simple ionization of Li in polar coordinates. The curves represent respectively the results obtained by the model potential, by all electron approach with f1 and g1, all electron with f1,f2, g1 and g2 and by the frozen core model, going from the exterior toward the center of the circle on the horizontal axis.

On Fig. 2, we show the variation of the TDCS in polar representation, where the scattering angle plays the role of the polar angle and the distance of a given point from the origin represents the magnitude of the TDCS. Here an incident electron of 250eV is considered and the angle between the ejection and scattering direction is taken to be equal to 140°. As expected all the curves possess the axis of symmetry at –70°. Here we have done a supplementary calculation in which we have neglected the inner electron exchange terms f2 and g2. We observe that the model potential and the frozen core models give very different magnitudes, and that the absence of f2 and g2 produces also a certain difference in the all electron calculation.

On Fig. 3 we present the variation of the asymmetry factor for a spin polarized target and projectile (e,2e) experiment defined by

$$A = \frac{\sigma^{(3)\uparrow\downarrow} - \sigma^{(3)\uparrow\uparrow}}{\sigma^{(3)\uparrow\downarrow} + \sigma^{(3)\uparrow\uparrow}} \quad (13)$$

in terms of ejection angle in an equal energy sharing regime. In our four electron approach, the triple differential cross section for parallel spin polarization is given by:

$$\sigma^{(3)\uparrow\uparrow} = (2\pi)^4 \frac{k_s k_e}{k_i} \left\{ |f1 + g2 + h1 - f2 - g1 - h2|^2 \right\} \quad (14)$$

and for anti parallel polarization by

$$\sigma^{(3)\uparrow\downarrow} = (2\pi)^4 \frac{k_s k_e}{2k_i} \left\{ |f_1 + g_2 + h_1 - f_2 - g_1 - h_2|^2 + |f_1 + g_1 + 2h_3 - f_2 - g_2 - h_1 - h_2|^2 \right\} (15)$$

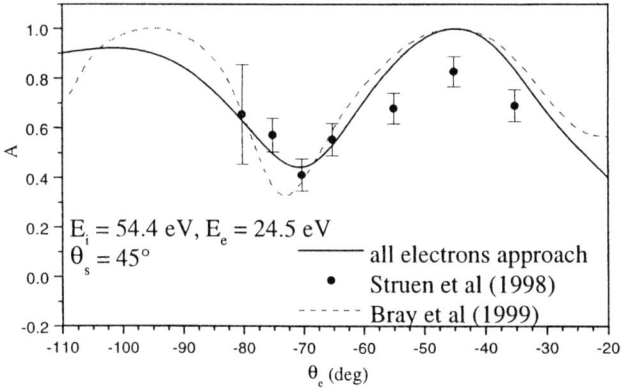

FIGURE 3. The variation of A in terms of the ejection angle in an equal energy sharing situation.

29

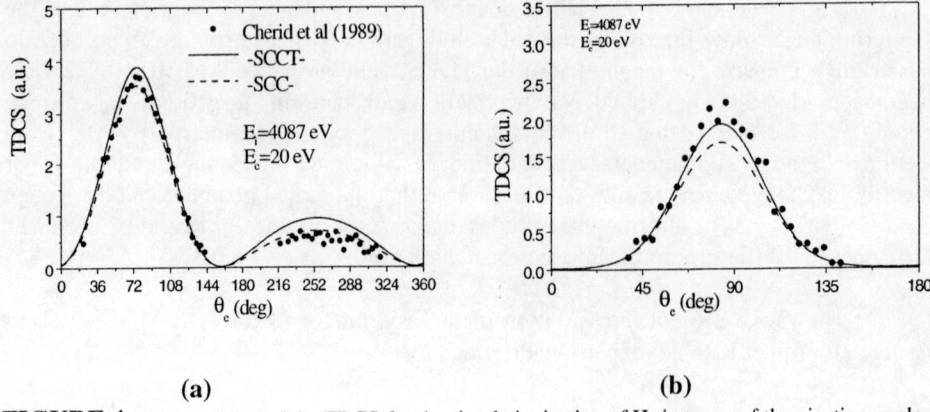

(a)   (b)

FIGURE 4. The variation of the TDCS for the simple ionization of H_2 in terms of the ejection angle. In (a) $\theta_s=1°$ and in (b) $\theta_s=3°$.

The comparison with recent experimental and theoretical results [10] and [11] shows that our approach improves the agreement with the experimental values.

In the diatomic case we have first of all verified our procedure by comparing our results to that of Cherid et al [12]. We applied two variants for the continuum function $\chi_e(\vec{k}_e,\vec{r})$ in Equation (10). In the first, which we will call separated centre continuum (SCC), we consider that the wave is centred on the nuclei (a or b). In the second, we use the same function but we introduce a phase, which corresponds to a translation of the plane wave part of the wave function to the centre of the molecule and we designate it by (SCCT). On Figures 4a and b we show that our procedure gives an excellent agreement with that of the experimental values.

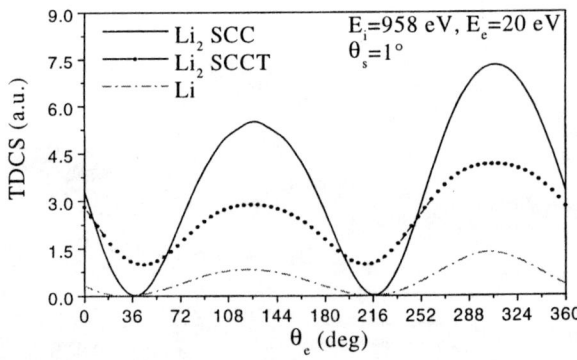

FIGURE 5. The variation of the TDCS of the simple ionization of Li and Li_2 in terms of the ejection angle.

30

Once our procedure was verified, we have performed some preliminary calculations on Li_2 to find the optimal conditions of this process. On Fig. 5 we present a typical situation for a small scattering angle and compare it to that of atomic lithium. We observe, as expected, that the TDCS of the diatomic target is much larger than that of the atomic one. This has also been observed between the TDCS of the (e,2e) ionization of diatomic and atomic hydrogen and can be explained by the presence of two valence electrons in the diatomic case and their larger dispersion in space compared to atomic targets. We observe also that in contrast to the H_2 case the SCCT results are quite different from that of SCC approach. This is due to the large inter nuclear distance of the Li_2 system (~5a.u.)

ACKNOWLEDGMENTS

We want to thank the CINES (Centre Informatique National de l'Enseignement Supérieur) for the attribution of computational facilities

REFERENCES

1. Dal Cappello C., Tavard C., Lahmam-Bennani A. and Dal Cappello M.C. *J. Phys. B: At. Mol. Opt. Phys* **17**, 4557-4564 (1984)
2. Zhang X., Whelan C.T. and Walters H.R.J. . *J. Phys. B: At. Mol. Opt. Phys* **25**, L457-L462 (1992)
3. Hafid H. and Joulakian B., Z. Phys D, **31**, 49-52 (1994)
4. Whelan C.T., Allan R.J., Walters H.R. and Zhang X., "(e,2e) effective charges, distorted waves and all that" in *e,2ᵉ and Related Processes*, edited by Whelan C.T., Walters H. R., Lahmam-Bennani A. and Ehrhardt H. Academic Publishers Kluwer 1993 pp. 1-28
5. Schulz M., *J. Phys. B: At. Mol. Phys* **6** ,2580-2599 (1973)
6. Jones S. and Madison D. H., *J. Phys. B: At. Mol. Phys* **27**, 14231428 (1973)
7. Boudali F., Najjari B. and Joulakian B. *J. Phys. B: At. Mol. Opt. Phys* **33**, 2383-2393 (2000)
8. Weck P., Joulakian B., Hanssen J., Fojon O. A. and Rivarola R. D., *Phys. Rev. A* **62**, 0114701-1-3 (2000)
9. Harris F.E., *J. Chem. Phys.* **32**, 3-18 (1960)
10. Streun M., Baum G., Blask W., Rasch J., Bray I., Fursa D. V., Jones S., Madison D. H., Walters H. R. J. and Whelan C. T. *J. Phys. B: At. Mol. Opt. Phys.* **31** 4401-4411 (1998)
11. Bray I., Beck J. and Plottke C. , *J. Phys. B: At. Mol. Opt. Phys* **32**, 4309-4320 (1999)
12. Chérid M., Lahmam-Bennani A., Zurales R.W., Lucchese R.R., Duguet A., Dal Cappello M.C. and Dal Cappello C., *J. Phys. B: At. Mol. Opt. Phys* **22**, 3483-3499 (1989)

(e,2e) collisions with polarized electrons and excited, oriented and spin polarized targets

E. Weigold*, J. Lower*, J. Berakdar† and S. Mazevet**

*Atomic and Molecular Physics Laboratories, Research School of Physical Sciences and
Engineering, Australian National University, Canberra ACT 0200, Australia
†Max-Planck Institute for Microstructure Physics, Weinberg 2, 06120 Halle, Germany
**T-division, Los Alamos National Laboratory, Los Alamos, NM87545, USA.

Abstract. An experiment has been performed in which a polarized electron beam ionizes an orbitally oriented and/or spin polarized valence electron of sodium. The cross section for this reaction is measured for well-resolved vector momenta of the two electrons in the final channel. A tensorial re-coupling scheme has been developed in which the measured quantities are expressed in terms of independent, irreducible spherical tensor components. For a comparison with experiment we performed numerical ionization-cross section calculations within the Distorted Wave Born Approximation (DWBA) and the Dynamically Screened Three Coulomb Waves (DS3C) theory.

INTRODUCTION

Experimental results are presented which probe the spin, the orbital and the charge dependence of electron-atom ionizing collisions. This is achieved by performing ionization coincidence measurements on laser excited sodium atoms in which both the spin and the orbital projection quantum states of the electron-atom system are determined prior to the collision. To disentangle the spin-dependent from the orbital orientation effect a tensorial re-coupling scheme has been developed in which the cross sections are expressed in terms of independent, irreducible spherical tensor components. For comparison with experiment, numerical values for the tensorial components are calculated using the the Distorted Wave Born Approximation (DWBA) [1, 2] and the Dynamically Screened Three Coulomb Waves (DS3C) theory [3].

EXPERIMENTAL METHODOLOGY

As a detailed description of the apparatus has appeared in previous publications [4, 5], only a brief description will be given here. The primary polarized electron beam used to induce the ionization process (degree of Polarization $P_e = 24\%$) is generated by photo-emission from a cesium and oxygen coated GaAs crystal under illumination by 810nm circularly polarized laser radiation. Inversion of the electron beam polarization from into, to out of the scattering plane (defined by the axes of the sodium and primary electron beams) is achieved by reversing the helicity of the diode laser radiation field through rotation of a quarter wave plate.

CP604, *Correlations, Polarization, and Ionization in Atomic Systems*
edited by D. H. Madison and M. Schulz
© 2002 American Institute of Physics 0-7354-0048-2/02/$19.00

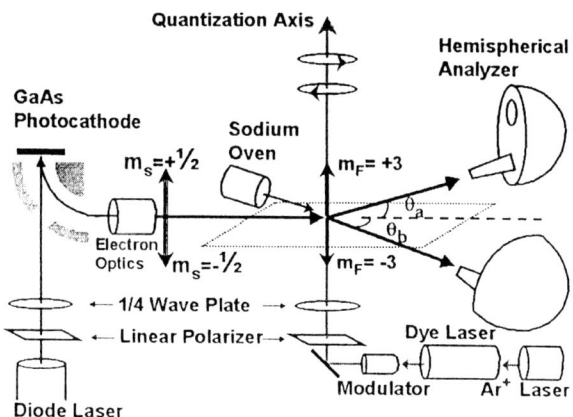

FIGURE 1. Schematic representation of the (e,2e) experimental apparatus. See text for details.

A frequency modulated 589nm circularly polarized laser beam is used to excite, spin polarize and in the case of the excited state atoms, orient the sodium target ensemble through pumping the 3s-3p transition with circularly polarized light. After a few excitation/decay cycles the target atoms gather exclusively in the two state system

$$3s^1 \, {}^2S_{1/2} \, (F = 2, \, m_F = +2 \, (-2))$$
$$\leftrightarrow 3p^1 \, {}^2P_{3/2} \, (F = 3, \, m_F = +3 \, (-3)) \tag{1}$$

for pumping by left-hand σ^+ (right-hand σ^-) circularly polarized radiation.

Scattered electrons emitted in the scattering plane are measured in two electrostatic hemispherical analyzers, incorporating position sensitive detectors, located on opposite sides of the incident beam. Each analyzer is independently rotatable about the dye laser beam axis which defines the quantization axis in the present measurements.

The experiments consisted of measuring the (e,2e) count rates as a function of the emission angle θ_b of one of the two final state electrons, for a fixed emission angle θ_a of the other, for each of the four combinations of atomic and electron beam polarization directions. For excited state ionization, the reactions considered are:

$$e(\uparrow) + Na(m_F = +3), e(\uparrow) + Na(m_F = -3),$$
$$e(\downarrow) + Na(m_F = +3), e(\downarrow) + Na(m_F = -3). \tag{2}$$

where $\uparrow$ ($\downarrow$) represent spin up (spin down) incident electrons.

THEORETICAL FORMULATION

In an earlier publication [6] we showed that the ionization cross section can be written in terms of independent, irreducible tensor components in the following manner:

$$\sigma(\Omega_a,\Omega_b,E_b) = \sum_{K=0}^{2J}\sum_{m_K=-K}^{K}\sum_{k=0}^{2s}\sum_{m_k=-k}^{k} c_k c_K \langle\{\mathbf{J}\}_{Km_K}\rangle\langle\{\mathbf{s}\}_{km_k}\rangle \sum_{SM_S} \Lambda^{Kk}_{m_K m_k}(SM_S)$$

(3)

where $\Lambda^{Kk}_{m_K m_k}(SM_S)$ is a spherical tensor of rank K with spherical components m_K. The constants c_K, c_k are given by the formula

$$c_j = 2^j\sqrt{\frac{(2j+1)!!(2J-j)!}{j!(2J+j+1)!}}.$$

$\langle\{\mathbf{J}\}_{Km_K}\rangle$ denotes an averaged value of the tensor product of angular momentum operators $\mathbf{J}$. That $\Lambda^{Kk}_{m_K m_k}$ is a spherical tensor has immediate consequences as far as the rotational transformation properties are concerned. Tensors with rank $K=0$ and/or $k=0$ are scalar with respect to rotations generated by $\mathbf{J}$ and/or $\mathbf{s}$. The tensors with rank $K=odd$ ($k=odd$) are orientation parameters whereas for $K=even$ ($k=even$) the tensors can be regarded as alignment tensors.

The relation (3) is valid for an arbitrary mutual angle between the natural quantization axes of the incoming electron beam and the polarized atomic target. If the polarized electron beam and the polarized target have a common quantization axis (as is the case in the present experiment) the density matrices become diagonal and Eqs.(3) reduces to

$$\sigma(\Omega_a,\Omega_b,E_b) = \quad \Lambda^{0,0}_{0,0}\left[P_{00}p_{00} + P_{00}p_{10}\frac{\Lambda^{0,1}_{0,0}}{\Lambda^{0,0}_{0,0}}\right.$$

$$+ \sum_{K=1}^{2J}\left(P_{(K=odd)0}p_{00}\frac{\Lambda^{(K=odd),0}_{0,0}}{\Lambda^{0,0}_{0,0}} + P_{(K=odd)0}p_{10}\frac{\Lambda^{K=odd,1}_{0,0}}{\Lambda^{0,0}_{0,0}}\right)$$

$$\left.+ \sum_{K=2}^{2J-1}\left(P_{(K=even)0}p_{00}\frac{\Lambda^{(K=even),0}_{0,0}}{\Lambda^{0,0}_{0,0}} + P_{(K=even)0}p_{10}\frac{\Lambda^{(K=even),1}_{0,0}}{\Lambda^{0,0}_{0,0}}\right)\right].$$

(4)

For the experimental arrangement shown in Fig.(1) the ionization cross section for the orbital $m_L=0$ is zero [7]. Therefore (and due to the neglect of any spin-flip reactions) only four (out of eight) parameters in Eqs.(4) are independent. These are Λ^{00}_{00}, Λ^{10}_{00}, Λ^{01}_{00}, Λ^{11}_{00}. To relate the measured count rates with the tensorial parameters we have introduced above, we group them in the following way:

$$\sigma_{av} = \mathcal{K}'\left[N^{\uparrow\Uparrow} + N^{\uparrow\Downarrow} + N^{\downarrow\Uparrow} + N^{\downarrow\Downarrow}\right] = \mathcal{K}'N_\Sigma$$

(5)

$$A_{orb} = \frac{1}{N_\Sigma}\left[N^{\uparrow\Uparrow} + N^{\downarrow\Uparrow} - N^{\uparrow\Downarrow} - N^{\downarrow\Downarrow}\right] \qquad (6)$$

$$A_{mag} = \frac{1}{N_\Sigma P_e}\left[N^{\uparrow\Uparrow} + N^{\uparrow\Downarrow} - N^{\downarrow\Uparrow} - N^{\downarrow\Downarrow}\right] \qquad (7)$$

$$A_{m,o} = \frac{1}{N_\Sigma P_e}\left[N^{\uparrow\Downarrow} + N^{\downarrow\Uparrow} - N^{\uparrow\Uparrow} - N^{\downarrow\Downarrow}\right]. \qquad (8)$$

Here $N^{\uparrow\Uparrow}$ ($N^{\downarrow\Uparrow}$) is used to describe the *measured* count rates for ionization when the target volume is pumped by left hand $\Uparrow$ circularly polarized radiation and ionized by an electron beam whose polarization vector is out of $\uparrow$ (or into $\downarrow$) the scattering plane. In the same manner $N^{\downarrow\Downarrow}$ ($N^{\uparrow\Downarrow}$) represents count rates when the target atoms are pumped by right hand $\Downarrow$ circularly polarized laser light and ionized by an electron beam whose polarization vector is into $\downarrow$ (or out of $\uparrow$) the scattering plane. $\mathcal{K}'$ is a normalization constant arising from the fact that the present measurements are relative and not absolute.

The above quantities are related to the tensorial parameters as follows: $\sigma_{av} = \sqrt{2}\Lambda_{00}^{00}$, $A_{orb} = -\frac{\sqrt{5}}{2}\Lambda_{00}^{10}/\Lambda_{00}^{00}$, $A_{mag} = -\frac{1}{2}\Lambda_{00}^{01}/\Lambda_{00}^{00}$ and $A_{m,o} = \frac{\sqrt{5}}{4}\Lambda_{00}^{11}/\Lambda_{00}^{00}$. The parameter σ_{av} is a scalar which describes the ionization cross section averaged over the projections of the electrons' spins and the sense of orbital rotation and is independent of the helicity of the laser light. The quantity A_{orb}, defined for a beam of *unpolarized* electrons, is proportional to the spin averaged *orbital dichroism*. It results from the dependence of the ionization cross section on the *orientation* of the atomic target ensemble. In contrast the tensorial parameter A_{mag}, hereafter referred to as the *magnetic dichroism*, changes sign when the polarization of the incoming electron beam is inverted but remains invariant under a change of the helicity of the photon (cf.Eq.7). It describes a spin up-down asymmetry for a polarized beam of electrons from an *aligned* ensemble of target atoms. The fourth independent tensorial component $A_{m,o}$ is an exchange induced antiparallel/parallel spin asymmetry and as such changes sign if the helicity of the photon is flipped or if the polarization of the incoming beam is inverted.

EXPERIMENTAL DATA AND COMPARISON WITH THEORY

The experimental results presented here are compared with distorted-wave Born approximation (DWBA) [1, 2] and the dynamically screened three-Coulomb wave model (DS3C) [3] calculations. Both these approximations reduce the scattering from the Na atom to a three-body problem by considering only the active (valence) electron of the Na atom [4]. The DWBA approach accounts for the short and long range interactions of both of the final-state continuum electrons with the field of the ion [1, 2], however their mutual electron-electron interaction is discarded from the treatment. In contrast, the DS3C method treats the three-body system in the final state as the sum of three decoupled two-body subsystems (the electron-electron, the electron-Na^+ and the electron-Na^+ two-body subsystems). The coupling of these three two-body subsystems is included in the theory via dynamical screening of the interaction strength of each of the three individual two-body subsystems [3].

35

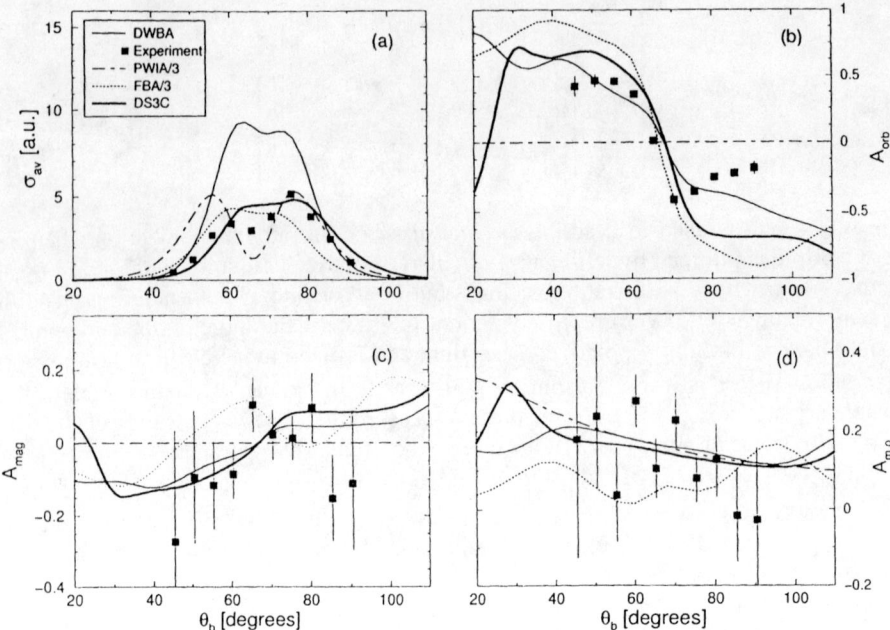

FIGURE 2. Comparison of the measured and calculated cross section parameters σ_{av} (a), A_{orb} (b), A_{mag} (c), and $A_{m,o}$ (d) (see Eqs. (5-8)), for ionization from the oriented and excited 3p state of Na with polarized electrons. The incident beam energy $E_0 = 151eV$ and the mean energy of the detected fast scattered electrons $\bar{E}_a = 127eV$ with corresponding scattering angle $\theta_a = 20^o$. The cross sections are plotted as a function of the slow electron scattering angle θ_b. The experimental average cross section (here seen in (a)) has been normalized to the DS3C theory. Solid and light lines are respectively the DS3C and DWBA calculations. Also shown by short and long dashes respectively are first Born approximation FBA (times 1/3) and plane wave impulse approximation PWIA (times 1/3) calculations.

In Fig.(2) we compare the results of measurement with theory for the quantities (5-8). Four calculations are shown (see figure caption), however only the more sophisticated DS3C and DWBA calculations will be discussed here. The incident beam energy is 151eV and the mean energy $\bar{E}$ of the detected fast electrons is 127eV with corresponding scattering angle $\theta_a = 20°$. In Fig.2(a) the averaged cross section data is presented and the experimental results are normalized to the DS3C theory. Clearly neither the DS3C or DWBA calculation is able to accurately describe the double peak structure. For the parameter A_{orb} in Fig.2(b) the DWBA provides the better description, suggesting that final state electron-electron correlation may not play a significant role under these kinematics.

In Fig. 2(c) the parameter A_{mag} is shown. The physical origin of the structures revealed by A_{mag} are made clearer by expressing it in terms of the direct and exchange amplitudes $A_{mag} \propto \left\{ \Re(f_{m_L=+1} g^*_{m_L=+1}) - \Re(f_{m_L=-1} g^*_{m_L=-1}) \right\} / \sigma_{av}$ where f_{m_L} and g_{m_L} are the state-resolved direct and the exchange amplitudes. This relation makes clear that A_{mag} is in fact an exchange induced quantity and it diminishes if an interference between g_{m_L}

and f_{m_L} is unlikely, e.g. if $|g_{m_l}|/|f_{m_l}| \to 0$. When the direction of the ejected electron coincides with the direction of the momentum transfer (i.e. when $\theta_b \approx 64^o$) the direct scattering amplitude $|f_{m_l}|$ predominates [8] and hence A_{mag} becomes small, and it increases for larger deviations from $\theta_b \approx 64^o$ where exchange scattering can become significant. Fig. 2(d) shows the results for the spin asymmetry $A_{m,o}$. This parameter can as well be written in terms of the direct and exchange amplitudes f_{m_L} and g_{m_L}.
$A_{m,o} \propto \left\{ \Re(f_{m_L=+1} g^*_{m_L=+1}) + \Re(f_{m_L=-1} g^*_{m_L=-1}) \right\} / \sigma_{av}$ (note in the present geometry the scattering from the state $m_L = 0$ does not contribute). In the binary collision regime, which is encompassed by the present kinematics, we can expect that in general $|f|$ will dominate (over $|g|$) so that $A_{m,o}$ is also generally small [8]. Both theories perform satisfactorily in comparison with experiment, although the large error bars preclude more definitive statements being made.

CONCLUSION AND FUTURE PERSPECTIVES

We have carried out (e,2e) cross sections measurements on sodium where the angular momentum projection state of the projectile and target are determined prior to the collision. To provide a general description we have developed a tensorial re-coupling scheme that factorizes the cross sections into components characterized by their spherical transformation properties. For a comparison with the experimental results we performed calculations including the Distorted Wave Born Approximation (DWBA) and the Dynamically Screened Three Coulomb Waves (DS3C) model. The results show that the initial state resolved ionization cross section depends both on the relative spin projections of the incident and bound state electrons and on the orientation of the initial atomic state. Reasonable agreement is found between theory and experiment. The theories can be improved by using improved descriptions of the initial state. Improvement in the experimental apparatus are underway by introducing new-generation toroidal electron analyzers and by employing an electron source of improved degree of polarization.

REFERENCES

1. McCarthy, I., and Weigold, E., *Electron-atom Collisions*, Cambridge University Press, Cambridge, 1995.
2. McCarthy, I., *Aust. J. Phys.*, **48**, 1 (1995).
3. Berakdar, J., *Phys. Rev. A*, **53**, 2314 (1996).
4. Lower, J., Elliott, A., Weigold, E., Mazevet, S., and Berakdar, J., *Phys. Rev. A*, **62**, 12706 (2000).
5. Dorn, A., Elliott, A., Lower, J., Mazevet, S., McEachran, R., McCarthy, I., and Weigold, E., *J. Phys. B: At. Mol. Opt. Phys.*, **31**, 547 (1998).
6. Lower, J., Weigold, E., Berakdar, J., and Mazevet, S., *Phys. Rev. A* (Accepted for Publication, 2001).
7. Berakdar, J., Engelns, A., and Klar, H., *J. Phys. B*, **29**, 1109 (1996).
8. Joachain, C., *At. Mol. Phys.*, **17**, 261 (1986).

Chemical Applications of Binary (e,2e) Spectroscopy

C.E.Brion[*], G.Cooper[*], R.Feng[*], S.Tixier[*], Y.Zheng[*], I.E.McCarthy[#], Z.Shi[+] and S.Wolfe[+]

[*] Department of Chemistry, University of British Columbia, Vancouver, BC V6T1Z1, Canada

[#] Department of Physics, The Flinders University of South Australia, Adelaide, SA 5001, Australia

[+] Department of Chemistry, Simon Fraser University, Burnaby, BC V5A 1S6, Canada

Abstract. Binary (e,2e) spectroscopy (also known as EMS) is applied to problems of chemical interest including imaging of orbital electron densities, electron transfer processes, investigation of different orbital models, the evaluation of wavefunctions and computational methods for larger molecules, frontier orbital electron densities in pharmaceuticals, and distorted wave effects at low momenta for π^* type MOs.

INTRODUCTION

Chemical effects are generally considered to be associated with the detailed behaviour of the valence (frontier) electrons, which are the thermodynamically most accessible and thus the most easily transferred electrons since they have the lowest IPs. Such ideas are embodied in Fukui's Frontier Orbital Theory of chemical reactivity [1]. The remaining, more tightly bound electrons can be thought of as mainly doing "physics". Binary (e,2e) spectroscopy at large momentum transfer, using symmetric non-coplanar geometry, is the basis of Electron Momentum Spectroscopy (EMS). EMS has been demonstrated to be a unique chemical probe [2] in that it can effectively image valence (frontier) orbital electron densities and provide direct experimental information on the nature of electron transfer processes which are important in bonding and reactivity. As such, EMS can provide new insights into understanding and predicting chemical behaviour at the fundamental electronic level.

ELECTRON MOMENTUM SPECTROSCOPY, ELECTRON TRANSFER AND ORBITALS

In the PWIA the EMS cross-section, σ_{EMS}, is proportional to the spherically averaged Dyson orbital momentum space electron density associated with the fast electron impact induced removal of an electron in going from an N electron initial state Ψ_N to the N-1 electron final state Ψ_{N-1}. The Dyson orbital may be calculated using MRSD-CI or by Many Body Green's Function methods. In this context it should be noted that the poles of the N to N-1 Green's function correspond [3] to ionization potentials (IP) while those for N+1 to N processes give electron affinities (EA). The IP

CP604, *Correlations, Polarization, and Ionization in Atomic Systems*
edited by D. H. Madison and M. Schulz
© 2002 American Institute of Physics 0-7354-0048-2/02/$19.00

$$\sigma_{EMS} = C \int d\Omega \left| \langle \mathbf{p}\Psi_{N-1} | \Psi_N \rangle \right|^2 \approx C \int d\Omega \left| \psi_i(\mathbf{p}) \right|^2 \qquad (1)$$

Dyson orbital density   orbital density

and EA (and the corresponding Dyson orbitals) thus relate directly to electron transfer processes and it is of interest that these quantities define the key chemical concept of the Mulliken electronegativity. The Dyson orbital density in equation (1) can be very well approximated by appropriate one electron orbital densities $\left| \psi_i(\mathbf{p}) \right|^2$ where $\psi_i(\mathbf{p})$ is typically either the Hartree-Fock or the Kohn-Sham i^{th} orbital momentum space wavefunction given by the Fourier transform of the familiar position space wavefunction $\psi_i(\mathbf{r})$. The results of a recent experimental and computational investigation [2] of equation (1) for the frontier orbitals of the hydrides HF, H_2O, NH_3, CH_4 and H_2S are summarised in figure 1. It can be seen that the MRSD-CI Dyson orbital densities quantitatively reproduce the respective valence orbital momentum distributions while the HF (independent particle) canonical molecular orbitals (CMOs) and especially the DFT (b3lyp and b3pw91) Kohn-Sham orbitals (KSOs) [4] provide good quantitative modeling of the Dyson orbitals and thus of the EMS measurements. KSOs are orbitals with "attitude" – i.e. they include correlation and are thus similar to natural orbitals. It should also be noted that CMOs and KSOs are intrinsically delocalised (i.e. canonical).

An alternative and very different type of orbital division (model) involves the localised molecular orbitals (LMOs) which were extensively promoted in the form of hybrid type localised Valence Bond (VB) orbitals by Linus Pauling [5,6] as an

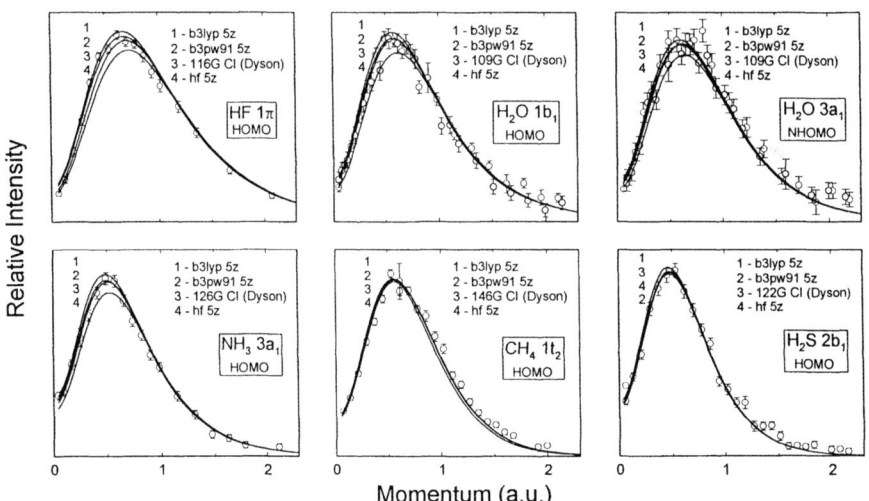

FIGURE 1. EMS cross sections and MRSD-CI Dyson orbital, Hartree-Fock orbital (HF) and DFT (b3lyp and b3pw91) Kohn-Sham orbital, density calculations using very large, diffuse and saturated basis sets.

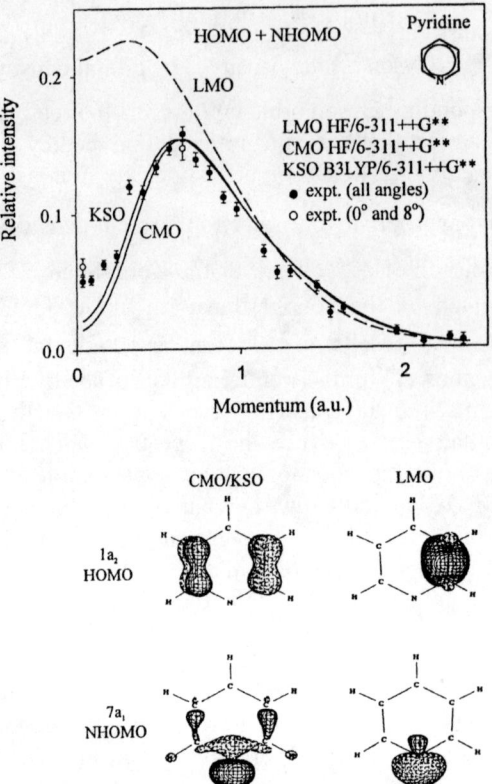

FIGURE 2. EMS cross sections and CMO, KSO and LMO (VB Hybrid) orbital momentum distributions for pyridine. Position space 3D simulations of the CMO/KSO and LMO orbital densities are also shown in the lower part of the figure.

alternative to Mulliken's delocalised CMOs [7]. Pauling's hybrid VB (i.e. LMO) terminology is intuitively simpler and continues to dominate much of the "language" of chemistry used in teaching and research. EMS experiments should be able to determine which, if any, of the Mulliken (CMO), Kohn-Sham (KSO) or Pauling hybrid VB (LMO) orbital models is appropriate for describing valence electron behaviour. For example figure 2 shows EMS measurements and CMO, KSO and LMO (VB hybrid) calculations (6-311++G**) for the frontier electrons of pyridine [8]. Similar measurements and calculations have been carried out for other molecules including CH_3OH [9], C_2H_4 [10], C_2H_2 [10], and CH_4, NH_3 and H_2O [11]. Results for the aromatic molecule pyridine are shown in figure 2. From the results in figure 2 and those for the other molecules [9-11], it seems that, while the intrinsically delocalised KSO and CMO models give very good and quite good quantitative descriptions, respectively, of experiment, the LMO (VB hybrid) localised description is not correct.

The excellence of the fit of the KSO calculations to the EMS experimental data in figures 1 and 2 suggests that the traditional view, that Kohn-Sham orbitals are purely mathematical constructs without physical significance [4], is not the case. The present work also strongly suggests that electron transfer in bonding and reactivity is likely to be well modeled by using delocalised KSOs (or CMOs) rather than by LMOs (i.e. localized VB hybrids). These findings are supported [2] by observations in the rather different situations pertaining to STM experiments on adsorbed C_{60} molecules [12] and Frontier orbital theory studies of reactivity [1]. In both phenomena canonical (i.e. delocalised) orbital treatments with KSOs or CMOs correspond to the experimental observations. The present findings have profound implications for the language of chemistry and for molecular modeling. It is also apparent that EMS should provide useful electronic insights into details of valence (frontier) electron behaviour in molecular systems, electron transfer and chemical behaviour. Some examples are discussed below.

EVALUATION OF BASIS SETS AND QUANTUM MECHANICAL METHODS FOR LARGER MOLECULES

Electron momentum distribution measurements from EMS provide an experimental test-bed for the testing of wavefunctions (basis sets), computational methods and also for the role of electron correlation effects in describing electronic structure [2,9]. A good example is given by recent studies [13] of the methylamines, ammonia and NF_3

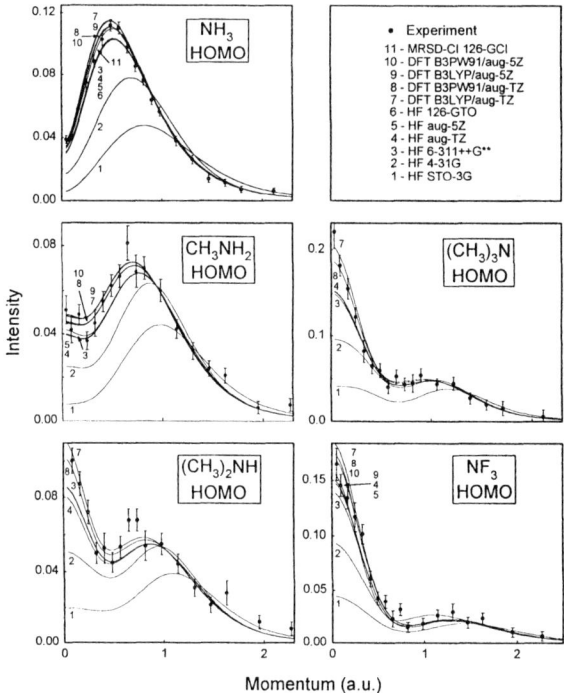

FIGURE 3. EMS measurements, and Hartree-Fock and Density Functional Theory calculations for NH_3, the methylamines and NF_3.

shown in figure 3. When the original EMS measurements [14,15] were made 15 years ago only simple basis set (STO-3G and 4-31G) Hartree-Fock level calculations were feasible for the molecules larger than NH_3 (see curves 1 and 2 on figure 3) with the resulting poor agreement. Large basis set CI calculations were, and still are, only possible with high accuracy for NH_3. It can be seen from figure 3 that good agreement is now obtained for all molecules using large basis set DFT calculations and this shows the need for including electron correlation when describing the chemically sensitive frontier orbitals.

STUDIES OF THE FRONTIER ORBITAL DENSITIES OF PHARMACEUTICALS

Larger organic molecules such as pharmaceuticals pose challenging computational problems for accurate modeling of the reactive frontier electron densities by quantum theoretical methods because of the large number of atoms and electrons involved. In such situations EMS studies can provide an experimental test-bed for the often drastic but necessary approximations (e.g. reduction in basis set size) which must be made. Recently EMS studies [16,17] have been made of the moderately large pharmaceuticals amantadine ($C_{10}H_{17}N$, an antiviral agent) and urotropine ($C_6H_{12}N_4$, an antibacterial agent). Results for urotropine [17] are shown in figure 4. The good agreement with experiment of the DFT (b3lyp and b3pw91) calculations and the large improvement over the Hartree-Fock (hf) treatments show that electron correlation is important in modeling the chemically sensitive low p (larger r) regions of the frontier orbital electron density of urotropine.

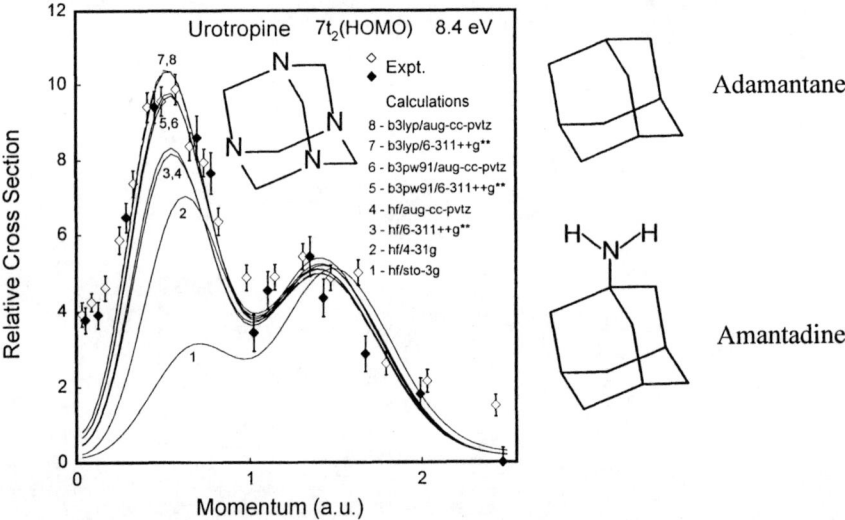

FIGURE 4. EMS cross section and Hartree-Fock (hf) and DFT (b3lyp and b3pw91) electron density calculations for the HOMO of Urotropine.

DISTORTED WAVE EFFECTS AT LOW MOMENTUM IN π* TYPE MOLECULAR ORBITALS

While distorted wave (DW) effects have often been observed at high p (> 2au) and are well understood it is now known that such effects can also dramatically increase EMS cross-sections at low p (< 1 au) in the particular case of atomic d orbitals [18]. From DWIA calculations as a function of energy, PWIA calculations and EMS experiments at 1200eV on Xe 4d, Cd 3d, Zn 3d etc., a rational model has suggested [18] that these low p DW effects arise from the gerade (even) nature of d (ℓ=2) atomic orbitals. Results for Xe 4d are shown in figure 5. Subsequently we have found that similar large departures from high level PWIA theory occur at low p in EMS cross-sections at 1200 eV in the case of certain types of molecular orbitals (they are all π* like in character) at 1200eV. Since π* molecular orbitals are similar to d-type atomic orbitals with regard to symmetry, it is possible that the observed effects at low p are also due to DW effects in molecules. These effects have been noticed for momentum profiles of π* type orbitals in a range of molecules including C_2H_4, CH_3COCH_3, HCHO, CH_3CONH_2, $HCONH_2$, CH_3OH, pyridine, o-xylene, adamantane, amantadine and, to a lesser extent, for the HOMOs of F_2, O_2 and NO. Typical results, for C_2H_4 and $HCONH_2$, including orbital wavefunction plots, are also shown on figure 5. Confirmation that the observed discrepancies are due to DW effects would require a wide range of experimental impact energies (difficult in EMS) or a molecular DWIA theory (not achieved thus far for multi-centre, i.e. molecular targets).

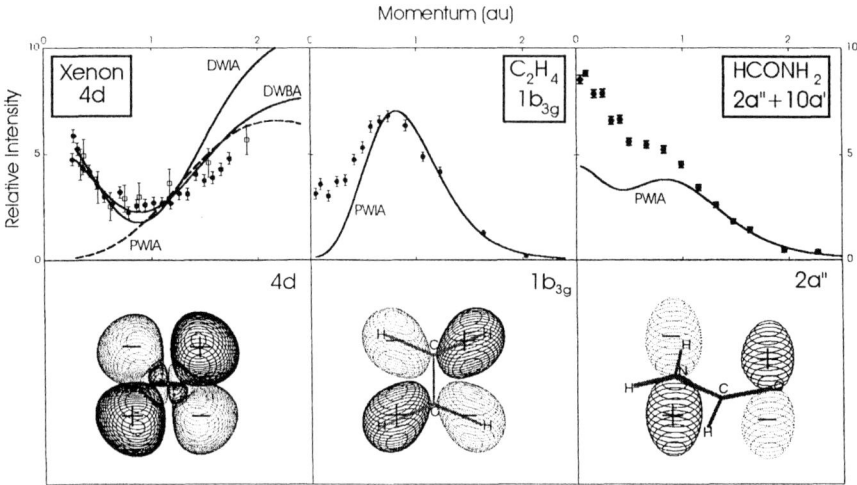

FIGURE 5. EMS cross sections (points), PWIA/DWIA calculations (lines) and orbital wavefunctions for Xe 4d, C_2H_4 1b3g and $HCONH_2$ 2a''.

ACKNOWLEDGEMENTS

Financial support for this work was provided by NSERC (Canada) and by The Peter Wall Institute for Advanced Studies (UBC).

REFERENCES

1. Fukui, K., Yonezawa, T., and Shingu, H., *J. Chem. Phys.* **20**, 722 (1952).
2. Brion, C.E., Cooper, G., Zheng, Y., Litvinyuk, I.V. and McCarthy, I.E., *Chem Phys* **270**, 13 (2001) and references therein.
3. von Niessen, W., Schirmer, J., and Cederbaum, L.S., *Comp. Phys. Reports* **1**, 57 (1984).
4. Kohn, W., Becke, A.D., and Parr, R.G., *J. Phys. Chem.* **100**, 12974 (1996).
5. Pauling, L., in *The Nature of the Chemical Bond,* Cornell Univ., Ithaca, New York,1st Ed. (1939), 2nd Ed. (1940), 3rd Ed. (1960).
6. Pauling, L., *J. Amer. Chem. Soc.* **53**, 1367 (1931); *J. Chem. Ed.* **69**, 519 (1992).
7. Mulliken, R., *Phys. Rev.* **40**, 55 (1932); *Phys. Rev.* **41**, 49 (1932)
8. Tixier, S., Shapley, W., Zheng., Y., Chong, D.P., Brion, C.E., Shi, Z., and Wolfe, S., *Chem. Phys.*, **270**, 263 (2001).
9. Rolke, J., Zheng, Y., Brion, C.E., Shi, Z., Wolfe, S., and Davidson, E.R., *Chem. Phys.*, **244**, 1 (1999).
10. Brion, C.E., Cooper, G., Zheng, Y., Shi, Z., and Wolfe, S., to be published.
11. Wolfe, S., Brion, C. E., Cooper, G., Zheng, Y., and Shi, Z., to be published.
12. Pascual, J.I., Gomez-Herrero, J., Rogero, C., Baro, A.M., Sanchez-Portal, D., Artacho, E., Ordejon, P., and Soler, J.M., *Chem. Phys. Letters* **321**, 78 (2000).
13. Brion, C.E., Young, J.B., Litvinyuk, I.V., and Cooper, G., *Chem.Phys.* **269**, 101 (2001).
14. Bawagan, A.O., and Brion, C.E., *Chem.Phys. Lett.* **137**, 573 (1987).
15. Bawagan, A.O., and Brion, C.E., *Chem. Phys.* **123**, 51 (1988).
16. Litvinyuk, I.V., Zheng, Y., and Brion, C.E., *Chem. Phys.* **261**, 289 (2000).
17. Litvinyuk, I.V., Young, J.B., Zheng, Y., and Brion, C.E., *Chem. Phys.* **263**, 195 (2001).
18. Brion, C.E., Zheng, Y., Rolke, J., Neville, J.J., McCarthy, I.E., and Wang, J., *J Phys B* **31**, L223 (1998).

Vector Correlations in Dissociative Photoionization of Diatomic Molecules in the VUV Energy Range

D. Dowek*, A. Lafosse*, M. Lebech*, J.C. Brenot*, P.M. Guyon*§,
J.C. Houver*, R. Lucchese#, P. Lin#

*Laboratoire des Collisions Atomiques et Moléculaires (UMR 8625)
Bât. 351,Université Paris Sud F-91405 Orsay Cedex FRANCE
§ Laboratoire pour l'Utilisation du Rayonnement Electromagnetique CNRS (UMR 130) Bât. 209D
Université Paris Sud F-91405 Orsay Cedex FRANCE
#Department of Chemistry, Texas A&M University, College Station, Texas 77843-3255, USA

Abstract. The (V_{A+}, V_e, **P**) vector correlation method for the study of dissociative photoionization (DPI) of a diatomic molecule AB, induced by linearly polarized synchrotron radiation (**P**), is presented. On the experimental side, imaging and time resolved coincidence techniques are combined to measure the V_{A+} and V_e velocity vectors for each DPI event. The information derived from the ion-electron kinetic energy correlation, and from the spatial analysis of the vector correlation, is illustrated on specific examples of dissociative photoionization of NO and O_2 following excitation of inner-valence electrons in the 20-25 eV photon energy range. Based on the general form of the molecular frame photoelectron angular distributions (MFPAD) reported here for linear molecules excited by linearly polarized light, we discuss in particular the new information gained from the azimuthal dependence of the $I(\theta,\phi)$ distribution, for each orientation of the molecular axis with respect to the polarization. The measured MFPADs are compared with recent calculations using the multichannel Schwinger configuration interaction method.

INTRODUCTION

The ionization continuum of molecules reached by absorption of a photon in the VUV range is structured by shape resonances and autoionizing states, and the subsequent evolution of the photoexcited molecule AB** can follow different schemes. In most cases ionization is fast with respect to the dissociation: electron emission takes place in the Franck-Condon (FC) region, and subsequently, the formed molecular ion may dissociate. When an autoionizing state is populated, depending on its lifetime, the electron emission may alternatively happen during or after the dissociation into neutral fragments. The study of vector properties in such fragmentation processes is a powerful method for the characterization of the photoejection dynamics [1]. In particular a number of experiments have been designed

CP604, Correlations, Polarization, and Ionization in Atomic Systems
edited by D. H. Madison and M. Schulz
© 2002 American Institute of Physics 0-7354-0048-2/02/$19.00

to probe molecular frame photoelectron angular distributions (MFPAD) which reveal the most detailed dynamical information through the determination of the dipole matrix elements, in moduli and relative phases [2]. For this purpose two different approaches may be considered. If the molecular state is aligned, either by optical pumping with polarized light [3] or by absorption on a surface [4], the MFPADs can be obtained from the photoelectron distributions measured in the laboratory frame. An alternative is to take advantage of dissociative ionization, and determine a posteriori the molecular orientation from the recoil direction of the fragment ion, in the frame of the recoil axial approximation [5]. In particular, the $I(\theta)$ polar dependence of the photoelectron angular distribution for inner-shell [6] and valence-shell [7,8] excitation of specific molecules aligned parallel or perpendicular to the linear polarization of the exciting light has been measured in recent angular resolved photoelectron-photoion coincidence (ARPEPICO) experiments, which most commonly use position sensitive detection. The vector correlation approach [9] presented here belongs to this second approach. We have investigated dissociative photoionization of a series of diatomic molecules (NO, O_2, CO, H_2, N_2) following excitation of inner-valence electrons. The $I(\theta,\phi)$ MFPADs have been determined for any orientation of the molecular axis with respect to the polarization, for a series of DPI processes identified by their reaction pathway. We illustrate the type of information obtained by considering the DPI of NO and O_2, and emphasize the relevance of the ϕ azimuthal dependence. This analysis is based on the general form of the MFPAD which is reported here for linear molecules ionized by linearly polarized light [10]. The measured MFPAD are compared with recent calculations using the multichannel Schwinger configuration interaction (MCSCI) method [11].

EXPERIMENTAL APPROACH

The ($\mathbf{V}_{A+}$, $\mathbf{V}_e$, $\mathbf{P}$) vector correlation method developed for the investigation of dissociative photoionization (DPI) of diatomic molecules induced by linearly polarized VUV synchrotron radiation ($\mathbf{P}$) (Super ACO, LURE, Orsay):

$$AB + h\nu(\mathbf{P}) \rightarrow (AB)^{**} \rightarrow A^+ + B^* + e \qquad (1)$$

consists in measuring for each event the $\mathbf{V}_{A+}$ and $\mathbf{V}_e$ velocity vectors of the nascent ion and electron. The velocity spectrometer described in details previously [12] combines time-of-flight resolved ion-electron coincidence detection and imaging techniques with two delay-line anode position sensitive detectors [13]. For each particle, ion and electron, the time-of-flight (TOF) and the (x,y) positions on the detector are determined, providing the three components of the correlated $\mathbf{V}_{A+}$ and $\mathbf{V}_e$ initial velocities. Related set-ups have been reported e.g. for the study of double ionization of D_2 [14], and time-resolved multiphoton dissociative ionization of NO_2 [15]. The interaction region is defined at the intersection of the molecular beam (x) and the light beam (y) linearly polarized along the x axis, and pulsed with a period of 120 ns in the two bunch mode. Ions and electrons are extracted by a DC electric field orthogonal to the (x, y) plane, whose magnitude (few tens of V/cm) ensures a 4π collection of both

46

particles, then travel through a field free region before reaching the detector. The main instrumental widths which govern the apparatus function are the time width of the light pulse ($\delta t \approx 0.8$ ns) and the width of the molecular beam ($\delta y \approx 4$ mm). After a proper filtering of the ion TOF spectra, the (A^+, e) coincident events are analyzed in two steps: (i) the ion-electron kinetic energy correlation is derived from the magnitudes of the ($\mathbf{V}_{A^+}$, $\mathbf{V}_e$) velocities (ii) the angular distributions are derived from the spatial analysis of the ($\mathbf{V}_{A^+}$, $\mathbf{V}_e$, $\mathbf{P}$) correlation.

ION-ELECTRON KINETIC ENERGY CORRELATION

In this section we discuss the example of dissociative photoionization of O_2 [12] for a photon excitation energy of 23.15 eV. The (E_{0^+}, E_e) kinetic energy correlation diagram (KECD) shown in Fig. 1(a) represents a bidimensional histogram of the (O^+, e) events. In this diagram, a process appears as an accumulation of data points forming a peak or a ridge, characterized by the set of kinetic energies of the ion and electron (E_{0^+}, E_e).

The KECD shows three distinct structures and illustrates the resolving power of the energy correlation. The dominant peak I is identified by the set of kinetic energies at its center (E_{0^+}, E_e) $\approx$ (0.86 eV, 2.6 eV), and corresponds to the formation of the ionic molecular state B $^2\Sigma_g^-$ in the FC region, followed by its predissociation to the limit L_1 :

$$O_2(^3\Sigma_g^-) + h\nu \rightarrow O_2^+(B\ ^2\Sigma_g^-) + e \rightarrow O^+(^4S) + O(^3P) + e \qquad (2)$$

On the other hand, we attribute peaks II (a and b), which remain at fixed $E_e(a) \approx 0.16$ eV and $E_e(b) \approx 0.8$ eV electron energies when the photon energy is scanned, to the excitation of the $O_2^*(3\ ^2\Pi_u, 4s\sigma_g)$ and $O_2^*(3\ ^2\Pi_u, 5s\sigma_g)$ discrete neutral Rydberg states, which dissociate into the $O(^3P) + O^*(^2P, 3d'')$ [8] and $O(^3P) + O^*(^2P, 4d'')$ neutral limits and then encounter atomic autoionization to the ionic limit L_1 [16]. The reaction pathways associated with these three processes are schematized on Fig. 1(b).

When the extraction field magnitude is chosen such that a 4π collection of electrons and ions is achieved for the processes of interest, the integration of the number of events in each structure directly provides the branching ratios for the different processes (here 70%, 10% and 8% for processes I, II(a) and (b) respectively).

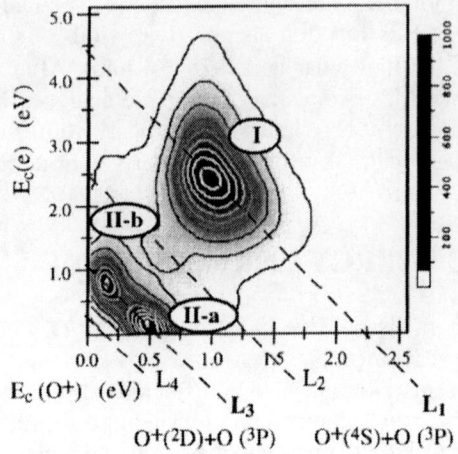
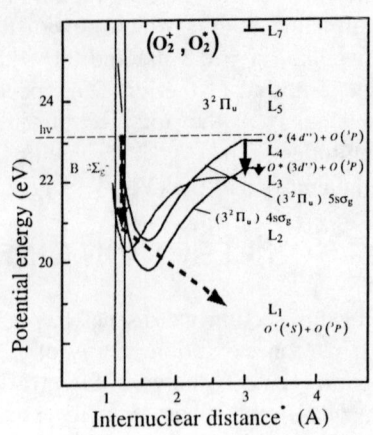

(O_2^+, O_2^{**})

$3^2\Pi_u$

$B\ ^2\Sigma_g^-$

$o^*(4d'') + o(^1P)$
$o^*(3d'') + o(^1P)$
$(3^2\Pi_u)\ 5s\sigma_g$
$(3^2\Pi_u)\ 4s\sigma_g$

$o^*(^1s) + o(^1P)$

Internuclear distance* (A)

$E_c(e)$ (eV)

$E_c(O^+)$ (eV)

$O^+(^2D) + O\ (^3P)$ $O^+(^4S) + O\ (^3P)$

FIGURE 1. (a) Kinetic Energy Correlation Diagram for DPI of the O_2 molecule at $h\nu \approx 23.15$ eV; The dashed straight lines of slope -2 correspond to the accessible dissociation limits L_1, L_2 ...
(b) Scheme of the relevant potential energy curves for O_2 and O_2^+ molecular states: identification of the reaction pathways for processes I and II of Fig(a) using dashed (I) and full (II) arrows.

Finally, the kinetic energy correlation diagram enables a proper selection of each process in order to perform the complete angular analysis, as discussed in the next section.

ANGULAR CORRELATIONS

After selection of a process in the KECD, the spatial analysis of the $(\mathbf{V}_{A+}, \mathbf{V}_e, \mathbf{P})$ vector correlation provides the complete $I(\theta_e, \phi_e, \chi_{A+})$ angular distribution, where χ_{A+} is the polar angle of the molecular axis with respect to the polarization direction, θ_e and ϕ_e are the polar and azimuthal emission angles of the electron in the molecular frame. After integration over the electron emission angles, the complete angular distribution $I(\theta_e, \phi_e, \chi_{A+})$ naturally contains the angular distribution of the ionic fragment with respect to the polarization of the form $(1 + \beta P_2(\cos\chi_{A+})$, where the asymmetry parameter β characterizes the symmetry of the AB** state in the continuum. It includes the $I_{//}(\theta_e)$ and $I_\perp(\theta_e, \phi_e)$ MFPADs for a molecule orientation parallel and perpendicular to the polarization axis respectively, but it contains also the $I_{\chi_{A+}}(\theta_e, \phi_e)$ MFPAD for any orientation of the molecular axis.

The general expression of the MFPAD for linear molecules ionized by linearly polarized light, where the molecular frame is defined with $\mathbf{z}$ oriented along the molecular axis $\mathbf{m}$, $\mathbf{x}$ lies in the plane defined by the molecular $\mathbf{m}$ and polarization $\mathbf{n}$ directions, oriented in the $\mathbf{n} - \mathbf{m}(\mathbf{n}\cdot\mathbf{m})$ direction, takes the remarkably simple form [10]:

$$I\left(\theta_{e}, \phi_{e}, \chi_{A^+}\right) = F_{00}\left(\theta_{e}\right) + F_{20}\left(\theta_{e}\right)P_{2}^{0}\left(\cos\chi_{A^+}\right) + F_{21}\left(\theta_{e}\right)P_{2}^{1}\left(\cos\chi_{A^+}\right)\cos(\phi_{e})$$
$$+F_{22}\left(\theta_{e}\right)P_{2}^{2}\left(\cos\chi_{A^+}\right)\cos(2\phi_{e}) \tag{3}$$

The F_{LN} functions are partial-wave expanded in terms of the dipole matrix elements. One interesting feature of the MFPAD form (3) is the simple dependence on the angle ϕ_{e}. We use this property in the extraction of the four $F_{LN}(\theta_{e})$ functions, which contain the maximum information derived from an experiment with linearly polarized light. The data analysis consists in a Fourier analysis in ϕ_{e} of the $I_{[\chi_{A^+}]}(\theta_{e}, \phi_{e})$ bidimensional histograms for chosen $[\chi_{A^+}]$ ranges of the molecular axis orientations [10]. We note that determination of the F_{21} function, which gives access to the relative phase between the dipole moments for the parallel and the perpendicular transition, requires the measurement of the MFPAD for a molecule which is neither parallel, nor perpendicular to the polarization axis. The possibility of obtaining this phase information was discussed recently for inner shell excitation of CO [17]. The azimuthal dependence of the MFPAD for a molecular oriented perpendicular to the polarization provides direct insight into the symmetry of the initial (neutral) and final (ionic) molecular states involved. For example, if these states both have Σ^+ or Σ^- symmetry the MFPAD reduces to the form $I_{\perp}(\theta_{e}, \phi_{e}) \propto \cos^2(\phi_{e})$, although it reduces to the form $I_{\perp}(\theta_{e}, \phi_{e}) \propto \sin^2(\phi_{e})$ if the initial and final states have a Σ symmetry of opposite reflection symmetry.

Here we illustrate the angular analysis on the example of the dissociative photoionization of NO leading to the c $^3\Pi$ state of NO$^+$ [9]:

$$NO(^2\Pi) + h\nu \rightarrow NO^+(c\ ^3\Pi) + e \rightarrow N^+(^3P) + O(^3P) + e \tag{4}$$

For this process the four $F_{LN}(\theta_{e})$ functions have been measured in the 22-25 eV photon energy range, and calculated using the MCSCI method [11]. The measured β

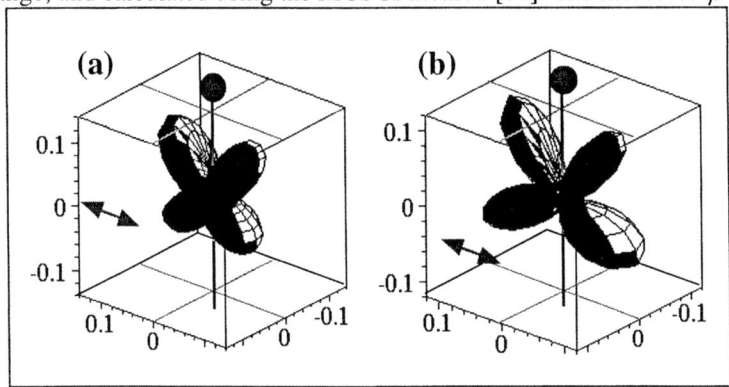

FIGURE 2. Measured (a) and calculated (b) MFPADs for reaction (4), for the geometry where the molecule is oriented perpendicular to the polarization axis.

49

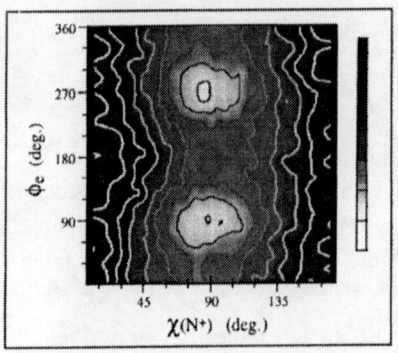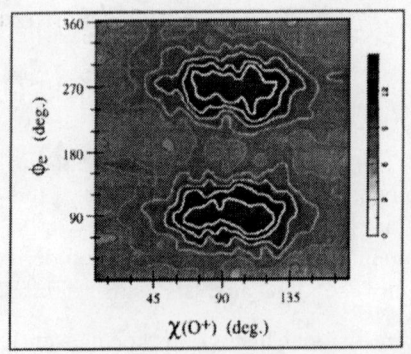

FIGURE 3. $[\chi_{A+}, \phi_e]$ angular correlation diagrams illustrating the different azimuthal dependence of the MFPADs for reaction (4) (left) and for reaction (5) (right) as function of the orientation of the molecular axis with respect to the polarization axis.

asymmetry parameter for this process is $\beta \approx 1$ indicating that reaction (4) corresponds to a dominant parallel transition. Figure 2 displays the experimental and theoretical results in the form of the MFPADs for the geometry where the molecule is oriented perpendicular to the polarization axis. The MFPADs are found to depend strongly on the polarization direction in the molecular frame. For the perpendicular transition shown in Fig.2 the dominant shape of the angular distribution is that of a $d\pi$ wave where only the lobes in the plane defined by the molecular and the polarization axes are populated, whereas the parallel transition shows an angular distribution characteristic of a $d\sigma$ wave. The shapes of the measured MFPADs are qualitatively very well predicted by the calculations for the three significant geometries, parallel, at the magic angle and perpendicular to the polarization. We point out that the shape of the MFPAD for the perpendicular transition is very sensitive to electronic correlation due to the coupling between different channels: the good agreement between the measured and calculated angular distribution was only achieved with a 17 channel calculation. The azimuthal dependence of the MFPAD for the perpendicular transition is close to a $\cos^2(\phi_e)$ distribution, indicating that the $NO(^2\Pi) \rightarrow NO^+(c^3\Pi)$ electronic transition, which corresponds to the creation of a hole in the 4σ orbital, behaves like a transition between Σ states of same reflection symmetry.

Finally we emphasized on Figure 3 the information derived from the azimuthal dependence of the MFPADs, by comparing the measured $[\chi_{A+}, \phi_e]$ angular correlation diagrams for reaction (4) and reaction (5) below:

$$O_2(^3\Sigma_g^-) + h\nu \rightarrow O_2^+(3\ ^2\Pi_u) + e \rightarrow O^+(^2P) + O(^3P) + e \tag{5}$$

Reaction (5), which corresponds now to the creation of a hole in the $1\pi_u$ orbital of $O_2(^3\Sigma_g^-)$, is dominated for $\chi_{A+} \approx 90°$ by a $\sin^2(\phi_e)$ distribution, in contrast with the finding for reaction (4), behaving like a transition between Σ states of opposite reflection symmetry. The theorctical analysis in progress finds very good qualitative agreement with this behavior.

ACKNOWLEDGMENTS

The support of the National Science Foundation (USA), through grant INT-0089831, and of the Centre National de la Recherche Scientifique (France) is gratefully acknowledged. We are grateful to M. Vervloet (LURE, Orsay), L. Spielberger and O. Jagutzki (Frankfurt University), A. Golovin (St Petersburg State University) and G. Raseev (LPPM, Orsay) for their collaboration.

REFERENCES

1. e.g. Greene, C.H., and Zare, R.N., *Annu. Rev. Phys. Chem.* **33**, 119 (1982); Siebbeles, L.D.A., *et al*, *J. Chem. Phys.* **100**, 3610 (1994).
2. Dill, D., *J. Chem. Phys.* **65**, 1130 (1976); Cherepkov, N.A., and Raseev, G., *J. Chem. Phys.* **103**, 8283 (1995).
3. Reid, K.L., Leahy, D.J., and Zare, R.N., *Phys. Rev. Lett.* **68**, 3527 (1992).
4. Kuznetsov, V.V., Cherepkov, N.A., and Raseev, G., *J. Phys.: Condens. Matter* **8**, 10327 (1996).
5. Zare, R.N., *J. Chem. Phys.* **47**, 204 (1967).
6. Shigemaza, E. *et al*, *Phys. Rev. Lett.* **74**, 359 (1995); Ito, K. *et al*, *Phys. Rev. Lett.* **85**, 46 (2000); Heiser, F. *et al*, *Phys. Rev. Lett.* **79**, 2435 (1997).
7. Downie, P. and Powis, I., *Phys. Rev. Lett.* **82**, 2864 (1999); Hikosaka, Y. and Eland, J.H.D., *Phys. Chem. Chem. Phys.* **2**, 4663 (2000).
8. Golovin, A.V. *et al*, *Phys. Rev. Lett.* **79**, 4554 (1997).
9. Lafosse, A. *et al*, *Phys. Rev. Lett.* **84**, 5987 (2000).
10. Lucchese, R. *et al*, submitted to *Phys. Rev. Lett.*
11. Stratmann, R.E., and Lucchese, R.R., *J. Chem. Phys.* **102**, 8493 (1995); Stratmann, R.E., Zurales. R.W., and Lucchese, R.R., *J. Chem. Phys.* **104**, 8989 (1996).
12. Lafosse, A. *et al*, *J. Chem. Phys.* **114**, 6605 (2001).
13. Jagutzki, *et al*, *Imaging Spectrometry IV*, in Proceedings of SPIE, edited by M.R. Descour and S.S. Shen **3438**, 322 (1998).
14. Dörner, R., *et al*, Phys. Rev. Lett. **81**, 5776 (1998).
15. Davies, J.A., *et al*, Phys. Rev. Lett. **84**, 5983 (2000).
16. Wills, A.A., Cafolla, A.A., and Comer, J., *J. Phys. B* **24**, 3989 (1991).
17. Gessner, O., et al, J. Electron. Spectrosc. Rel. Phenom. **101-103**, 113 (1999).

Electron-electron coincidence experiments: from atoms to solids

R. Gotter *, S. Iacobucci ♣, A. Ruocco ♠, G. Stefani ♠

♠ *Dipartimento di Fisica and Unita' INFM Universita' di Roma Tre*
Via della Vasca Navale 84, I-00146 Roma

♣ *Istituto Metodologie Avanzate Inorganiche CNR and Unita' INFM Roma 3*
Area della ricerca di Roma, CP 16, I-00016 Monterotondo

* *Laboratorio Nazionale TASC INFM*
Area Di Ricerca, Basovizza 34012 Trieste

Abstract. In atoms and molecules alike, electron-electron coincidence spectroscopies (e-e) have been extensively applied to study single particle as well as correlated properties of the target. Recent years have witnessed the first successful (e-e) on solids and surfaces.

The experimental effort undertaken in order to apply (e-e) to solids is twofold. On one hand the realization of photon induced resonant double photoionization , i.e. core ionization followed by autionization (Auger) process in which Auger and photoelectron are detected in cincidence (APECS). On the other hand the realization of direct single electron ionization of valence states where both final electrons are detected in coincidence and discriminated in energy and momentum. In the first case, value and limitations of the so called two step mechanism for core ionisation/relaxation processes is under scrutiny. In the second case, the possibility of direct measurement of binding energy and momentum distribution is under test.

Although many aspects of these experiments are common to gas phase and condensed matter, some noticeable differences do exist; most of them amounting to final state interaction with the periodic potential that sustains the electronic states of the solid target.

INTRODUCTION

Break-up ionization reactions have been extensively used to investigate single particle as well as many-body properties of isolated atoms and molecules. Among the various experimental methods adopted, time coincidence experiments have attracted, over the past quarter of a century, a steadily increasing interest because of their capability to approach complete determination of the kinematics of the process. Currently used electron spectroscopies detect and analyze only one of the final state particles (two at least in a break-up reaction) and the measured cross section is then averaged over the dynamical variables of the undetected particles. In this way, the most of the delicate characteristics of the process are overlooked. This shadowing

CP604, *Correlations, Polarization, and Ionization in Atomic Systems*
edited by D. H. Madison and M. Schulz
© 2002 American Institute of Physics 0-7354-0048-2/02/$19.00

effect can be eliminated, at least partially, by measuring multiple differential cross-sections. Till recently, the full (spin excluded) differential cross-section has been investigated only for photodouble ionization and electron impact ionization [1]. In both cases, the experiment consists of collecting time correlated pairs of electrons at selected emission angles and kinetic energies. The probability distribution of the pairs is then measured as a funcion of the energy and/or the momentum balance of the ionizing reaction under study. By selecting valence or core ionization, different classes of experiments are built. They are usually termed (hv,2e) for photoelectron-photoelectron, (APECS) for Auger-photoelectron and (e,2e) for scattered-ejected electrons. In recent years, these experiments have been applied to solids, namely to surfaces and adsorbates and this report will focus on progresses done by APECS and (e,2e) at surfaces, either resolved or integrated in angle. On the one hand, to extend these spectroscopies to solids asks for introducing substantial complexity to the process and it might jeopardize the validity of the experimental methods and theoretical models developed for gases. On the other hand, it allows to gather new, otherwise unreachable, information on the electronic structure and the ionization dynamics of complex matter.

This paper will highlight these differences and similarities by the help of a few selected experiments. Angle integrated APECS will be discussed in the second section while the third will be devoted to the angle resolved one. The forth section will deal with grazing angle (e,2e) experiments and some concluding remarks will be drawn in the last section.

ANGLE INTEGRATED APECS

Although core photoemission (XPS) and Auger spectroscopy have been since long used to investigate surfaces and adsorbates, it was only in the late seventies that Haak et al. [2] succeeded in correlating in time the two final electrons of the L shell ionization in Copper, thus demonstrating the feasibility of such an experiment and highlighting its capability to physically separate overlapping components of an Auger spectrum. The photoemitted electron selects the individual atomic site while the time coincident Auger spectrum detects only electrons generated by the decay of the selected hole state. This method was particularly effective in pin pointing, for instance, Auger decays originating from an L_1 satellite embedded in the L_2VV spectrum. The first theoretical interpretation of those processes was made shortly afterwards by Hono and Wendin [3] and Gunnarson and Schönhammer [4]. These works pointed to the band like valence-electron states as the main difference between solids and atom/molecule. It was also speculated [3] that in solids coincidence experiments bring little extra information when the core hole intrinsic width is small compared to the energy interval over which the local density of states is distributed. In other words, whenever all the excitations created during the core photoemission propagate away before the Auger decay actually takes place, the APECS spectrum doesn't contain more information than the individual XPS and Auger spectra already do. As a consequence, in solids the most evident advantage of APECS over non coincidence spectroscopies is discrimination of overlapping Auger features, that is not a negligible one.

Following those early papers the field has been almost dormant till advent of the high brilliance synchrotron radiation sources. The unique capability of APECS to disentangle overlapping spectra while enhancing surface sensitivity, makes it very attractive for surface analysis [5]. A recent investigation on surface alloys is an excellent example of the capability of APECS at physically discriminating overlapping signals [6]. Surface alloys are metallic systems that intermix only in the first layer. Systems such as Ag/Cu are immiscible in the bulk, so it is unusual that they form an alloy at all, unless the thermodynamics at surface make this possible. Ag/Cu surface alloy has been observed by STM for coverages below 0.1ML. The surface alloying process is expected to result in a systematic shift of the centroid of the impurity d-band. The associated change in the electronic structure will, of course, influences chemical properties, magnetic properties, etc. It would then be relevant to experimentally test it. Unfortunately, the overlap of the Ag 4d band and the Cu 3d band makes it essentially impossible to conclude whether the Ag levels shift or not. By using APECS is possible: (1) eliminate background from the substrate, (2) isolate the Ag signal, (3) have higher surface sensitivity, and (4) measure the intrinsic line shape of the Auger spectrum. Therefore, it was possible to look for shifts in the MVV Auger spectrum of the impurity that would reflect shifts in its d-bands. By APECS the impurity Auger spectra of the Ag/Cu(100) surface alloy has been isolated and the centroid of the impurity d-levels appears shifted away from the Fermi level, as expected, but the amplitude of the effect is much smaller than predicted, a few hundred meV instead of a few eV [6]. It has been possible to accurately measure this shift because of the strong discrimination between silver and copper originated Auger decay channels provided by APECS.

Recently, the question of what can be learned from an APECS experiment was newly addressed by Hono [7] who suggested this as a suitable method to shed more light on the "backbonding " metal-ligand charge transfer structures present in Auger and photoelectron spectra of adsorbed molecules, such as CO on Ni. He also pointed out that APECS might provide valuable insights on initial to final state relaxation time of 3d metal compounds for which it is not known whether the shake-up/off states can relax to the lowest energy state before their decay starts.

Already during early days of APECS Sawatzky [8] suggested that if the core hole intermediate state was short lived, it would have then be possible to perform XPS with energy resolution not any more limited by the photoline natural width. This prediction was confirmed by Jensen et al. [9] with an experiment on Cu (001) that measured the $M_{2,3}$ photolines in coincidence with $M_{2,3}VV$ Auger line. In this experiment, the coincidence photoline was narrower than the non coincidence one and the energy position of the peak maximum displayed a linear dispersion versus the energy selected for the partner coincident Auger electron. This clearly demonstrates that the energy balance of the process is satisfied jointly by the two final electrons rather than individually by each of them within its own step of generation, i.e. photoemission and subsequent Auger decay.

In summary, APECS promise to be an effective method in studying charge transfer and delocalization time in complex matter as it has the ability to separately probe the local electronic structure of individual atoms in a multicomponent system. APECS can then be used to investigate the electronic structure at a particular site in the solid and to monitor how that site changes as the solid is perturbed.

ANGLE RESOLVED APECS

Individual photoelectrons and Auger electrons are emitted with an angle distribution that results from sampling all possible final states that conserve energy and momentum. Therefore, in a conventional non coincidence experiment, the intensity of Auger (photoelectron) current into a given solid angle is the result of a suitable averaging over all energies and angle allowed to the partner photoelectron (Auger).

In a coincidence experiment, only the subset of events for which both electrons (auger and photoelectron) fall within the angle and energy windows accepted by the detectors will contribute to the measured current of probability for detecting time correlated pairs. The ground breaking works performed by Kammerling and Schmidt on noble gases [10] had already shown that the enhanced discrimination capability inherent in these experiments gives a deeper insight into the alignment of the intermediate core hole state and on the matrix elements governing the following non-radiative decay process [11].

To elucidate value and limitation of such a coincidence experiment on solids, let's start discussing its atomic analogous. For linearly polarized light and within the dipole approximation, the electric field vector provides a convenient reference axis in order to describe the ionization pattern. When two electrons are ejected, as it happens in core ionization followed by Auger decay, momenta and spin polarization of the electron pair are connected and the fragmentation picture becomes more complex. Two different approaches can be followed in describing the core excitation-Auger relaxation process. The first one, usually termed *two step*, amounts to factorizing out the Auger event from the photoionization. It treats the core hole state as a well defined real state, as opposite to the second approach, usually termed the *one-step*, in which the core hole is described by virtual intermediate states spanning over the full series of single electron excited states, including the continuum. In this latter case, the Coulomb operator responsible for the Auger decay operates over the full Z particle system and both final electrons are involved and their pair wavefunction has to fulfil the correct boundary conditions. The *one step* will then be suitable of describing inner shell excitation/ionization with non radiative decay as well as post collisional and other final state interactions.

Early AR-APECS experiments on free atoms, satisfactorily interpreted within the two step model and dipole approximation frame, broke the ground towards achievement of the so called *complete* experiment where alignment of the core hole intermediate state and phase and amplitude of the Auger matrix elements are determined [12]. Inadequacy of the two step model has been recently pointed out by studying the resonant double photoionization of Ne via an AR-APECS experiment in which the kinetic energy of the photoelectron matches exactly the energy of the Auger electron [13]. The overall experimental energy resolution, narrower than the natural linewidth of the intermediate state, has allowed to observe the combined effects of post-collision interaction in the final state and interference due to the indistinguishability of the two electrons. The measured pair probability energy distribution, as measured at small mutual angle, exhibit a strong interplay of post-collision interaction and interference effects, in agreement with the theoretical prediction of Sheinerman and Schmidt [14].

In summary, for the *two step* model to be valid, the following assumption must be hold:
1. Finite lifetime of the core hole intermediate state.
2. Well defined parity and angular momentum of the states involved.
3. Absence of overlap with neighboring states.
4. Negligible post collisional and final state interactions.

The additional information provided by APECS is very promising for its application to solids, particularly in understanding several effects connected with the dynamics of core hole excitation and Auger decay in the condensed phase. The relevant parameters of this reaction can be finely tuned in a coincidence experiment. For instance, the core hole intermediate state lifetime changes by choosing deep or shallow core levels; the kinetic energy of the two outgoing electrons is a function of the chosen decay and the photoelectron energy can be tuned inside the same decay; different screening effects are expected when going from metals to semiconductors; in magnetic systems the correlation between the core hole and the valence band is expected to be more relevant in virtue of the linear magnetic dichroism effect present in the core hole photoemission. All these aspects can switch on/off or can tune the radius of the correlation sphere (a pictorial view of the interaction described by the Thomas-Fermi potential) inside which the behavior of the two electrons is expected to be strongly correlated.

The first step towards attainment of AR-APECS in solids was made by a pioneering experiment performed on the Cu(111) surface at ELETTRA with the ALOISA beam line [15]. By this experiment it has been measured the angular distribution of the Cu $L_3 M_{45} M_{45}$ Auger electrons detected in coincidence with Cu $2p_{3/2}$ photoemission line tuned at a diffraction maximum and minimum alternatively [16]. The ALOISA apparatus, with its multicoincidence system, is at present the only one capable of efficiently performing such an experiment fully resolved in angle and a detailed description is given elsewhere [17]. Briefly, the radiation impinges on the sample at grazing incidence and the surface normal lie in the plane determined by the photon beam and its polarization vector (nearly p-polarization). Seven electron analyzers are devoted to simultaneously acquire ten different coincident pairs. Two of them (termed bimodal or scanning) rotate around the photon beam axis and around an axis normal to it. These two analyzers are usually employed to perform a polar scan in the scattering plane defined by the surface normal and the photon beam axis. The other five analyzers (18° apart and termed axial or fixed) are positioned on a plane containing the photon beam axis that can rotate around it. In this way the two bimodal analyzers measure an angular distribution in coincidence with five different values of the momentum wavevector selected by the five axial analyzers. In the following we will refer as *integrated signal* the angular distribution obtained by summing the coincidence events detected by the five axial analyzers together with any one of the bimodal analyzers. We will instead refer as *single pair signal* to the angular

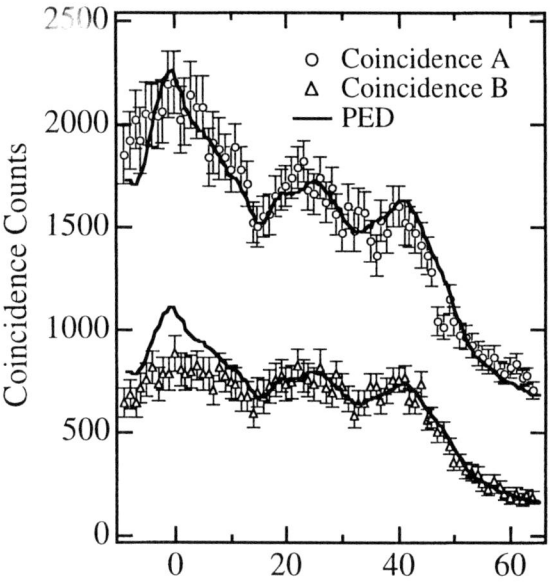

Figure 1. AR-APECS on Cu (111). *Integrated* angular distribution of coincidence M_3 photoelectrons measured correlated with M_3VV Auger electrons emitted in a plane containing the photon beam direction and forming an included angle of 35° (A) and 37° (B) with the surface normal. The monochromatic light beam (250 eV) impinges onto the surface with 3° (A) and 6° (B) grazing angle and nearly P polarization. Zero bimodal angle corresponds to direction perpendicular to the surface. Error bars represent statistical uncertainty on the coincidence events accumulated summing the contribution of the five axial analyzers.

distribution of electron pairs detected by one of the bimodal analyzers in coincidence with only one axial analyzer. It is clear that the former coincidence probability is a sum above the five axial electron momentum values selected by the axial analyzers that, because of its reduced statistical uncertainty, highlights differences between coincidence and non coincidence measurements. The latter type of experiment is more sensitive to correlation effects that show up when comparing coincidence rates relative to different orientations of the axial electron wavevector.

As a first example we consider the AR-APECS performed on the Cu(111) surface, measuring the angular distribution of Cu $3p_{3/2}$ photoelectrons in coincidence with the M_3VV Auger electrons. In fig.1 the *integrated signal* of the photoelectron polar scan performed in two slightly different geometric conditions are displayed and compared with the conventional non coincidence photoelectron diffraction (PED) spectrum. The kinematics relative to figure 1A and 1B differ only by 3° in the grazing incidence angle of the radiation and 3° in the position of the axial analyzer's plane. The coincidence angular distribution closely follows the behavior of the PED in the case of fig. 1A, while a sizeable decrease of the intensity along the surface normal is displayed in the second spectrum (fig 1B, less grazing geometry). Two reasons may be responsible for this difference.

The observed change in the near normal AR- APECS yield could be a consequence of the sharp change in reflectivity experienced by the 250 eV incident radiation when the grazing angle changes from 6° to 3°. The enhanced surface sensitivity of AR-APECS with respect to PED is the most probable explanation for the findings reported in fig. 1.

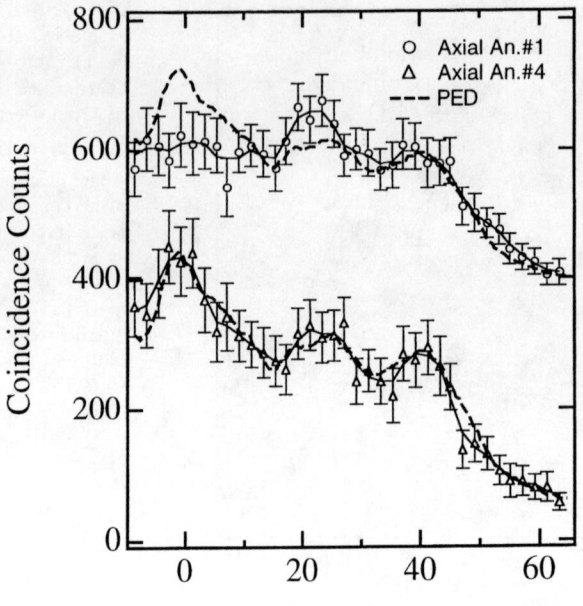

Figure 2 AR-APECS on Cu (111). *Single pair* signal angular distribution of coincidence M_3 photoelectrons measured correlated with M_3VV Auger electrons detected by the axial analyzers located in a plane containing the photon beam direction and forming an angle of 35° with the surface normal. Analyzer #1 and #4 are at an angle of 54° and 126°, respectively, from the propagation direction of the light beam. The monochromatic light beam (250 eV) impinges onto the surface with a 3° grazing angle and nearly P polarization. Zero bimodal angle corresponds to the surface normal. Error bars are statistical uncertainties on the coincidence rate, the line is a best fit to the PED.

In figure 2 the *single pair* angular distribution, corresponding to the integrated signal reported in fig. 1A, obtained in the more grazing incidence case and in coincidence with two different axial analyzers (#1 and #4 that are 54° apart) are compared. In the upper curve a decrease of the intensity along the surface normal and an increase at 22° from the surface normal is evident, while the AR-APECS reported in the lower curve agrees fairly well with the non coincidence PED angular distribution.

This difference cannot be explained in terms of different surface sensitivity (is almost the same in the two cases) so a correlation effect between the Auger electron and the parent photoelectron detected in coincidence can be hypothesized.

A second example of correlated behavior of Auger and photoelectron is given by the angular distribution of $L_3M_{45}M_{45}$ Auger electrons in coincidence with $2p_{3/2}$ photoelectron measured on the Ge(100) surface with a 1450 eV photon beam [18]. In this case, a comparison between the conventional Auger electron diffraction (AED) and the *integrated signal* of the coincident angular distribution displays sizeable differences. Namely, the coincidence angular distribution is less modulated than the AED polar scan. These findings suggest that the coincidence signal is more surface sensitive and thus displays a reduction in the modulation associate to escape emission parallel to atomic rows in the crystal lattice (forward focussing). The *single pair* signals confirm the reduction of the AR-APECS angular distribution modulation amplitude and, furthermore, exhibit remarkable differences in the angle at which the intensity maximum is found for the different Auger-photoelectron pairs. It is interesting to note that the for axial analyzers located at angles that are roughly symmetric with respect to the normal to the surface (which is coincident with the radiation electric field versor if the 3° grazing angle is neglected), the AR-APECS angular distributions are very similar. These latter findings are not merely reducible to enhanced surface sensitivity effects and, to be interpreted, they require a realistic model for excitation and relaxation dynamics to be developed. The examples so far discussed deal with quite different cases. In the case of copper we are dealing with core-valence-valence Auger transitions in a metal and the energy width of the intermediate state is large compared to the final 2-hole state. As a consequence, none of the aforementioned conditions for validity of the two step model is satisfied, the measured probability current for coincidence electron pair is, to a large extent, merely the product of the photo- and Auger-electron probability current. In other words, strong final state interactions (diffraction from the lattice) overwhelm possible correlation of the final electron pair originated by the intermediate state alignment. In the case of Ge lifetime and delocalization of the states involved should not be such that the two step model is to be ruled out. In this case, the AR-APECS *integrated signal* displays noticeable differences from the non coincidence AED, i.e. the electron pair is ejected in a correlated manner, though not simply reducible to a two independent step model and, once more, diffraction of the electron pair form the crystal lattice plays a relevant role. Similar investigations on Ag, Fe/Ni, and Cu LMM have shown that the simple two step model is unable to account for experimental findings whose only common denominator is final state diffraction from the crystal lattice.

GRAZING ANGLE e,2e

A clear evidence for relevance of diffraction to the angular distribution of electron pairs ejected from a surface is given by (e,2e) at grazing angle. In these experiments a monochromatic electron beam is brought to collide at grazing angle with a solid surface and the final, unbound, electron resulting from a single ionization process is detected coincident in time and discriminated in energy and momentum. Similarly to what has happened for atoms and molecules, these experiments can be used to study the dynamics of the electron interaction with and within a solid. We shall concentrate on two examples of (e,2e) experiments in reflection geometry. In these break-up experiments the kinematics of the ionizing event is fully determined and two main issues are and have always been at stake :

a) which is the dominant mechanism that leads to ejection of electron pairs from a solid surface

b) which information, not already available from currently used spectroscopies, is yielded by this coincidence spectroscopy.

Depending on the kinematics chosen, the correlated behaviour of the electron pairs might prevail on the independent particle one.

The experimental aim is to determine the kinematics of the electron collision as fully as possible and the goal is achieved by detecting coincident in time the two final electrons. The reaction is initiated by a monochromatic electron impinging onto the surface, momenta and energies of the two final electrons are measured as well. The first experiment of this kind was performed few years ago [19]. A more recent study [20] shows that, at least at low energy, the pair of final electrons, ejected from the surface in a reflection (e,2e) event, can be described by a quasi-particle with an internal degree of freedom. The energy distribution of the pair reveals structures that might be interpreted as resulting from diffraction of the quasi-particle from the crystal lattice. This gives the unique chance to study the dynamics of direct electron collisions a few electron volts above the vacuum level.

For energies way above the vacuum level, the (e,2e) event is dominated by direct impact processes and provide a "portrait" of the electrons moving in the initial bound state that can not be given by any other kind of electron spectroscopy. Based on the experience gained in atomic physics, it is now well established that (e,2e) has the unique capability of measuring the momentum distribution of the initial state directly.

The potential capability of high energy (e,2e) to measure surface momentum densities was recognised as long ago as the late 1960's, but it was not until recently that the Rome's group has succeeded in performing the first momentum density determination in grazing angle reflection geometry [21]. The experiment, performed on a sample of pyrolitic graphite (HOPG) using 300 eV energy incident electrons and a grazing angle of 7°, established the feasibility of a spectroscopy that could shed light on binding energy and momentum density of solid surfaces. In order to measure the momentum density the collisional model must be reliable.

The model adopted and successfully applied till now for the grazing angle reflection geometry, is based on the First Born approximation. The projectile is treated as a plane wave whereas the target electron initial and final states are described by one-electron wave functions in the momentum space representation. Similarly to the three step model of bulk photoionisation in solids, the ejected electron wave function within the solid matches the energy and the parallel component of the momentum of the plane wave in the vacuum

Various dynamical models have been suggested to describe the grazing reflection (e,2e) experiment, among them those in which the projectile reflects specularly from the target and then scatters inelastically from the bound electron has been demonstrated to be particularly accurate [22]. This assumption is corroborate by recent experimental evidences obtained by measuring excitation spectra of HOPG by electron impact at intermediate energies and in specular reflection geometry [23]. This work shows that the dominant collision mechanism consist of two independent interactions: an elastic one with the crystal lattice that is followed or preceded by an inelastic one with the valence electrons.

Results of the (e,2e) experiments on HOPG are shown in figure 3 where the energy separation spectrum for the valence band (difference between initial and final state energy), as interpreted within the aforementioned double scattering collision model, is shown in conjunction with a calculation performed in the plane wave first Born approximation and within the LMTO approximation [21]. In this approximation, after a nearly vertical transition in the target momentum space, the kinematically determined

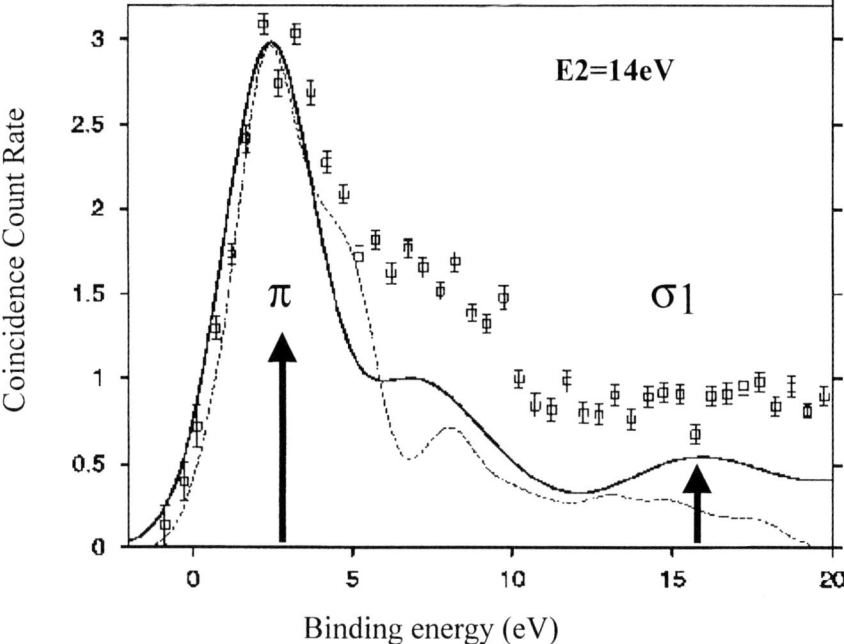

Figure 3 Measured and calculated (e,2e) energy separation spectrum on HOPG. The full line shows the experiment (error bars) deconvoluted with respect to the multiple scattering. The calculated energy separation spectrum shown by the dashed line takes into account $G_{//}$ assisted transitions (see text for details) [22]. Arrows represent ionizing vertical transitions originated in the 1st Brillouin zone.

vacuum state of the ejected electron is projected onto a well defined region in the valence electron momentum space. However, this direct mapping of the valence band is modified by a surface reciprocal lattice vector ($G_{//}$) exchanged while the ejected electron propagates through the crystal interface.

In the case of the experiment on highly oriented pirolitic graphite (HOPG) reported in figure 3, all the vertical transitions kinematically allowed originate in the firs Brillouin zone where only the σ_1 and π bands have significant presence. Because of the $G_{//}$ exchange, contributes from the n-th Brillouin regions, i.e. from the σ_1 and σ_2 orbitals, must be expected in the measured energy separation spectrum. The spectra shown in figure 3 do confirm presence of contributions coming from non vertical transitions that originate from final state interactions with the crystal lattice and that show up as continuous contribution to the coincidence count rate in the binding energy region between the two vertical transitions. Only these $G_{//}$ assisted transitions explain the signature of all the four occupied HOPG bands in the measured energy separation spectrum.

CONCLUSIONS

The early phase of application of (e,2e) spectroscopies to solids and surfaces has progressed much over the past few years. Two processes have been mainly explored: the electron scattering driven single ionization (e,2e) and the core hole mediated double photoionization APECS.

The capability of (e,2e) to measure with the same experiment binding energy and momentum distribution of the valence bound states in solids is now well established. It is now common wisdom that, to properly interpret these results, it is crucial to account for momentum exchange between the two unbound electrons and the crystal lattice potential, i.e. final state interaction with the target potential.

The investigations so far conducted on angle resolved APECS have shown two main differences with respect to ordinary Auger or XPS. The coincidence experiments undoubtedly display an enhanced surface sensitivity as compared to non coincidence photoemission ones. Furthermore, any Auger (XPS) experiment implies an averaging process over the dynamical parameter of the undetected photoelectron (Auger). This averaging is avoided by AR-APECS, hence fine details of the Auger-photoelectron angular correlation can be highlighted. For isolated atoms, allignement of the core hole state and post collisional interaction (PCI) of the final electrons in the presence of the residual ion are recognized as the ruling processes for the Auger-photoelectron pair angular distribution. For AR-APECS on surfaces a similar model is valid if PCI is replaced by diffraction from the crystal lattice. In this perspective AR-APECS from solid surfaces can be suitable for deepening the knowledge of the core-hole/Auger decay mechanism in the solid matter, by means of the momentum selectivity of the two electrons, and to the fact that is less affected by the diffraction with the crystal lattice. A full understanding of momentum resolved APECS will be also useful in studying the double photo ionization (DPI) where an electron pair is directly created from the valence band, without the mediation of a core-hole decay.

From the experimental point of view, multicoincidence parallel detection apparata are mandatory in order to keep within acceptable limits the execution time of both kind of coincidence experiments.

In conclusion, for both coincidence spectroscopies, (e,2e) and AR-APECS on solids, most of the modeling already developed for atoms and molecules can be retained provided final state interaction of the unbound electrons from the crystal lattice (diffraction) is properly described.

ACKNOWLEDGEMENTS

The authors are grateful to the ALOISA beamline staff members for the valuable support provided during the AR-APECS experiments performed at the ELETTRA synchrotron radiation facility, they are also indebted to INFM for financial support provided through the "Supporto ELETTRA" programme.

BIBLIOGRAPHY

1. Stefani, G., "Recent advances in electron-electron coincidence experiments" *in New Directions in Atomic Physics*, Edited by Whelan et al., Kluwer Academic/Plenum Publishing, New York, 1999
2. Haak, H.W., Sawatsky, G.A., Thomas, T.D., Phys. Rev. Lett.: **41**, 1825 (1978)
3. Ohno, K., Wendin G., J. Phys, At. Mol. Phys.: **B12**, 1305 (1979)
4. Gunnarson, O., Schönhammer, K., Phys. Rev. Lett.: **46**, 859 (1981)
5. See, A.K.; Siu, W.K.; Bartynski, R.A.; Nangia, A.; Weiss, A.H.; Hulbert, S.L.; Wu, X.; Kao, C.C, Surface Science, **383**: L735 (1997)
6. Arena, D.A.; Bartynski, R.A.; Nayak, R.A.; Weiss, A.H.; Hulbert, S.L.,. Phys. Rev. B **63**, 155102/1 (2001)
7. Ohno, K., Journal of El. Spec. and Rel, Phenom., **104**: 109 (1999)
8. Sawatzky, G.A., Auger Photoelectron Coincidence Spectroscopy, in Auger electron spectroscopy, C.L. Briant and R.P. Messmer, Editors. 1988, Academic Press: Boston. p. 167-243.
9. Jensen, E.; Bartynski, R.A.; Hulbert, S.L.; Johnson, E.D.; Garrett, R.,. Physical Review Letters:. **62**: 71 (1989)
10. Kammerling, B. and V. Schmidt, Phys. Rev. Lett.,. **67, 1848** (1991)
11. Schmidt, V., Nuclear Instruments and Methods in Physics Research, B **87**, 241 (1994)
12. Kabachnik, N.M., J.Phys.B: At. Mol. Opt., **32**, 1769 (1999)
13. Rioual, S.; Rouvellou, B.; Avaldi, I.; Battera, G.; Camiloni, R.; Stefani, G.; Turri, G., Phys. Rev. Lett.:. **86**, 1470 (2001)
14. Sheinerman, A., Schmidt, V., J.Phys.B: At. Mol. Opt., **30**, 1677 (1997)].
15. Floreano, L.; Naletto, G.; Cvetko, D.; Gotter, R.; Malvezzi, M.; Marassi, L.; Morgante, A.; Santaniello, A.; Verdini, A.; Tommasini, Rev. Sci. Instr., **70**, 3855 (1999)
16. Gotter, R., Attili, A., Ruocco, A., Arena, D., Bartynski, R.A., Iacobucci, S., Marassi, L., Luches,P., Cvetko, D., Floreano, L., Morgante A, Tommasini, F., Stefani, G., J. Phys. IV, **9**, Pr6-161 (1999)
17. Gotter, R., Ruocco, A., Morgante, A., Cvetko, D., Floreano, L., Tommasini, F., Stefani, G.,Nuclear Instruments and Methods in Physics Research, A, **467-468**, 1468 (2001)
18. Gotter, R., et al., paper in preparation (2001)
19. Kirschner, J.; Artamonov, O.M.; Samarin, S.N. Phys Rev. Lett.**74**, 1462 (1995)
20. Berakdar, J.; Samarin, S.N.; Herrmann, R.; Kirschner, J. Phys Rev. Lett. **81**, 3535 (1998)
21. Rioual, S.; Iacobucci, S.; Neri, D.; Kheifets, A.S.; Stefani, G., Phys. Rev. B **57**, 2545 (1998)
22. A.S.Kheifets, S.Iacobucci, A.Ruocco, R.Camilloni, and G.Stefani, Phys. Rev. B. **57** 7360 (1998)
23. Ruocco, A., Milani, M., Nannarone, S., Stefani, G.; Phys. Rev. B **59**, 13359 (1999)

Manifestations of electronic correlation in finite and extended systems

J. Berakdar*, O. Kidun* and A. Ernst*

*Max-Planck-Institut für Mikrostrukturphysik, Weinberg 2, 06120 Halle, Germany

Abstract. In this work we discuss the influence of electronic correlation in finite and extended systems. In particular we stress the fact that the electronic correlation renormalizes the particle-particle interaction in a characteristic way and interpolate these ideas, well-known for extended systems, to small systems. We also sketch briefly how the thermodynamics and critical phenomena in finite systems may be treated.

INTRODUCTION

Over the past decade there has been an impressive progress in miniaturization techniques that aim ultimately at the fabrication of atomic-size devices whose features are controlled primarily by the quantal behaviour of a finite number of correlated particles [1]. Therefore, it is of interest to develop microscopic theoretical models that connect phenomena akin to few-body quantum systems to those occurring in the thermodynamic limit (large volume V, larger number of particle N and finite particle density $n = N/V$). In this work we explore differences and common features in the behaviour of small and extended systems. Two aspects are emphasized: the cooperative response of a system to an external perturbation and the treatment of critical phenomena in finite systems.

COLLECTIVE RESPONSE AND SHORT-RANGE DYNAMICS

The primary source of knowledge on a given system is provided by its characteristic response to external perturbations. In many cases this response is dependent on the collective behaviour of the constituents of the systems, as in the Faraday effect where the delocalized electrons in a metallic surface re-arrange among them self as to shield an external electric field. These correlated fluctuations of the density are determined by the so-called polarization operator $\Pi(\mathbf{q}, \omega)$ which depends on the momentum $\mathbf{q}$ and the frequency ω. On the other hand the polarization of the medium modifies the properties of the particle-particle interaction $U(\mathbf{q}, \omega)$. The modified potential U_{eff} is related to U and $\Pi(\mathbf{q}, \omega)$ through the integral equation [2, 3]

$$U_{eff} = U + U\Pi U_{eff}. \tag{1}$$

CP604, *Correlations, Polarization, and Ionization in Atomic Systems*
edited by D. H. Madison and M. Schulz
© 2002 American Institute of Physics 0-7354-0048-2/02/$19.00

This relation can be formally written as

$$U_{eff} = \frac{U}{1 - U\Pi}. \qquad (2)$$

The *screening* term $\kappa(\mathbf{q}, \omega) := 1/(1 - U\Pi)$ is usually called the generalized dielectric function [3] and plays a central role in a variety of phenomena. E.g. the electrical conductivity $\sigma(\mathbf{q}, \omega)$ of a plasma is obtained from $\kappa(\mathbf{q}, \omega)$ as $\sigma(\mathbf{q}, \omega) = i\omega(1 - \kappa)$. >From Eq.(2) it is clear that the determination of the renormalized interaction U_{eff} and of the dielectric function κ requires the knowledge of the polarization function Π. In essence Π is a two-point Green function that describes the particle-hole excitations. Its lowest order approximation Π_0 is provided by the random phase approximation (RPA) as

$$i\Pi_0(\mathbf{q}, \omega) = \frac{2}{(2\pi)^4} \int d\mathbf{p}\, d\xi\, G_0(\mathbf{q} + \mathbf{p}, \omega + \xi) G_0(\mathbf{p}, \xi). \qquad (3)$$

Here G_0 is the free, single particle Green function. The evaluation of the integrals (3) can be performed analytically for a homogeneous system [3]. In this work we concentrate on the long wave-length limit (long-range screening) in which case one obtains $\Pi_0 \approx -2N(\mu)$ where $N(\mu)$ is the density of states at the Fermi level μ. If we are dealing with an electronic system, like a metallic cluster, the naked interaction $U(\mathbf{q})$ is given $U(\mathbf{q}) = 4\pi/q^2$ and according to Eq.(2) the screened interaction reads in the long wave-length limit

$$U_{eff} = \frac{4\pi}{(q^2 + 8\pi N(\mu))}. \qquad (4)$$

The form of this potential in configuration space is obtained via a Fourier transform: $U_{eff} = \frac{e^{-r/\lambda}}{r}$. The advantage of this simplified form of the interaction is that it allows a transparent discussion of the nature of collisions from many-particle systems: In a charged two-particle scattering events with small momentum transfer (far collisions) dominate as deduced from the form factor of the naked potential $U \propto 1/q^2$. In scattering from a polarizable medium these events are cut out due to the finite range of the renormalized scattering potential U_{eff}, i.e. scattering occurs for close collisions where the medium is not able to screen the external field. For a detailed discussion of this point in the case of ionizing electron collisions from C_{60} we refer the interested reader to the work [5] of this volume. Here we would like to emphasize that in scattering processes from many electron systems it is important to account for the cooperative behaviour of the target electrons which results in screening. The reward for resolving the non-trivial task of evaluating the screening effects is that the interactions are then of a short-range and the description of the scattering dynamics can be done using standard methods of scattering theory, such as the first order Born approximation (see the discussion below). Thus, the real obstacle in describing low-energy collisions from many-particle systems is in obtaining an adequate expression for the polarization propagator.

For a uniform dense electron gas one can employ the RPA to obtain useful approximate expression for Π. For a real system such as a metal or semi-conductor surface one should however start with a realistic single particle Green function in Eq.(3) in order to derive the particle-hole excitations. Such a starting Green function can be obtained from

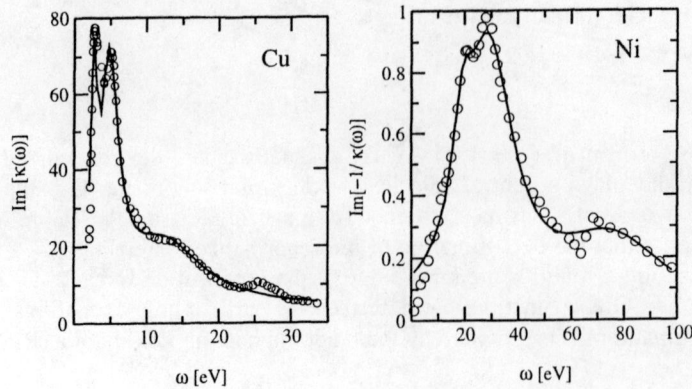

FIGURE 1. The calculated imaginary part of the (bulk) dielectric functions of Cu and the loss function of Ni are compared to the experimental values as deduced from the electron-energy loss spectra [6]. For Ni the experiment is done for a wave vector value of $q = (0.25,0,0)\frac{2\pi}{a}$ where the lattice constant a is $a = 6.65a_0$ (a_0 is the Bohr radius).

density-functional theory within the local-density approximation. The strategy adopted here is the so-called *GW* method [4] which is conceptually well-known for a long time but computationally still poses a challenge, in particular in the case of surfaces. According to this method one derives first the dynamical response χ of the system. From this we deduce the screened interaction U_{eff}, as done above which in turn determine the so-called self-energy Σ and, via the Dyson equation, the interacting single-particle Green function. In matrix form this can be written as

$$\chi(\mathbf{r},\mathbf{r}',\tau) = -iG(\mathbf{r},\mathbf{r}',\tau)G(\mathbf{r}',\mathbf{r},\tau) \tag{5}$$

$$U_{eff}(\mathbf{r},\mathbf{r}',\tau) = U(\mathbf{r},\mathbf{r}') + \int d^3x \int d^3x' U(\mathbf{r},\mathbf{x})\chi(\mathbf{x},\mathbf{x}',\tau)U_{eff}(\mathbf{x}',\mathbf{r}',\tau) \tag{6}$$

$$\Sigma(\mathbf{r},\mathbf{r}',\tau) = iG(\mathbf{r},\mathbf{r}',\tau)U_{eff}(\mathbf{r},\mathbf{r}',\tau) - \delta(\mathbf{r}-\mathbf{r}')V_{xc}(\mathbf{r}) \tag{7}$$

$$G(\mathbf{r},\mathbf{r}') = G_0(\mathbf{r},\mathbf{r}') + \int d^3x \int d^3x' G_0(\mathbf{r},\mathbf{x})\Sigma(\mathbf{x},\mathbf{x}')G(\mathbf{x}',\mathbf{r}'). \tag{8}$$

In these equations we operate in the configuration space and in the time domain. In principle, Eqs.(5-8) has to be solved self consistently starting from the Kohn-Sham Green function in Eq.(5) to arrive at the Green function (8) which is then inserted in Eq.(5). In this procedure, as done in Eq.(7), one should subtract the exchange and correlation potential to arrive at the correct self-energy.

Fig.1 shows the imaginary part of the dielectric function of Cu and the loss function of Ni as obtained from the above *GW* scheme [6]. As indicated by the results of Fig.1 the *GW* approach provides in some cases a reasonable description of the response of a many-body system.

The simple analytical form (3) of the polarization is based on the RPA expression for a uniform medium. The range of validity of such a treatment is estimated from its physical meaning: The interaction creates virtual electron-hole pairs. Within the RPA one considers these events to be incoherent, i.e. the phase of the electron-hole pair is lost right after it's creation and does not affect the creation of the next pair. This assumption is reasonable for systems with a large density of particles, for it is more probable to scatter from different particles consecutively than to undergo multiple scattering from the same scatterer. For a dilute system the RPA is no longer valid and one has to resort to other methods such as the ladder approximation for the polarization propagator [3] in which case the virtual electron-hole pair interacts repeatedly before it disappears. A similar situation of multiple coherent scattering is encountered for systems with few-interacting particles, say three or four electrons. The basic ideas concerning the renormalization of particle-particle interactions are still however valid. This can be seen from the following argument: The fundamental quantity that describes the dynamic of the system is the N-particle Green function $G^{(N)}$. Formally it satisfies the algebraic (Lippmann-Schwinger) relation

$$G^{(N)} = G_0 + G_0 U G^{(N)} = G_0 + G_0 U G_0 + G_0 U G_0 U G_0 + \cdots \qquad (9)$$

where G_0 is a reference (solvable) system and the interaction U is given by $U = G_0^{-1} - (G^{(N)})^{-1}$. Usually, one aims at evaluating a limited number of certain terms of the perturbation series (9) involving the naked interaction U. However, Eq.(9) can also be written formally as

$$G^{(N)} = G_0 + G_0 \frac{U}{1 - G_0 U} G_0 = G_0 + G_0 U_{eff} G_0. \qquad (10)$$

The operator $1/(1 - G_0 U)$ plays the role of the dielectric function in the RPA (cf. Eq.(2)). The essence of its effect is that it renormalizes the interaction U. As evident from Eq.(9) this renormalization procedure amounts to a sub-sum of all terms in the perturbation series up to the first order (Born) term. In practice G_0 is a diagonal many-body matrix, which is appropriately chosen. Analogous to the many-body case where one has to evaluate the polarization propagator, one needs to invert the matrix $1 - G_0 U$. Having done that the particle-particle interactions can be renormalized and a first-order (Born) treatment of the scattering dynamics is then sufficient.

THERMODYNAMICS PROPERTIES AND PHASE TRANSITIONS IN FINITE SYSTEMS

As stated above the properties of a system are encompassed in the Green function. In this section we point out that thermodynamic properties and critical phenomena and the cross over from the thermodynamic limit to confined small systems can also be described with the help Green function techniques. Strictly speaking, finite systems do not expose phase transitions [7]. However, one expect to observe the onset of a critical behaviour when the system approaches the thermodynamic limit. The traditional theory concerned with these questions is finite-size scaling theory [9]. Here we take another route, originally due to Yang and Lee[7] and Grossmann et al. [8] developed to treat critical phenomena in macroscopic systems.

The argument of Yang and Lee[7] is the following: Thermodynamical quantities, such as the specific heat C_V are obtained as a derivative with respect to the inverse temperature

β of the logarithm of the canonical partition function $Z(\beta)$,

$$C_V = \beta^2 \partial_\beta^2 \ln Z(\beta) = f(\beta, Z(\beta))/Z(\beta).$$

Here f is some analytical function and for the Boltzmann constant we assume $k = 1$. Therefore, divergences in the thermodynamic quantities, which signify phase transitions are connected to the zero points of $Z(\beta)$. These zero points are generally complex valued.

Grossman *et al.* [8] applied the concept of Yang and Lee to the canonical ensemble. In this case the inverse temperature $\beta = 1/T$ is continued analytically to $\beta = \Re(\beta) + i\Im(\beta)$. The phase transitions are then the crossings of the zero points line of $Z(\beta)$ with the real β axis. The crucial point is that in the thermodynamic limit $N \to \infty$, $V \to \infty$ and $v = V/N < \infty$ (V is the volume, N is the number of particles) the zero points approach to an infinitesimal small distance the real axis. For this reason, the characteristic phase-transition divergences appear in the thermodynamical quantities. For finite systems $Z(\beta)$ has only finite zero points which can not lie necessarily infinitely close the real axis. Therefore, the thermodynamic quantities show smooth peaks rather then divergences. The positions and widths of these peaks can be obtained from the real and imaginary parts of the zero points laying closest to the real axis [9].

To apply this approach to quantum finite systems, such as a Bose-Einstein condensate we consider a system of N interacting particles. The canonical partition function of a correlated system can be expressed in terms of the many-body Green function as

$$Z(\beta) = \int dE \, \Omega(E) \, e^{-\beta E}. \tag{11}$$

Here $\Omega(E)$ is the density of states which is related to the imaginary part of the trace of $G^{(N)}$ via

$$\Omega(E) = -\frac{1}{\pi} \Im \operatorname{Tr} G^{(N)}(E). \tag{12}$$

Therefore, as in the preceding sections the problem reduces to find appropriate expressions for the many-body Green functions. Recently it has been shown that the N-body Green function satisfies a recursion relation where the strength of interaction is successively reduced [10]:

$$G^{(N)} = \sum_{j=1}^{N} G_j^{(N-1)} - (N-1)G_0. \tag{13}$$

For a brief discussion of the limitations of this relation we refer to Ref.[10]. $G_j^{(N-1)}$ is the Green function of a system in which only $N-1$ particles are interacting while particle j is independent of all other particles. G_0 is a reference Green function of an independent-particle system. With the help of Eq.(13) one can construct flow equations to map the interacting system onto a non-interacting one in $N-1$ steps [11].

>From Eqs.(11,13) we deduce for the partition function the recursion relation

$$Z^{(N)} = \sum_{j=1}^{N} Z_j^{(N-1)} - (N-1)Z_0. \tag{14}$$

Here Z_0 is the partition function of the independent particle system (taken as a reference) while $Z_j^{(N-1)}$ is the canonical partition function of a system in which the interaction strength is diluted by cutting all interaction lines that connect to particle j.

Eq(14) allows to study the thermodynamic properties of finite systems on a microscopic level as well as to investigate the inter-relation between the thermodynamics and the strength of correlations. Critical phenomena can be studied using the idea put forward by Yang and Lee. For example if we are interested in the onset of condensation in a quantum Bose gas one should look at the ground-state occupation number $\eta_0(N, \beta)$. This is given by

$$
\begin{aligned}
\eta_0(N, \beta) &= -\frac{1}{\beta} \frac{\partial_{\varepsilon_0} Z^{(N)}(\beta)}{Z^{(N)}(\beta)} \\
&= -\frac{1}{\beta} \frac{\sum_{j=1}^{N} \partial_{\varepsilon_0} Z_j^{(N-1)} - (N-1)\partial_{\varepsilon_0} Z_0}{Z^{(N)}}.
\end{aligned}
\tag{15}
$$

Here ε_0 is the ground-state energy. By means of this equation one can study systematically the influence of the interaction on the onset of the critical regime or one may chose to find the roots of Eq.(14) in the complex β plane and to identify the zero point that systematically approach the real β axis signifying the transition point in the thermodynamic limit.

REFERENCES

1. Timp, G. L.,*Nanotechnology*, Springer, New York, 1999.
2. Gonis, A., *Theoretical materials science : tracing the electronic origins of materials behavior*, (Warrendale, Pa : Materials Research Society, 2000)
3. Fetter, A. L., Walecka J. D., *Quantum theory of many-particle systems*, (New York, McGraw-Hill, 1971)
4. Fulde, P., *Electron Correlation in Molecules and Solids*, Springer Series in Solid-State Sciences, Vol. 100, (Springer Verlag, Berlin, Heidelberg, New York, 1991); Hedin, L., *J. Phys. C* **11**, R489-R528 (1999).
5. Kidun, O., Berakdar, J., *this volume*.
6. Feldkamp, L. A., Stearns, M. B., and Shinozaki, S. S., *Phys. Rev. B* **20**, 1310 (1979).
7. Yang, C.N., Lee, T.D., *Phys. Rev.* **97**, 404; 1952, *ibid*, **87**, 410 (1952).
8. Grossmann, S., Rosenhauer, W., *Z. Phys.* **207**, 138 (1967); *ibid* **218**, 437 (1969).
9. Barber, M.N., *Phase Transitions and Critical Phenomena*, eds. Domb, C., Lebowity, J.L., pp. 145-266 (1983).
10. Berakdar, J., *Phys. Rev. Lett.* **85**, 4036 (2000).
11. Berakdar, J., *Surf. Rev. and Letters* **7**, 205 (2000).

Electron momentum spectroscopy of metals

M. Vos, A.S. Kheifets and E. Weigold

Atomic and Molecular Physics Laboratories, Research School of Physical Sciences and Engineering, Australian, National University Canberra 0200 Australia

Abstract. Electron momentum spectroscopy measurements of aluminum, copper and gold are presented. Data were taken for incoming electrons with an energy of 50 keV, and both outgoing electrons with an energy near 25 keV. In all cases we used an identical deconvolution procedure, which consistently removes all intensity at high energy loss values. However the intensity becomes vanishing small only for binding energies of about twice the bandwidth. The observed spectra compare well with many body calculations based on the cumulant expansion scheme

INTRODUCTION

Electron momentum spectroscopy (EMS) provides very direct information about the electron wave function in atoms, molecules and solids [1]. Of these materials the solid state is the most challenging for a quantitative understanding. On a one-particle level the discrete electron states of atoms and molecules are replaced by dispersing states, described by the band structure. The electron-electron interaction is modified by the response (screening) of the material. The resulting final states that can be created by the annihilation of a target electron is therefore qualitatively different from those in atoms and molecules. EMS is of course well suited to study these satellites but EMS of materials has the distinct disadvantage (compared to EMS of atoms and molecules) that multiple scattering can not be avoided, even for the thinnest free standing films one can prepare.

To minimize these problems as far as possible a new spectrometer was constructed at the ANU in which the incoming and outgoing electrons have a high energy (50 keV incoming energy, 25 keV outgoing energy of both electrons) [2, 3]. The low level of multiple scattering in this spectrometer, and hence the easy with which the many body effects are revealed was evident for extremely thin carbon films [4]. Here we want to investigate to what extent the same effects can be measured using this spectrometer for heavier elements.

RESULTS

Some results obtained on this spectrometer for various metals are shown in fig. 1. Near the Fermi level the momentum densities of all three metals (polycrystalline films of aluminum, copper and gold) show distinct peaks. The thickness of the films is not exactly known, as the films are sputter-thinning by argon ions, before the measurement. How-

CP604, Correlations, Polarization, and Ionization in Atomic Systems
edited by D. H. Madison and M. Schulz
© 2002 American Institute of Physics 0-7354-0048-2/02/$19.00

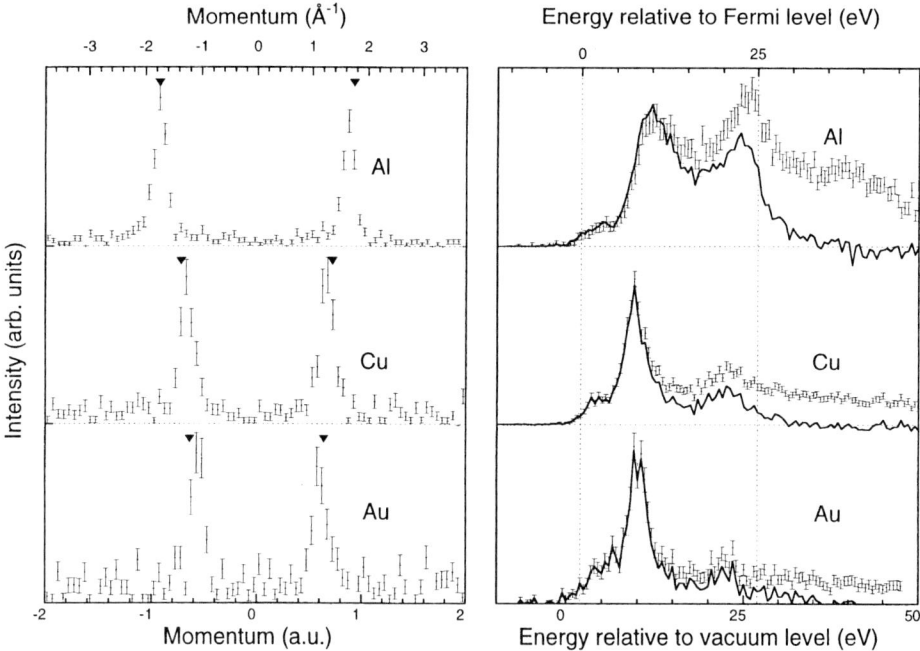

FIGURE 1. Momentum densities at the Fermi level (left) for Al, Cu and Au films. We indicate with triangles the Fermi wave vector as calculated from a free electron model. In the right panel we show energy spectra at zero momentum for these films. The full line is an estimate of the intensity with the effects of inelastic multiple scattering removed,as described in the text

ever we expect the thickness to be in the 50 Å range. Thus with the new spectrometer we can obtain data for the heaviest elements (although with a lower count rate). The quality of the momentum density is especially good for aluminum. It has very small intensity away from the two peaks. For Cu and especially Au, the intensity away from the peaks seems to comprise a larger fraction of the peak intensity. Thus elastic scattering effects are more severe for the heavier elements.

Near the Fermi level the main contribution to the electron density is in all three cases due to *sp*-electrons. These electrons have very extended wave functions and behave like free electrons in all cases. The magnitude of the Fermi vector, can, within the free electron model simply be calculated based on the *sp*-electron density [5] and the appropriate values are indicated in figure 1. The observed peak positions are at slightly smaller momentum values. This is probably due to finite energy resolution: also electrons, slightly away from the Fermi edge, and hence with smaller wave vectors will contribute to the measured intensity.

The densities right at the Fermi level are particular simple to interpret: elastic scattering can affect its shape, inelastic scattering will reduce its intensity, but not affect its shape. In the right panel we display the spectra near zero momentum of these metals.

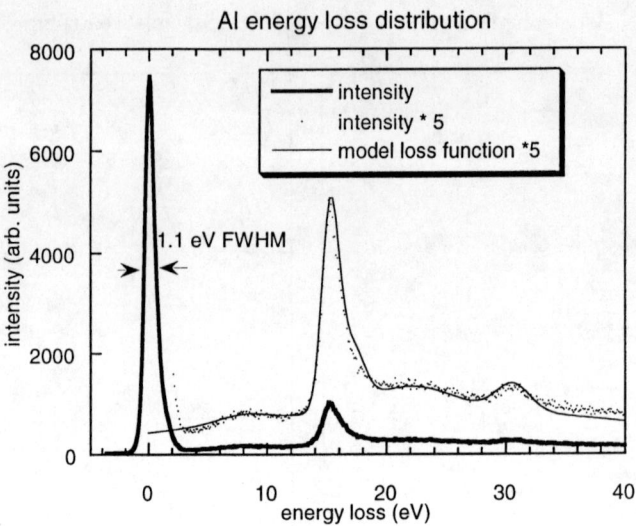

FIGURE 2. The measured energy loss distribution for an aluminum film and the empirical fit with a number of Gaussian line shapes used for deconvolution

These data are affected both by elastic and inelastic multiple scattering. Somewhat surprisingly the single particle peak at the bottom of the band stands out most clearly for the heavier elements:Cu and Au. Thus the quality of the energy spectra appears not to deteriorate with increasing atomic number. In none of the three cases does the intensity drop to zero for high binding energy, although the Au case the intensity is surprisingly small. We want to remove the effect of multiple scattering in all three cases by a deconvolution procedure, that has no free parameters. This procedure is outlined in the next section. An earlier deconvolution approach is described by Jones and Ritter [6]. Another approach is to 'add' the expected effects of multiple scattering to the theory, using Monte Carlo procedures [7]. The Monte Carlo procedure allows for the inclusion of elastic and inelastic effects simultaneously, but its results depend somewhat on the model to derive the elastic and inelastic cross sections, and the model we use for electronic structure calculations. The method described here requires less theoretical input.

The first stage is to determine the loss function. For this we do a non-coincident i.e. a *singles* experiment as illustrated in fig. 2 for aluminum. The energy of the incoming beam is reduced to 25 keV, so the analyzers measure the elastic peak and low loss part of the energy loss spectrum. Ideally we would do this at zero degrees, but in an (e,2e) spectrometer there is usually no analyzer at forward angles. Thus we measure the energy loss distribution at 45°. All electrons detected are thus deflected by a large-angle elastic scattering event. Some of the transmitted electrons are scattered inelastically as well(ie electronic excitations occurred along the incoming and/or outgoing trajectory). From the spectra we determine the ratio of the area of the zero-loss peak (i.e. the electrons that are only scattered elastically) to the area of the inelastic part of the spectrum (i.e. those electrons that suffered inelastic excitations as well). This loss function is characteristic for particles that have traversed a certain thickness of material. As is clear from fig.

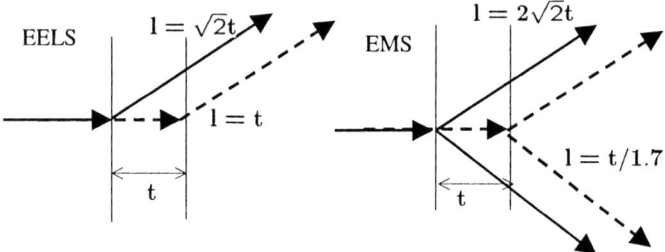

FIGURE 3. Schematic representation of the path length of electrons through a thin film for a singles energy loss experiment and an EMS experiment.

3 the effective thickness l will vary from t to $\sqrt{2}t$, depending on the depth at which the elastic scattering event occurred. On the average the length traversed (effective thickness) by the incoming electrons will be $\simeq 1.2t$. For very thin samples (when the chance of two inelastic scattering events is small) the shape of the energy loss part of the spectrum does not depend on the thickness. Only the ratio between the zero loss peak of the spectrum and the inelastic part will change with thickness. The ratio will be about one, if the effective thickness of the film is equal to one inelastic mean free path of 25 keV electrons. Note that for the Flinders University (e,2e) spectrometer [8], the approximation that only one inelastic scattering event occurs fails for the slow (1.2 keV) electron. Hence the procedure described here, can not be used in that case.

We now assume that this loss function can be used to describe the inelastic scattering of the (e,2e) event. This is no problem for the outgoing particles, as their energy is identical to that of the singles measurement, but somewhat questionable for the incoming beam. The incoming electron has twice the energy, and hence a smaller cross section for electronic excitations. The mean free path between inelastic collisions increases approximately as $E^{0.75}$ for keV electrons [9] i.e. the path length of the incoming particle is reduced by 1.68, as far as the probability of inelastic scattering is concerned. In fig. 3 we show again the extremes in possible path length for an (e,2e) event. It will vary from $t/1.68$ ((e,2e) event occurred at the exit surface), to $2\sqrt{2}t$ (event at the entrance surface). On average the path length is close to $1.7t$.

Thus the film appears thicker in a coincident experiment compared to a singles experiment by a factor $1.7/1.2 \simeq 1.4$. Now we subtract the effect of inelastic multiple scattering from the EMS data in the following way: We start at the Fermi level, here $E = 0$. All events here are *not* contaminated by inelastic multiple scattering. Some intensity at $E + \Delta$ is due to (e,2e) events at the Fermi level in combination with an energy loss event of magnitude Δ. If in the singles spectra, with the zero loss peak normalized to 1, we find an intensity of y an energy loss Δ, then, per unit intensity at E we can expect a background in the (e,2e) spectra, at energy $E + \Delta$ of $1.4y$. This is the amount that is subtracted in the deconvolution procedure. Next we consider the next energy $E + \delta$, slightly away from the Fermi level. It could only be contaminated from (e,2e) events at the Fermi level plus small energy loss. However this has been subtracted in the previous iteration. Hence we can consider the modified intensity as being free from contamination due to inelastic scattering. Now we use this modified intensity to

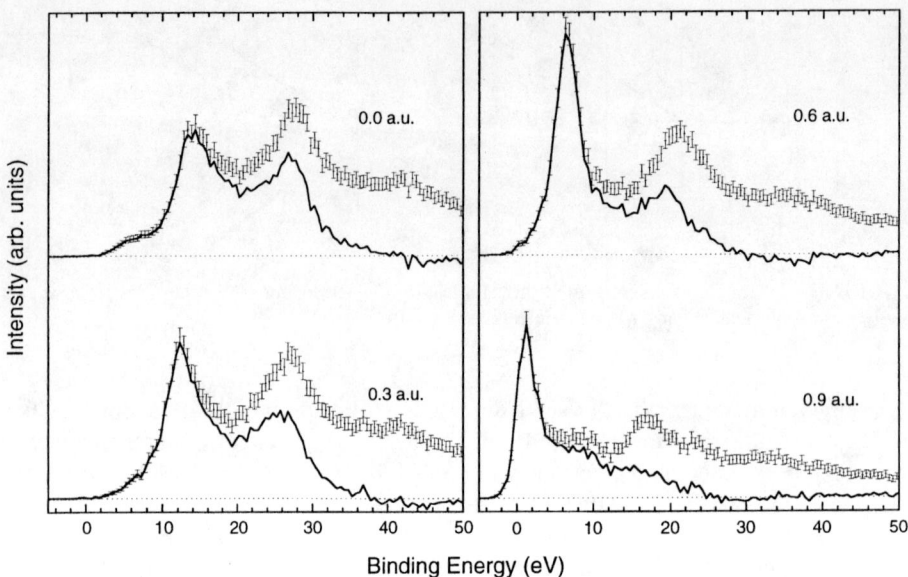

FIGURE 4. Raw and deconvoluted spectra for aluminium films at selected momentum values.

correct the larger binding energy intensity from contamination from (e,2e) events at energy $E + \delta$. This process continues until the high binding energy limit of the spectrum is reached. This deconvolution procedure depends on the measured loss spectrum only, without adjustable parameters. It is expected to work well for films of thickness less or equal to one inelastic mean free path at 25 keV.

The result of this procedure is shown in the spectra of fig. 1. In spite of the large difference in the initial shape the intensity for the three metals, the intensity decreases to zero at high binding energy as required for all three cases. This gives us some confidence that the procedure followed, captures at least the main part of the physics involved. Moreover the procedure does not only produce reasonable spectra at zero momentum but also for other parts of the dispersing structure. This is demonstrated in fig. 4 for the case of aluminum. The quasi-particle peak disperses to the Fermi level, but in no case does the intensity drop straight to zero at higher binding energy. Its width decreases, as lifetime broadening becomes vanishing small near the Fermi edge, and all width is due to the finite experimental resolution.

In fig 5 the theoretical spectra for Al, Cu and Au are displayed, based on the cumulant expansion scheme [10, 11]. The theoretical spectra are convoluted with the experimental resolution of 2 eV. Note that the general trend in these calculations is exactly as in the measured deconvoluted spectra in fig. 1. Relatively large satellites for Al, medium for Cu, and only an high-energy tail for the spectrum of Au.

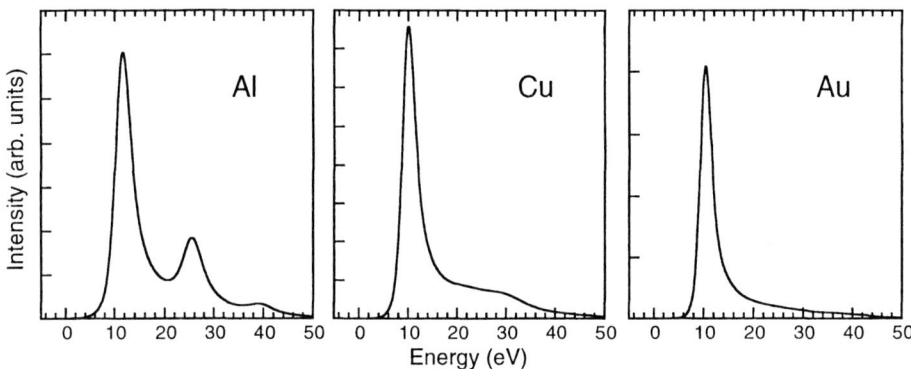

FIGURE 5. Theoretical spectra near zero momentum, based on the cumulant expansion scheme. The theory was convoluted with the estimated energy resolution of 2 eV.

CONCLUSION

we have demonstrated that the new spectrometer has dramatically increased the range of targets that can be studied. Previously the heaviest elements for which valence band spectra could be obtained, with great effort, were Ge [12] and Cu [13]. Here we demonstrate that, with the high-energy spectrometer we can get quite good spectra, even from Au samples. In none of the cases studied, does the measured intensity drop to zero abruptly below the quasi-particle peak. These tails are absent in band structure calculations based on density functional theory, but are obtained in calculations that take many-body effects explicitly into account.

REFERENCES

1. Weigold, E., and McCarthy, I., *Electron Momentum Spectroscopy*, Kluwer Academic/Plenum, New York, 1999.
2. Vos, M., Cornish, G., and Weigold, E., *Rev. Sci. Instrum.*, **71**, 3831 (2000).
3. Vos, M., and Weigold, E., *J. Electron Spectrosc. Relat. Phenom.*, **112**, 93 (2000).
4. Vos, M., Kheifets, A., and Weigold, E., *Phys. Rev. B*, **63**, 033108 (2001).
5. Ashcroft, N., and Mermin, N., *Solid State Physics*, Holt-Saunders, 1976.
6. Jones, R., and Ritter, A., *J. Electron Spectrosc. Relat. Phenom.*, **40**, 285 (1986).
7. Vos, M., and Bottema, M., *Phys. Rev. B*, **54**, 5946 (1996).
8. Storer, P., Caprari, R., Clark, S., Vos, M., and Weigold, E., *Rev. Sci. Instrum.*, **65**, 2214 (1994).
9. Tanuma, S., Powell, C., and Penn, D., *Surface and Interface Analysis*, **20**, 77 (1993).
10. Aryasetiawan, F., Hedin, L., and Karlsson, K., *Phys. Rev. Lett.*, **77**, 2268 (1996).
11. Vos, M., Kheifets, A., Weigold, E., Canney, S., Holm, B., Aryasetiawan, F., and Karlsson, K., *J. Phys.: Condens. Matter*, **11**, 3645 (1999).
12. Cai, Y., Storer, P., Kheifets, A., McCarthy, I., and Weigold, E., *Surf. Sci.*, **334**, 276 (1995).
13. Guo, X., Fang, Z., Kheifets, A., Canney, S., Vos, M., McCarthy, I., and Weigold, E., *Phys. Rev. B*, **57**, 6333 (1998).

Accurate Amplitudes for Electron-Impact Ionization from Fully Correlated Wavefunctions

Mark D. Baertschy

JILA, 440 UCB, Boulder, CO, 80309-0440

Abstract. Energies near the ionization threshold present a particular challenge for electron-impact ionization calculations. The effect of electron-electron correlation is stronger at lower energies where qualitatively different behavior in the differential cross sections has been predicted. Most of the predictions of the near-threshold behavior have come from semi-classical work. Besides the famous "Wannier threshold law" for the total ionization cross section there are predictions of a flattening of the energy-sharing cross section and a change in shape of the symmetric, co-planar, triply differential cross section. In the case of hydrogen, the near-threshold behavior is expected to be observable within a few eV of threshold. Using the method of exterior complex scaling we can produce accurate wave functions for electron-impact ionization of hydrogen. From these we can extract the complete ionization amplitude. To the extent that comparison is possible, our results agree well with experimental data and the predictions of semi-classical theory.

EXTERIOR COMPLEX SCALING

We solve the time-indepedent Schrödinger equation for electron-hydrogen scattering using the method of exterior complex scaling (ECS) to simplify the scattering boundary conditions. The total wave function Ψ^+ is partitioned into an appropriately symmetrized unperturbed state $\Psi^0_{k_i}$, describing a free electron with momentum k_i incident on a ground state hydrogen atom,

$$\Psi^0_{k_i} = \frac{1}{\sqrt{2}}\left[\Phi_{1s}(\mathbf{r_1})e^{i\mathbf{k_i}\cdot\mathbf{r_2}} + (-1)^S\Phi_{1s}(\mathbf{r_2})e^{i\mathbf{k_i}\cdot\mathbf{r_1}}\right], \qquad (1)$$

and a scattered wave term Ψ^+_{sc}. The wave functions are either symmetric (for total spin $S = 0$) or anti-symmetric ($S = 1$) with respect to interchange of the two electronic coordinates. The scattered wave function, Ψ^+_{sc}, is defined as the outgoing solution of the inhomogeneous differential equation

$$(E - H)\Psi^+_{sc}(\mathbf{r_1},\mathbf{r_2}) = (H - E)\Psi^0_{k_i}(\mathbf{r_1},\mathbf{r_2}) \qquad (2)$$

that comes from rearrangement of the Schrödinger equation. Details of how we solve Eq. (2) can be found elsewhere [1] and will be described only briefly here.

The scattered wave, Ψ^+_{sc}, is expanded in terms of coupled spherical harmonics, $\mathcal{Y}^{L0}_{l_1,l_2}(\hat{r}_1,\hat{r}_2)$,

$$\Psi^+_{sc}(\mathbf{r_1},\mathbf{r_2}) = \sum_{L,l_1,l_2} \frac{i^L}{r_1 r_2}\psi^L_{l_1 l_2}(r_1,r_2)\mathcal{Y}^{L0}_{l_1,l_2}(\hat{r}_1,\hat{r}_2). \qquad (3)$$

CP604, *Correlations, Polarization, and Ionization in Atomic Systems*
edited by D. H. Madison and M. Schulz
© 2002 American Institute of Physics 0-7354-0048-2/02/$19.00

Each of the radial functions, $\psi_{l_1 l_2}^L(r_1, r_2)$, has outgoing wave boundary conditions. In order to calculate the $\psi_{l_1 l_2}^L$ we use the ECS transformation where both radial coordinates are mapped to a contour,

$$r \rightarrow \begin{cases} r, & r < R_0, \\ R_0 + (r - R_0)e^{i\eta}, & r \geq R_0, \end{cases} \tag{4}$$

that is real for $r \leq R_0$, but rotated into the upper half of the complex plane beyond R_0. Any outgoing wave evaluated on this contour becomes exponentially damped beyond R_0. Under the ECS transformation the $\psi_{l_1 l_2}^L$ are square integrable functions that we calculate directly on a two-dimensional radial grid by solving sets of coupled, two-dimensional, complex differential equations for each value of L and S. The complex scaling point R_0 used ranges from $80a_0$ to $130a_0$. Within a box of length R_0 the radial coordinates are real and the calculated $\psi_{l_1 l_2}^L$ are equal to those of the physical (unscaled) scattered wave.

IONIZATION AMPLITUDES

Our formalism for calculating the ionization amplitude, $\mathcal{F}(\mathbf{k_1}, \mathbf{k_2})$, described in detail elsewhere [2], is based on a volume integral expression for $\mathcal{F}(\mathbf{k_1}, \mathbf{k_2})$ that is a function of the momentum vectors for both outgoing electrons,

$$\mathcal{F}(\mathbf{k_1}, \mathbf{k_2}) = \langle \Phi_{\mathbf{k_1}}^{(-)} \Phi_{\mathbf{k_2}}^{(-)} | E - T - V_1 | \Psi_{sc}^+ \rangle . \tag{5}$$

In Eq. (5), E is the total energy, T is the two-electron kinetic energy operator, V_1 is the sum of all one-electron potentials defined as

$$V_1(\mathbf{r_1}, \mathbf{r_2}) \equiv -\frac{1}{r_1} - \frac{1}{r_2} , \tag{6}$$

and the $\Phi_{\mathbf{k}}^{(-)}(\mathbf{r})$ are Coulomb functions with effective charge one, momentum $\mathbf{k}$, and incoming-wave boundary conditions. The integral in Eq. (5) must be evaluated over a finite volume that lies within a region where both electronic coordinates are real. We must choose this volume to be within the region where both radial coordinates are real since the calculated $\psi_{l_1 l_2}^L$ are not usable in the complex-scaled region. Equation (5) has an equivalent surface integral representation which appears upon application of Green's theorem and is more convenient for numerical calculations:

$$\mathcal{F}(\mathbf{k_1}, \mathbf{k_2}) = \frac{1}{2} \int_S (\Phi_{\mathbf{k_1}}^{(+)} \Phi_{\mathbf{k_2}}^{(+)} \nabla \Psi_{sc}^+ - \Psi_{sc}^+ \nabla \Phi_{\mathbf{k_1}}^{(+)} \Phi_{\mathbf{k_2}}^{(+)}) \cdot d\hat{\mathbf{S}} . \tag{7}$$

We evaluate Eq. (7) using the partial-wave expansion of Ψ_{sc}^+ in Eq. (3) to produce partial-wave amplitudes, $f_{l_1 l_2}^L$, so that

$$\mathcal{F}(\mathbf{k_1}, \mathbf{k_2}) = \sum_{L, l_1, l_2} i^{L - l_1 - l_2} e^{i(\eta_{l_1} + \eta_{l_2})} f_{l_1 l_2}^L(k_1, k_2) \mathcal{Y}_{l_1, l_2}^{L0}(\mathbf{k_1}, \mathbf{k_2}) . \tag{8}$$

77

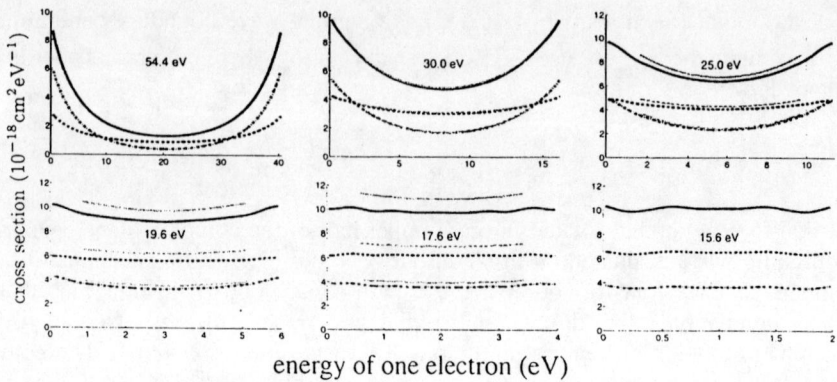

FIGURE 1. Singly differential cross sections for e-H scattering at the collision energies indicated. Individual components for singlet (dashed line) and triplet (dot-dash line) are shown. Where applicable, earlier results based on flux-extrapolation [1] are shown in light gray.

The $f_{l_1 l_2}^L$ are calculated for each $\psi_{l_1 l_2}^L$ by numerically integrating along a quarter-circle of some hyperradius $\rho = \sqrt{r_1^2 + r_2^2}$.

DIFFERENTIAL CROSS SECTIONS

The triply differential cross section (TDCS), which is differential in energy sharing and the directions of both outgoing electrons, is obtained directly from $\mathcal{F}$,

$$\sigma_{ion}(\mathbf{k_1}, \mathbf{k_2}) = \frac{16\pi^2}{k_i^2} |\mathcal{F}(\mathbf{k_1}, \mathbf{k_2})|^2 . \tag{9}$$

The singly differential cross section (SDCS) is differential with respect to the energy of one of the outgoing electrons and is obtained from the TDCS by integrating over the angular coordinates of both momenta $\mathbf{k_1}$ and $\mathbf{k_2}$. Because of orthonormality of the $\mathcal{Y}_{l_1, l_2}^{L0}$ in Eq. (8) the SDCS can be expressed as a simple sum of individual partial wave terms,

$$\sigma_{ion}(k_1, k_2) = \frac{16\pi^2}{k_i^2} \sum_{L, l_1, l_2} |f_{l_1 l_2}^L(k_1, k_2)|^2 . \tag{10}$$

The cross section definitions given in Eqs. (9) and (10) are normalized so that the total ionization cross section is obtained by integrating the SDCS over half of the total energy,

$$\sigma_{ion}(E) = \int_0^{E/2} \sigma_{ion}(\sqrt{2\varepsilon}, \sqrt{2(E-\varepsilon)})d\varepsilon . \tag{11}$$

Since the electron-electron repulsion is of infinite range, it is natural to inquire about the dependence of the cross sections on the size and shape of the volumes enclosed

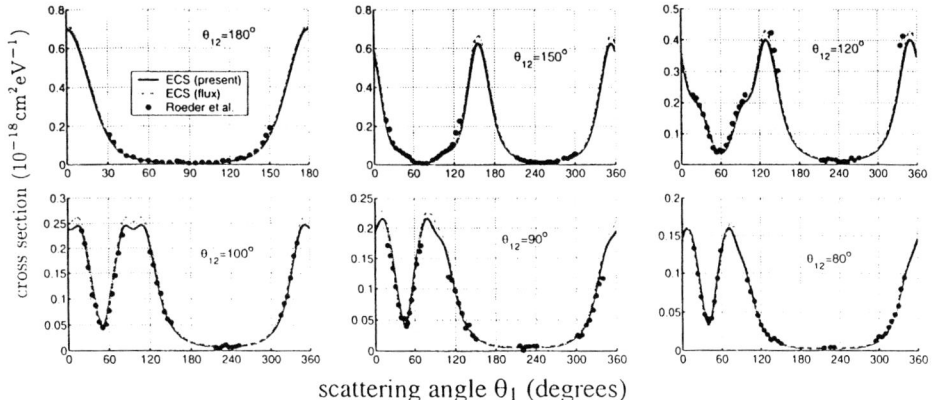

FIGURE 2. Equal-energy sharing, coplanar TDCS for 25 eV incident energy with θ_{12} fixed. Internormalized measurements [3], originally reported in arbitrary units, were multiplied by 0.16 to fit calculated cross section. Solid curves: present results. Broken curves: earlier results from ref. [1].

in computing the amplitudes. We have made the empirical observation that the volume integral in Eq. (5) should always be over a hypersphere. We found that evaluating the $f_{l_1 l_2}^L$ by integrating over other closed surfaces, for example over a hypercube defined by $r_1, r_2 \leq R_0$, produced spurious oscillations in the SDCS on the order of several percent. As a function of the hyperradius at which the underlying amplitudes are calculated the cross sections exhibit small amplitude oscillations generally on the order of a few percent. We average over a small range of hyperradius to obtain final results using the observed dependence upon the hyperradius to estimate their accuracy.

The SDCS calculated for several different collision energies, along with the individual singlet and triplet components, are shown in Fig. 1. We estimate that the SDCS are accurate to within 5% for 17.6 eV and 2% for 30 eV. For the four collision energies covered in ref. [1] we also show the SDCS derived from flux-extrapolation. Agreement in shape between results from the two methods is very good and confirms the flattening of the SDCS with decreasing energy predicted by semiclassical theory [4].

At 15.6 eV collision energy we were unable to obtain reliable results using flux-extrapolation. Even the $f_{l_1 l_2}^L$ exhibit a more pronounced dependence upon hyperradius than those calculated at the other energies, with the expected uncertainty of our SDCS being higher at this energy. The SDCS is found to have the same flattened shape first noticed at 17.6 eV. At both of these energies, the new procedure reveals subtle structure in the SDCS that could not be found by the flux-extrapolation method.

At 54.4 eV collision energy the $f_{l_1 l_2}^L$ should be more accurate. However, convergence of the angular momentum expansion requires more partial wave terms at this energy than we were able to include. This limits the accuracy of the 54.4 eV SDCS shown in Fig. 1 which we present here primarily to illustrate the dependence of the shape of the SDCS upon the total energy. Our *ab initio* calculation included angular momentum components up to $L = 13$. Beyond $L = 4$ the individual L components decay exponentially with increasing L. We estimated the contributions from $L > 13$ by assuming a simple ex-

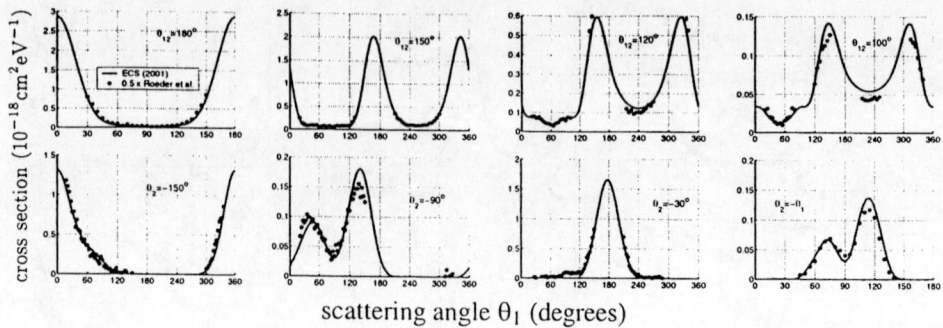

cross section $(10^{-18} \text{ cm}^2 \text{ eV}^{-1})$

scattering angle θ_1 (degrees)

FIGURE 3. Equal-energy sharing TDCS at 15.6 eV incident energy for various coplanar geometries. Absolute experimental data [3, 5] has been multiplied by 0.5.

ponential decay law for large L. The effect of the large L components that were estimated is to raise the SDCS by about 7% near the wings of the distribution.

Unlike the SDCS, the TDCS requires a construction of the complete ionization amplitude according to Eq. (8), so that any phase inconsistencies among the $f_{l_1 l_2}^L$ would adversely affect the calculated TDCS. In other words, the SDCS tests only the magnitudes of the calculated $f_{l_1 l_2}^L$ but the TDCS also tests the phases. In Figs. 2 and 3 we present the equal energy sharing TDCS at 25 and 15.6 eV collision energies for a variety of coplanar geometries. More results of this type are reported elsewhere [2]. In these geometries the trajectories of the incident and two outgoing electrons all lie in a single plane. For instance, each panel in Fig. 2 shows a one-dimensional slice of the TDCS for 25 eV collision energy corresponding to a particular fixed angular separation θ_{12} between the directions of the two outgoing electrons.

Figure 2 contains our previous results, based on flux extrapolation, along with the current results. The agreement between the TDCS results obtained from these two very different methods is a strong indicator of the correctness of both methods. For the flux extrapolation method this implies that error introduced by extrapolating from finite box-sizes was small. For our calculation of the ionization amplitude this implies that there is no phase inconsistency among the $f_{l_1 l_2}^L$ calculated at finite hyperradii. Thus, we have found that Eq. (5) (or, equivalently, Eq. (7)) leads to a viable method of calculating the complete ionization amplitude despite the existence of a formally divergent phase [6].

Experimental values due to Röder *et al.* [3] are also shown in Fig. 2. These results were originally presented in arbitrary units. We have normalized these values by choosing a single scaling factor to give the best overall fit between experiment and our calculations. Although comparison in absolute magnitude is not possible at these energies we see that agreement in the overall shape is quite good.

In Fig. 3 we show results for the TDCS at 15.6 eV collision energy. At this energy we were unable to produce reliable results using flux-extrapolation. The ionization amplitude expression is less sensitive to box-size and allows us to calculate cross sections at lower collision energies than what we were able to treat using flux-extrapolation. Absolute measurements at this energy [3, 5] are available. However, there is evidence that the normalization of the 15.6 eV data is incorrect [7]. Our calculations indicate that the absolute measurements are too high by a factor of 2. Multiplying the experimental values by 0.5 gives good overall agreement with our results, as shown in Fig. 3. Distorted partial-wave (DPW) calculations for $\theta_{12} = 180^o$ by Pan and Starace [5, 8] also indicate that the 15.6 eV data is too large by a factor of 2. In general, we have found that agreement between our results and DPW results (the latter available only for $\theta_{12} = 180^o$) is excellent. We show results for a variety of different geometries in Fig. 3. In all geometries, including those with θ_2 fixed, dividing the experimental values by two gives good agreement with our calculation.

ACKNOWLEDGMENTS

This work was supported by the U.S. DOE Office of Basic Energy Science, Division of Chemical Sciences. Computational work was carried out at the National Energy Research Scientific Computing Center at Lawrence Berkeley National Laboratory.

REFERENCES

1. M. Baertschy, T. N. Rescigno, W. A. Isaacs, X. Li, and C. W. McCurdy, Phys. Rev. A **63**, 022712 (2001).
2. M. Baertschy, T. N. Rescigno, and C. W. McCurdy, Phys. Rev. A **64**, 022709 (2001).
3. J. Röder, J. Rasch, K. Jung, C. T. Whelan, H. Ehrhardt, R. Allan, and H. Walters, Phys. Rev. A **53**, 225 (1996).
4. J. M. Rost, Phys. Rev. Lett **72**, 1998 (1994).
5. J. Röder, H. Ehrhardt, C. Pan, A. F. Starace, I. Bray, and D. Fursa, Phys. Rev. Lett. **79**, 1666 (1997).
6. R. K. Peterkop, *Theory of Ionization of Atoms by Electron Impact* (Colorado Associated University Press, Boulder, CO, 1977).
7. I. Bray, J. Phys. B **33** (2000).
8. C. Pan and A. F. Starace, Phys. Rev. A. **45**, 4588 (1992).

Electron impact ionization using an R-matrix approach

M. P. Scott[*][1], T. Still[†], N.S. Scott[†], and P.G. Burke[*]

[*]Department of Applied Mathematics and Theoretical Physics, Queen's University Belfast, Belfast BT7 1NN, U.K.

[†]Department of Computer Science, Queen's University Belfast, Belfast BT7 1NN, U.K.

Abstract. Two approaches based on the R-matrix method to study electron impact ionization at intermediate incident electron energies are discussed. These are the R-matrix with pseudo-states approach (RMPS) and the intermediate energy R-matrix method (IERM). Recent results to illustrate the key features of these methods are presented.

INTRODUCTION

Over recent years the low-energy atomic R-matrix method has been extended to enable the accurate treatment of electron-atom/ion scattering at intermediate energies. By the term 'intermediate energies' we mean incident electron energies from the ionization threshold to about three to four times this value. In this energy regime the incident electron has sufficient energy to ionize one of the electrons of the target atom or ion and therefore any accurate theoretical treatment of electron scattering must be able to account for the effects of the infinite number of continuum states of the ionized target plus ejected electron as well as the infinite number of bound states of the target atom or ion lying below the ionization threshold. Within the R-matrix framework two approaches have been developed which include carefully constructed pseudo-states in the close coupling expansion of the total wavefunction to represent, in some average way, the high-lying Rydberg states and continuum states of the target which cannot be included explicitly. These are the R-matrix with pseudo-states approach (RMPS) [1] and the intermediate energy R-matrix method (IERM) [2]. It is possible to use the RMPS, IERM and other close coupling approaches, where the target state expansion has been augmented by the inclusion of pseudo-states to account for the target continuum, to calculate an estimate of the cross section for electron impact ionization. This can be obtained

---

[1] e-mail: m.p.scottqub.ac.uk

CP604, *Correlations, Polarization, and Ionization in Atomic Systems*
edited by D. H. Madison and M. Schulz
© 2002 American Institute of Physics 0-7354-0048-2/02/$19.00

by summing the excitation cross sections to pseudo-states whose energy levels lie above the ionization threshold, or by determining the overlap of the pseudo-states with the discrete and continuum spectra of the target atom or ion. In section 1 we will briefly present the basic R-matrix theory common to both the RMPS and IERM approaches; in section 2 we will present some recent applications using these approaches and in section 3 we will discuss future work in this area.

THEORY

The R-matrix method proceeds by partitioning configuration space into two regions by a sphere of radius $r = a$, where r is the relative coordinate of the scattering electron and the centre of gravity of the target atom or ion. This sphere is chosen to completely envelope the electronic orbitals of the target atom or ion. Hence, in the internal region ($r \leq a$) exchange and correlation effects between the scattering electron and the target electrons must be included, whereas in the external region exchange effects can be neglected and the problem simplifies considerably.

In the internal region the $(N+1)$-electron wavefunction at energy E is expanded in terms of an energy-independent basis set, ψ_k, as

$$\Psi_E = \sum_k A_{Ek}\psi_k \ . \tag{1}$$

The basis states, ψ_k, are expanded in the form

$$\psi_k(\mathbf{x}_1,\ldots,\mathbf{x}_{N+1}) = \mathcal{A}\sum_{ij}\tilde{\Phi}_i(\mathbf{x}_1,\ldots,\mathbf{x}_N;\hat{\mathbf{r}}_{N+1}\sigma_{N+1})r_{N+1}^{-1}u_{ij}(r_{N+1})c_{ijk}$$

$$+ \sum_j \chi_j(\mathbf{x}_1,\ldots,\mathbf{x}_{N+1})d_{jk} \ , \tag{2}$$

where the channel functions $\tilde{\Phi}_i$ are obtained by coupling the orbital and spin angular momenta of the target states Φ_i with those of the scattered electron to form eigenstates of the total orbital and spin angular momenta L and S, their z components M_L and M_S and the parity π. The set of states Φ_i will include target eigenstates and pseudo-states to allow for the effect of the infinite number of highly excited bound states and continuum states of the target atom or ion which cannot be explicitly included in the calculation. The u_{ij} are members of a complete set of numerical orbitals used to describe the radial motion of the scattered electron and the χ_j are $(N + 1)$-electron configurations included to allow for short range correlation effects between the scattered and target electrons. The coefficients c_{ijk} and d_{ij} are obtained by diagonalizing the $(N + 1)$-electron Hamiltonian matrix in the internal region.

Each of the target eigenstates and pseudo-states is expanded in terms of a sum of orthonormal configurations

$$\Phi_i(\mathbf{x}_1,\ldots,\mathbf{x}_N) = \sum_j \phi_j(\mathbf{x}_1,\ldots,\mathbf{x}_N)c_{ij} \ , \tag{3}$$

83

where the ϕ_j are constructed from a set of orthonormal one-electron orbitals which can be either bound physical orbitals or pseudo-orbitals, included to represent electron correlation effects or to represent the target continuum.

As discussed in the introduction, we have developed two different R-matrix methods to study electron scattering at intermediate energies the R-matrix with pseudo-states (RMPS) and the intermediate energy R-matrix method (IERM). In both approaches the effect of the high lying bound states and the target continuum are represented by a discrete set of pseudo-states. However, the pseudo-orbital bases used in these two approaches are different. For the RMPS basis the pseudo-orbitals are constructed as the minimum Sturmian-type orthonormal set of functions $r^i e^{-\alpha r}$ that are also orthogonal to the physical orbitals included in the calculation. The choice of α is arbitrary, although it will affect the convergence pattern of a finite pseudo-state expansion. We therefore have a large degree of flexibility in the formation of the RMPS pseudo-state basis and it is possible to construct a basis which can be used over a wide range of incident electron energies. The RMPS approach has the advantage that it is easy to implement into existing codes and is therefore readily applicable to any N-electron target atom or ion.

To date the IERM method has only been developed for 'one-electron' targets. The pseudo-orbitals in this approach are members of the same numerical continuum basis which is used to describe the motion of the scattered electron. The radial functions of these orbitals are solutions of the differential equation

$$\left(\frac{d^2}{dr^2} - \frac{l(l+1)}{r^2} + \frac{2Z}{r} + k_{nl}^2 \right) u_{nl}(r) = 0 \tag{4}$$

for each angular momentum l, subject to the boundary conditions

$$u_{nl}(0) = 0 \ ,$$

$$\frac{a}{u_{nl}(a)} \left. \frac{du_{nl}(r)}{dr} \right|_{r=a} = b \ , \tag{5}$$

where b is a constant, usually taken as zero. The IERM basis produced from these functions is a more densely packed pseudo-state basis, with respect to the target state energy levels, than the RMPS pseudo-state basis. It is dependent on the R-matrix boundary, with the density increasing with increasing radius. It is therefore more appropriate in the study of scattering processes such as electron impact ionization close to threshold. As an example, the IERM target state basis used in the study of electron impact ionization of atomic hydrogen using an R-matrix boundary radius of 400 a.u. is given in Figure 1. The basis consists of up to $n = 14$ physical states for each target angular momentum, $l = 0, 1$ and 2, augmented by 66 additional pseudo-states per angular momentum.

In conjuction with the IERM method, a 2-dimensional R-matrix propagator has also been developed [2,3]. The 2-dimensional R-matrix propagator is an extension of the IERM method in which the (r_1, r_2) space in the internal region is subdivided

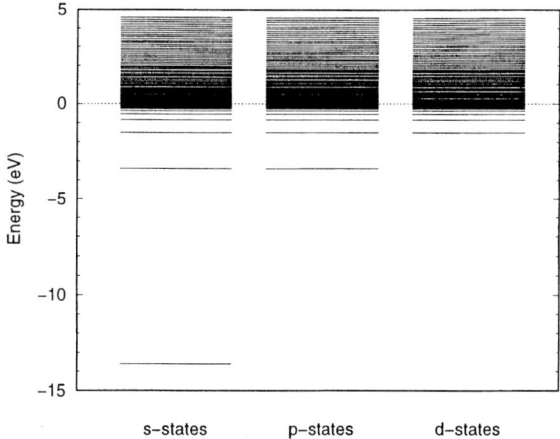

FIGURE 1. Distribution of IERM basis target states for electron impact ionization of atomic hydrogen near threshold

into a number of sectors. The Schrödinger equation is solved using a basis-state expansion in each sector and elementary R-matrices relating the radial functions to their derivatives on the boundaries of each sector are evaluated. These are then used to propagate a global R-matrix, $\mathcal{R}$, across the internal region. On the boundary of the internal region we transform to an IERM target state basis of the form discussed above, transforming $\mathcal{R}$ accordingly. The 2-dimensional R-matrix propagator thus allows us to extend the R-matrix boundary far beyond that possible with the traditional 'one-sector' codes.

APPLICATIONS

R-matrix with pseudo-states (RMPS)

The RMPS pseudo-state basis enables us to study electron impact ionization over a wide range of incident electron energies. The method as been applied successfully by a number of authors to study electron impact ionization from targets including H [4], He [5], Li$^+$ [6], Mg$^+$, Al^{2+} and Si^{3+} [7]. By careful construction of the pseudo-state basis it is possible to include both direct and indirect ionization mechanisms in the calculation. As an example we consider electron impact ionization of C^{3+}. In this case, in addition to direct ionization, which can be written as

$$ e^- + C^{3+}(1s^2 2s) \rightarrow C^{4+}(1s^2) + e^- + e^-, \tag{6} $$

three important indirect ionization processes can occur which proceed through the formation of autoionizing states. These are excitation autoionization (EA) given by

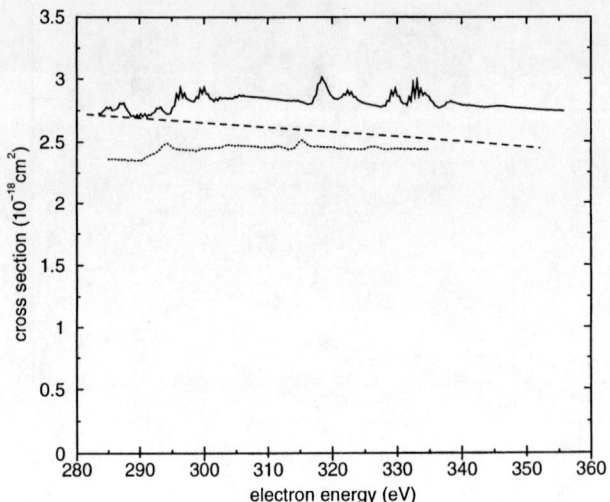

FIGURE 2. Electron impact ionization of C^{3+}. ——, convoluted RMPS results, including direct and indirect ionization processes and interference effects; – – –, RMPS results including direct ionization process only (Scott *et al* [9]); $\cdots$, experimental data of Müller *et al* [10].

$$e^- + C^{3+}(1s^2 2s) \to C^{3+*}(1s2snl) + e^-$$
$$\downarrow$$
$$C^{4+}(1s^2) + e^- + e^-, \tag{7}$$

resonant excitation double autoionization (REDA) given by

$$e^- + C^{3+}(1s^2 2s) \to C^{2+*}(1s2snln'l')$$
$$\downarrow$$
$$C^{3+*}(1s2sn''l'') + e^-$$
$$\downarrow$$
$$C^{4+}(1s^2) + e^- + e^- \tag{8}$$

and resonant excitation auto double ionization (READI) given by

$$e^- + C^{3+}(1s^2 2s) \to C^{2+*}(1s2snln'l')$$
$$\downarrow$$
$$C^{4+}(1s^2) + e^- + e^-, \tag{9}$$

where the stars in these equations refer to autoionizing states. Processes (7), (8) and (9) give rise to resonance structure on top of a smooth direct ionization background due to process (6).

In Figure 2 we compare the results of an RMPS calculation by Scott *et al* [9], which included the five lowest lying physical states of C^{3+}, 21 pseudo-states representing the continuum and 16 autoionizing states, with experimental measurements

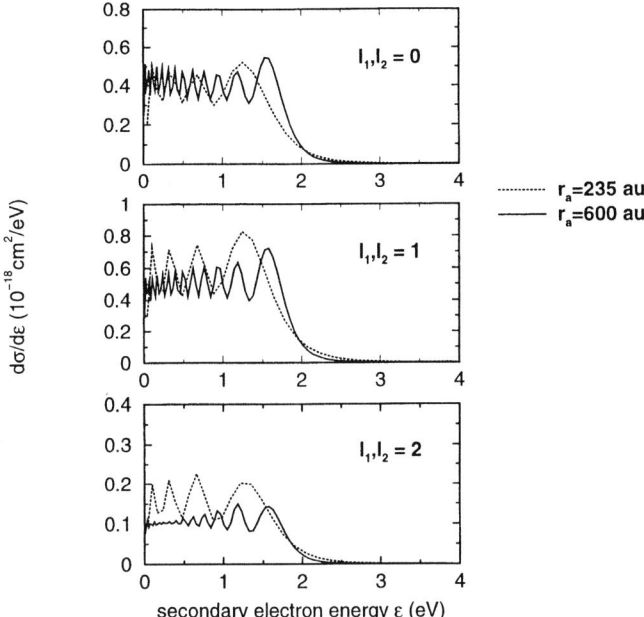

FIGURE 3. ^1S singly differential cross section for electron impact ionization of atomic hydrogen at 17.6 eV incident electron energy. 'Raw' IERM/2-dimensional R-matrix propagator results for different R-matrix boundary radii (see text).

of Müller *et al* [10]. The theoretical data for the direct ionization process are also included. The good agreement between the RMPS calculations and the experimental measurements in Figure 2 illustrates the validity of the pseudo-state approach for both direct and indirect ionization processes.

Intermediate energy R-matrix method (IERM)

As an example of the application of the IERM method we consider electron impact ionization of atomic hydrogen and concentrate particularly on the ^1S singly differential cross section (sdcs) at incident electron energy 4 eV above threshold. It is possible to obtain an estimate of the sdcs from the cross sections for excitation to the individual pseudo-states using the numerical procedure adopted by Konovalov *et al* [8]. The 'raw' sdcs results produced exhibit numerous oscillations which increase in frequency as the R-matrix boundary is increased. A step function in this data at $E/2$ has been predicted by Bray [11]. In Figure 3 we present results for the 'raw' sdcs data for IERM calculations with different R-matrix boundary radii, where the increasing number of oscillations with increasing R-matrix radius is demonstrated. The contributions from different target angular momenta, $l = 0, 1, 2,$

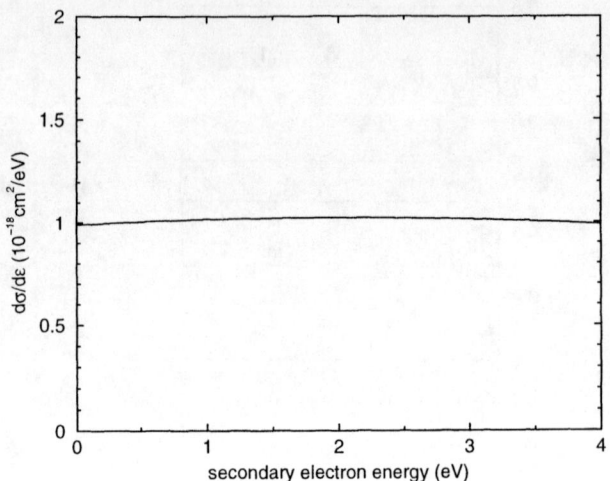

FIGURE 4. ^1S singly differential cross section for electron impact ionization of atomic hydrogen at 17.6 eV incident electron energy. Smoothed IERM/2-dimensional R-matrix propagator results for R-matrix boundary radius of 600 a.u. (see text).

are considered separately. To obtain an estimate of the actual sdcs results from the raw data we have assumed symmetry of the results about $E/2$ and fitted the results to the form $a + b(\epsilon - E/2)^2$ where E is the energy of the incident electron with respect to the ionization threshold and ϵ is the secondary electron energy. Data in the region of the step function has been omitted and we have calculated a and b so that the integral of the sdcs between 0 and $E/2$ equals the total ionization cross section at E. In order to extend the results to these large radii we have used the 2-dimensional R-matrix propagator which has recently been adapted to run on the Cray T3E. In Figure 4 we show the smoothed results using a boundary radius of 600 a.u.. 110 states were included per angular momentum, with 83 lying between the ionization threshold and 4 eV above threshold. These results are in good agreement with the TDCC and ECS results reported by Baertschy *et al* [12]. However, other close coupling approximation results reported by Bray [13] and Bartschat [14], using the CCC and RMPS methods respectively, tend to favour a more concave curve.

FUTURE WORK

While the current results using the IERM method with the 2-dimensional R-matrix propagator are encouraging, we are at present extending this method to fit the R-matrix defined along the boundary $r_1 = a$, $0 \le r_2 \le a$ to an asymptotic form analogous to that used by Poet [15] and by Jones and Stelbovics [16]. This should remove the oscillations seen in our present approach.

ACKNOWLEDGMENTS

This work has been financed, in part, by the EPSRC. TS also acknowledges the receipt of an ESF/Queen's University postgraduate studentship.

REFERENCES

1. Bartschat, K., Hudson, E.T., Scott, M.P., Burke, P.G., and Burke, V.M., *J. Phys. B: At. Mol. Opt. Phys.* **29**, 115–123 (1996)
2. Burke, P.G., Noble, C.J., and Scott, M.P., *Proc. Roy. Soc. A* **410**, 287–310 (1987)
3. Dunseath, K.M., LeDourneuf, M., Terao-Dunseath, M., and Launay, J.M., *Phys. Rev. A* **54** 561–572 (1996)
4. Bartschat, K., and Bray, I., *J. Phys. B: At. Mol. Opt. Phys.* **29**, L577–L583 (1996)
5. Hudson, E.T., Bartschat, K., Scott, M.P., Burke, P.G., and Burke, V.M., *J. Phys. B: At. Mol. Opt. Phys.* **29**, 5513–5526 (1996)
6. Brown, G.J.N., Scott, M.P., and Berrington, K.A., *J. Phys. B: At. Mol. Opt. Phys.* **32**, 737–748 (1999)
7. Badnell, N.R., Pindzola, M.S., Bray, I., and Griffin, D.C., *J. Phys. B: At. Mol. Opt. Phys.* **31**, 911–924 (1998)
8. Konovalov, D.A., Bray, I., and McCarthy, I.E., *J. Phys. B: At. Mol. Opt. Phys.* **27**, L413–L419 (1994)
9. Scott, M.P., Huaguo, T., and Burke, P.G., *J. Phys. B: At. Mol. Opt. Phys.* **33**, L63–L70 (2000)
10. Müller, A., Hofmann, G., Tinschert, K., and Salzborn, E., *Phys. Rev. Lett.* **61** 1352–1355 (1988)
11. Bray, I., *Phys. Rev. Lett.* **78** 4721–4724 (1997)
12. Baertschy, M., Rescigno, T.N., McCurdy, W.C., Colgan, J., and Pindzola, M.S., *Phys. Rev. A* **63** R50701 (2001)
13. Bray, I., *private communication* (2001)
14. Bartschat, K., *private communication* (2001)
15. Poet, R., *J. Phys. B: At. Mol. Phys.* **13**, 2995–3008 (1980)
16. Jones, S., and Stelbovics, A., *Phys. Rev. Lett.* **84** 1878–1881 (2000)

Close-coupling approach to ionization processes

I. Bray[*], D. V. Fursa[†] and A. T. Stelbovics[*]

[*]Centre for Atomic, Molecular and Surface Physics, School of Mathematical and Physical Sciences, Murdoch University, Perth 6150, Australia
[†]Electronic Structure of Materials Centre, The Flinders University of South Australia, G.P.O. Box 2100, Adelaide 5001, Australia

Abstract. We briefly review recent progress in the field of electron-impact ionization of light atoms concentrating on those theories which attempt to fully solve the underlying scattering problem. Comparison between competing theories and experiment shows up some unexpected discrepancies.

INTRODUCTION

The continued growth in computational power has allowed the emergence of highly computationally intensive techniques for solving ionization problems. The most spectacular example is the work of Rescigno et al. [1] who claimed to have solved the electron-hydrogen ionization problem. They utilise the exterior complex scaling (ECS) method which requires a two-dimensional direct numerical integration out to large distances. Careful usage of the transition of the coordinates from real to complex numbers enables the evaluation of the total wavefunction of the system, without recourse to three-body boundary conditions. Having a numerical wavefunction then allowed the extraction of the scattering information, first via a flux method [2], and then more accurately utilising amplitude formulations [3]. The resulting cross sections are in best overall agreement with available e-H experiments to date.

Another example of substantial progress made possible by modern computational resources is the development of time-dependent techniques [4, 5]. Application to double photoionization of helium [6], a near equivalent of electron-impact ionization of He^+, has shown excellent agreement with experiment [7], as well as other computer-intensive approaches including the hyperspherical R-matrix method [8] and the convergent close-coupling (CCC) theory [7].

It is the latter approach that has been pursued by the present authors. Though initially the close-coupling method [9] was designed for elastic scattering and discrete excitation, a simple extension to ionizing processes is possible [10]. In this paper we consider application of the CCC method to low energy e-H ionization with equal-energy outgoing electrons. We compare with experiment and the ECS theory.

CP604, *Correlations, Polarization, and Ionization in Atomic Systems*
edited by D. H. Madison and M. Schulz
© 2002 American Institute of Physics 0-7354-0048-2/02/$19.00

CCC THEORY

The details of the CCC approach to ionization have been given by Bray and Fursa [10], and subsequently, following the work of Stelbovics [11], slightly modified for the case of equal-energy outgoing electrons [12]. Briefly, we first obtain N square-integrable target states by diagonalising the target Hamiltonian using a Laguerre basis

$$\langle \phi_f^N | H_T | \phi_i^N \rangle = \varepsilon_f^N \delta_{fi}. \tag{1}$$

The idea relies on

$$\lim_{N \to \infty} \sum_{n=1}^{N} |\phi_n^N\rangle\langle\phi_n^N| = \sum_{n} |\phi_n\rangle\langle\phi_n| = I. \tag{2}$$

The states are used to expand the total electron-atom wavefunction

$$|\Psi_i^{(+)}\rangle = \mathcal{A}|\psi_i^{(+)}\rangle \approx \mathcal{A}\sum_{n=1}^{N} |\phi_n^N\rangle\langle\phi_n^N|\psi_i^{(+)}\rangle. \tag{3}$$

Close-coupling equations are formed in momentum space for the T matrix at a total energy $E = \varepsilon_i^N + k_i^2/2 = \varepsilon_f^N + k_f^2/2$

$$
\begin{aligned}
\langle \mathbf{k}_f \phi_f^N | T | \phi_i^N \mathbf{k}_i \rangle &= \langle \mathbf{k}_f \phi_f^N | V | \phi_i^N \mathbf{k}_i \rangle \\
&+ \sum_{n=1}^{N} \int d^3 k \frac{\langle \mathbf{k}_f \phi_f^N | V | \phi_n^N \mathbf{k} \rangle \langle \mathbf{k} \phi_n^N | T | \phi_i^N \mathbf{k}_i \rangle}{E + i0 - \varepsilon_n^N - k^2/2}.
\end{aligned}
\tag{4}
$$

Upon solution of (4) the discrete amplitudes show step-function behaviour [13]

$$\lim_{N \to \infty} \langle \mathbf{k}_f \phi_f^N | T | \phi_i^N \mathbf{k}_i \rangle = 0, \text{ for } k_f^2/2 < \varepsilon_f^N. \tag{5}$$

The ionization, or (e,2e) amplitude is defined by

$$f^N(\mathbf{k}_f, \mathbf{q}_f, \mathbf{k}_i) = \langle \mathbf{q}_f^{(-)} | \phi_f^N \rangle \langle \mathbf{k}_f \phi_f^N | T | \phi_i^N \mathbf{k}_i \rangle, \tag{6}$$

where $\mathbf{q}_f^{(-)}$ is a continuum eigenstate of H_T with energy $q_f^2/2 = \varepsilon_f^N \le k_f^2/2$. Following the work of Stelbovics [11] it follows that solving (4) is like taking a finite Fourier expansion of a step-function. Accordingly, at the step the amplitudes converge to half the required values. Hence for $q_f^2/2 = k_f^2/2 = E/2$ we use $2f^N(\mathbf{k}_f, \mathbf{q}_f, \mathbf{k}_i)$.

COMPARISON OF THEORY AND EXPERIMENT

With such an approach the CCC theory yields absolute agreement with all e-He ionization measurements where the two outgoing electrons share the excess energy E equally [see 14, and references therein]. Surprisingly, the situation for the simpler atomic hydrogen target is less clear.

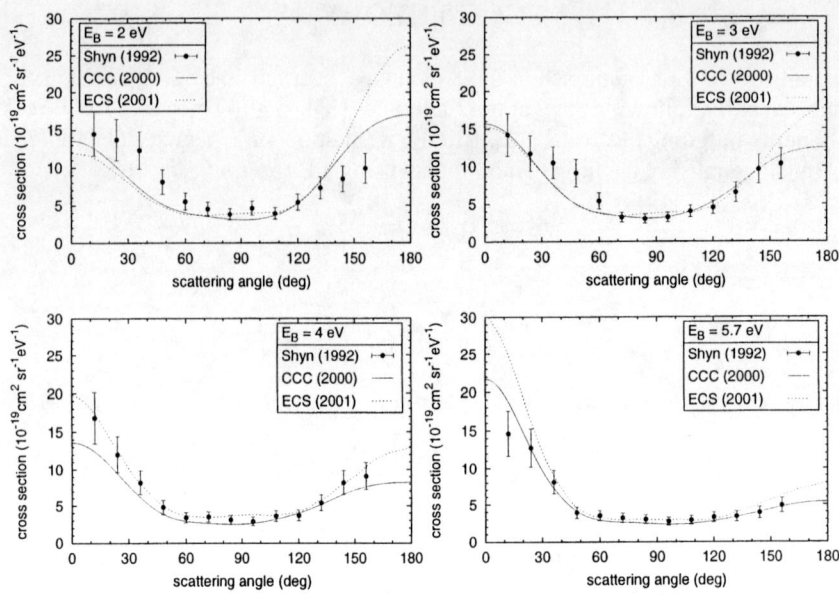

FIGURE 1. Doubly differential cross sections for 25 eV e-H ionization. The experiment is due to Shyn [15], the ECS theory is due to Isaacs et al. [16] and the CCC calculations are due to Bray [17].

We begin by looking at e-H ionization by 25 eV electrons. Here we have absolute doubly differential cross sections (DDCS) measured by Shyn [15]. These describe the angular distribution of the electron ejected with energy E_B. In Fig. 1 we present the experiment and the ECS [16] and CCC [17] theories. We see generally good agreement between the two theories and experiment. The biggest discrepancies between the theories are for the small and large scattering angles, which contribute least to the singly differential, and hence the total ionization, cross section owing to the $\sin(\theta)$ factor in the integration. Of particular interest is the case of $E_A = E_B = 5.7$ eV, for which relative triply differential cross section (TDCS) data are available [18].

The comparison of the 25 eV equal-energy-sharing TDCS is given in Fig. 2. The data are only available for the fixed separation angle θ_{AB} of the two outgoing electrons. For brevity of presentation we take the two smallest available θ_{AB} and two largest. As the data are relative we are free to move it collectively up and down by a single factor. We choose to normalise the experiment at the small θ_{AB}, where ECS and CCC are in good agreement. In doing so we see that at the larger θ_{AB} ECS agrees well with experiment, but the CCC theory is substantially too low. The fact that CCC is much lower than ECS is related to the discrepancy at the small and large scattering angles of the corresponding DDCS of Fig. 1. We are at a loss to explain why CCC would agree with ECS for the smallest θ_{AB}, which yield the smallest cross sections, yet be so different for the largest θ_{AB} with the largest cross sections. Apparently, upon integration over the solid angle $d\Omega_A$ the resultant DDCS only disagrees at the extreme angles.

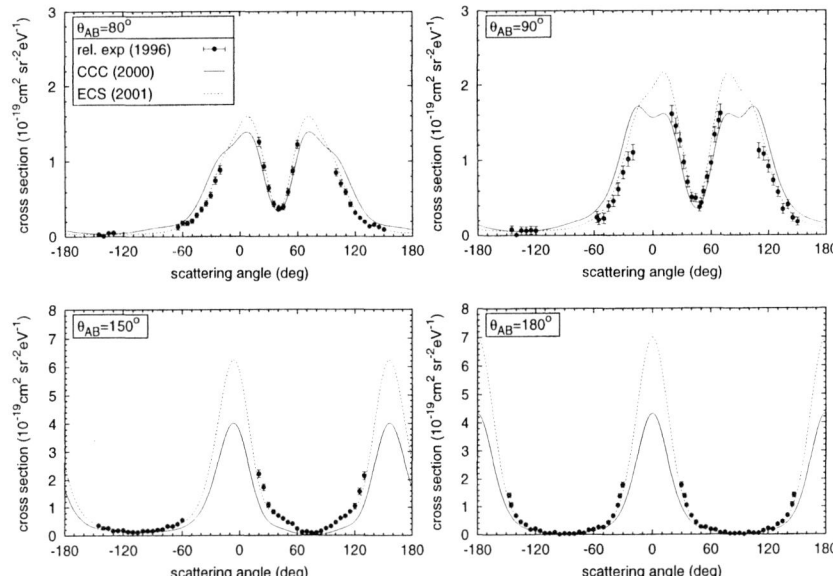

FIGURE 2. Triply differential cross sections for 25 eV e-H ionization with equal-energy outgoing electrons. The experiment is due to Röder et al. [18], the ECS theory is due to Baertschy et al. [3] and the CCC calculations are due to Bray [12].

We next turn our attention to the two energies where absolute TDCS data exist, that of 15.6 and 17.6 eV. In both cases the outgoing electrons have equal energy, $E_A = E_B = 1$ eV and $E_A = E_B = 2$ eV, respectively.

The 15.6 eV data are presented in Fig. 3. The experiment [19] is absolute, but has uncertainty of $\pm 35\%$ in the overall normalisation. However, internormalisation is claimed to be accurate to within 10%. At this energy there are not only fixed θ_{AB} data, but also fixed θ_A data. The latter are particularly important because the cross sections are often large, and also, they allow for internal consistency checks. Whenever the two sets of data intersect they must have a common point, as is generally the case with the 15.6 eV data [12].

From the figure we see remarkable agreement between the two theories and experiment so long as uniform scaling factors are applied. Experiment is a factor of two larger than the ECS theory, and a factor of three larger than the CCC theory. It is particularly surprising that the two theories disagree with each other only in the overall magnitude, with 3 CCC$\approx$ 2 ECS.

A similar situation occurs at 17.6 eV, presented in Fig. 4. Though at this energy the experimental data show some internal inconsistency [12] this is likely to affect the smallest cross sections measured. The geometries presented are similar to those of Fig. 3, and once more show excellent agreement between theory and experiment except for overall normalisation factors. Once again experiment is a factor of two greater than the ECS theory and a factor of three than the CCC theory.

93

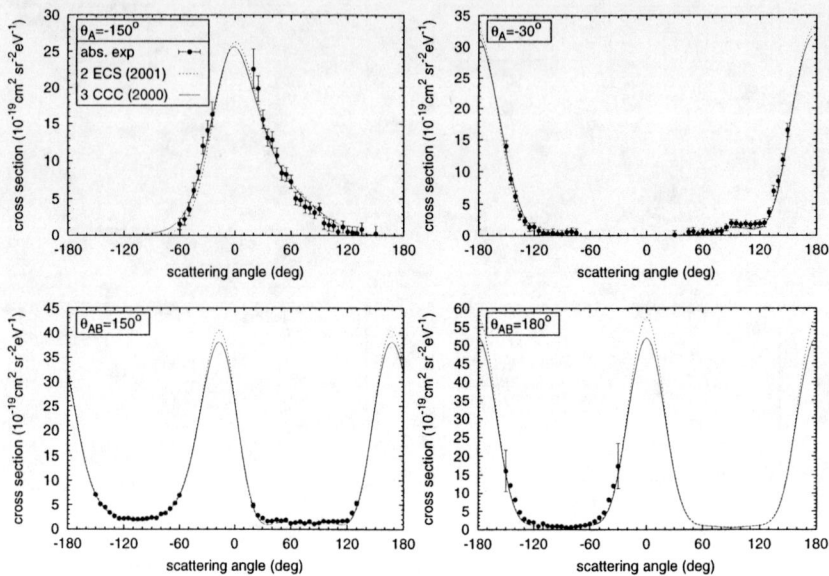

FIGURE 3. Triply differential cross sections for 15.6 eV e-H ionization with equal-energy (1 eV) outgoing electrons. The experiment is due to Röder et al. [19] and references therein, the ECS theory is due to Baertschy et al. [3] and the CCC calculations are due to Bray [12].

CONCLUSIONS

There has been much progress in the last few years in the ability of theory to reproduce measurements of electron-impact ionization fully differential cross sections. In the process some astonishing and unexpected discrepancies between competing theories have been found. The ECS theory yields the most accurate e-H angular distributions at the higher energies suggesting that something is going wrong with the CCC theory for these cases. At the lower energies both theories yield comparable angular distributions which, however, are around a factor of 2/3 apart in overall magnitude. Nevertheless, both theories yield accurate total ionization cross sections.

While we are presently investigating the CCC implementation at the higher energies, we would be grateful for the application of the time-dependent close-coupling (TDCC) theory [4] to this problem. Most importantly, new accurate absolute experimental observations would be very welcome. In particular, measurements of absolute double differential cross sections at the lower energies would be helpful in establishing the required magnitudes.

REFERENCES

1. Rescigno, T. N., Baertschy, M., Isaacs, W. A., and McCurdy, C. W., *Science*, **286**, 2474–2479 (1999).

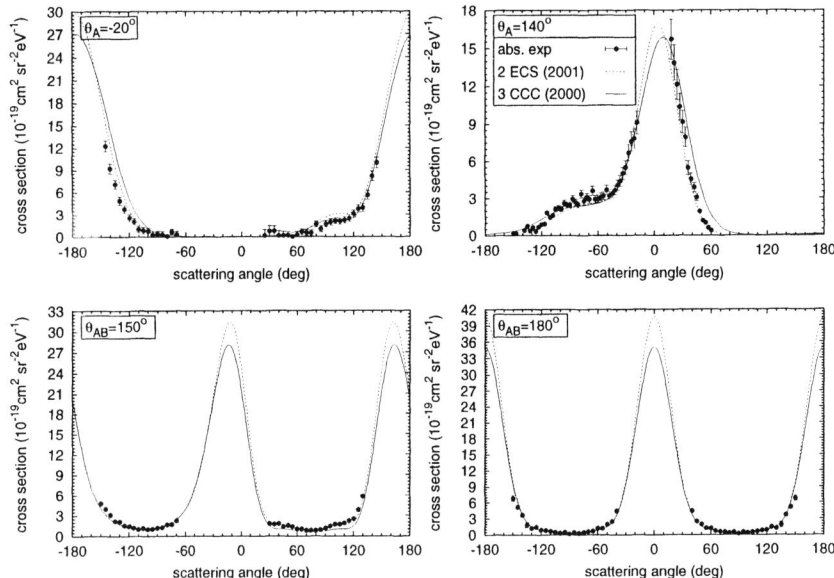

FIGURE 4. Triply differential cross sections for 17.6 eV e-H ionization with equal-energy (2 eV) outgoing electrons. The experiment is due to Röder et al. [19] and references therein, the ECS theory is due to Baertschy et al. [3] and the CCC calculations are due to Bray [12].

2. Baertschy, M., Rescigno, T. N., Isaacs, W. A., Li, X., and McCurdy, C. W., *Phys. Rev. A*, **63**, 022712 (2001).
3. Baertschy, M., Rescigno, T. N., and McCurdy, C. W., *Phys. Rev. A*, **64**, 022709 (2001).
4. Pindzola, M. S., and Schultz, D. R., *Phys. Rev. A*, **53**, 1525–1536 (1996).
5. Madison, D., Odero, D., and Peacher, J., *J. Phys. B*, **33**, 4409–4423 (2000).
6. Colgan, J., Pindzola, M., and Robicheaux, F., *J. Phys. B* (2001).
7. Bräuning, H., Dörner, R., Cocke, C. L., Prior, M. H., Krässig, B., Kheifets, A., Bray, I., Bräuning-Demian, A., Carnes, K., Dreuil, S., Mergel, V., Richard, P., Ullrich, J., and Schmidt-Böcking, H., *J. Phys. B*, **31**, 5149–5160 (1998).
8. Malegat, L., Selles, P., and Kazansky, A., *Phys. Rev. Lett.*, **85**, 4450–4453 (2000).
9. Massey, H. S. W., and Mohr, C. B. O., *Proc. Roy. Soc. A*, **136**, 289–311 (1932).
10. Bray, I., and Fursa, D. V., *Phys. Rev. A*, **54**, 2991–3004 (1996).
11. Stelbovics, A. T., *Phys. Rev. Lett.*, **83**, 1570–1573 (1999).
12. Bray, I., *J. Phys. B*, **33**, 581–595 (2000).
13. Bray, I., *Phys. Rev. Lett.*, **78**, 4721–4724 (1997).
14. Bray, I., Fursa, D. V., and Stelbovics, A. T., *Phys. Rev. A*, **63**, 040702 (2001).
15. Shyn, T. W., *Phys. Rev. A*, **45**, 2951–2956 (1992).
16. Isaacs, W. A., Baertschy, M., McCurdy, C. W., and Rescigno, T. N., *Phys. Rev. A*, **63**, 030704 (2001).
17. Bray, I., *Aust. J. Phys.*, **53**, 355–398 (2000).
18. Röder, J., Rasch, J., Jung, K., Whelan, C. T., Ehrhardt, H., Allan, R. J., and Walters, H. R. J., *Phys. Rev. A*, **53**, 225–233 (1996).
19. Röder, J., Ehrhardt, H., Pan, C., Starace, A. F., Bray, I., and Fursa, D. V., *Phys. Rev. Lett.*, **79**, 1666–1669 (1997).

Coplanar (e,3-1e) and (e,3e) Experiments : Importance of Non-First Order Effects in the Double Ionization of Helium

A. Lahmam-Bennani and A. Duguet

Laboratoire des Collisions Atomiques et Moléculaires (UMR 8625),
Université de Paris-Sud XI, F-91405 Orsay cedex

Abstract. New coplanar (e,3e) and (e,3-1e) experiments on helium at about 600 eV incident energy are presented. The results are discussed in comparison with state-of-the-art first order theories and with former (e,3e) experiments at higher energy. The whole set is analysed in terms of double ionization mechanisms. Clear evidence is given for the increasing importance of second or higher order effects in the projectile target interaction when decreasing the incident energy.

INTRODUCTION

(e,3e) experiments are electron impact kinematically complete double ionization (DI) experiments in which energies and angles of all participating particles are determined in the final state, and therefore fully differential cross sections are measured [1].

The ideal target for such studies is helium as it yields a pure 4-body problem in the final state. However, the He experiments proved to be difficult due to the low DI cross section and to the need for using triple coincidence techniques. Therefore, experiments were conducted on heavier gases [2,3]. It is only since ~3 years that their feasibility on He was demonstrated [4] using a multi-angle detection system based on a Double Toroidal Analyzer (DTA), and subsequently [5] using the COLTRIMS method. Detailed experiments were reported since then by our group [6-8] and by the Freiburg group [9].

In these papers, we presented (e,3e) results on He at 5.5 keV and 1.1 keV impact energy, for equal energy of the two ejected electrons and small momentum transfer, K to the target. The emphasis was there put on the comparison with state-of-the-art, first order theoretical models: the 3C, the Correlated 4-body Final State (C4FS) and the Convergent Close Coupling (CCC) models. Each of these models successfully describes the photo-double ionization process, so that the question was legitimately rised: are the observed differences between experiment and theory to be attributed to deviations from the optical limit (reached at infinite incident energy and vanishing momentum transfer)? Or should they be attributed to non-first order effects in the projectile-target interaction, not included in these models, such as the two-step2 (TS2) mechanism [2] involving two successive collisions of the projectile with the target?

CP604, *Correlations, Polarization, and Ionization in Atomic Systems*
edited by D. H. Madison and M. Schulz
© 2002 American Institute of Physics 0-7354-0048-2/02/$19.00

Starting from our first experiments at 5.5 kV, two possible directions could be followed, by extending the measurements either to larger incident energy and smaller K to probe when and how is the optical limit reached; or to lower incident energy and larger K to enhance and study the second or higher order effects. It is this second path that we decided to follow first, by performing two new experiments on He: an (e,3e) one under kinematics similar to the 5.5 and 1.1 keV experiments, but at ~ 600 eV impact energy and equal energy sharing, and an (e,3-1e) one, also at 0.6 keV but with unequal energy sharing. Results from these experiments are presented below, putting the emphasis on non-first order effects and their evolution with decreasing incident energy.

RESULTS AND DISCUSSION

The experimental set-up is described in detail in [2,10]. Briefly, the incident electron (energy E_0, momentum $\mathbf{k_0}$) is scattered under the angle θ_a with energy E_a and momentum $\mathbf{k_a}$, while two electrons are ejected from the target in the directions θ_b and θ_c (energies E_b, E_c). The central element of the system is the DTA, which allows the 2 ejected electrons to be energy-selected and multi-angle detected in a coplanar configuration (all electrons lie in the same plane). In an (e,3e) experiment, all three angles and energies are measured and the final electrons are detected in a triple coincidence a-b-c. One may also measure double coincidences a-b or a-c, hence a (e,3-1e) DI experiment, or a (e,2e) single ionization (SI) experiment, depending on the chosen energy balance.

(e,2e) Single Ionization

As noted above, the central idea here is non-first order effects in DI processes, i.e. deviations from first order, first Born (FB) theoretical models. We first recall some well established findings for (e,2e) SI, under kinematical conditions which are very close to those used in (e,3e) experiments. Recent (e,2e) experimental results obtained at 5.5 keV impact energy, and a 10 eV electron ejected from He show excellent agreement [7] with various FB theoretical models, the latter being practically undistinguishable from each other. Similar, new (e,2e) experiments at a 10 times smaller incident energy, 535 eV [11] are well reproduced by three theoretical models at the binary peak, among which the FB calculations show a recoil peak substantially smaller than that which is experimentally observed. Such agreement at the binary and disagreement at the recoil is typical of all FB calculations. The disagreement is largely removed when one compares with the full CCC calculations [12], which include the scattered - ejected e-e interaction, and hence go beyond FB.

Hence the conclusion that at 5 keV incident energy the SI process is largely a first order process in the interaction of the projectile with the target, whereas at 500 eV, small but yet sizeable deviations to FB can be observed, which are largely taken into account by the full CCC. This conclusion was already pointed out and quantified in Fig.1 of [13], where the (e,2e) binary-to-recoil intensity ratio at constant ejected energy and momentum transfer is plotted versus $1/\sqrt{E_0}$. The FB prediction for this ratio is independent of the incident energy. In contrast, the experimental data show a dependence upon E_0. Very clearly, the FB limit for SI is practically reached at 4-8 keV within some 5-10 %, whereas at 600 eV deviations from the FB of the order of 30% are observed.

The same point can be illustrated for the intermediate case of SI accompanied with excitation. This (e,2e) process bears some analogy with an (e,3e) one as it implies two active target electrons. Hence, second or higher order effects are expected to be significantly more important than in the direct ionization to the ion ground state. Indeed, experiments at 5 and 1 keV [14] as well as calculations [15] which include interactions up to second order of the projectile with the target show deviations from FB substantially larger than in (e,2e) SI to the ion ground state. Thus, one should expect these deviations to be very likely even larger in the (e,3e) DI, indicating stronger contributions of second and possibly higher order terms in the projectile target interaction.

(e,3e) Double Ionization

To this end, we compare our (e,3e) results successively obtained at 5, 1 and 0.6 keV, presented in Fig. 6 of [11] as contour plots.

At 5 keV, the experimental and CCC theoretical cross section angular distributions are in satisfactory agreement with each other, as far as the most prominent characteristics of these distributions are concerned: the 2-lobe structure is present in both, and the diagonal nodal line, which is reminiscent of the zero intensity corresponding to the forbidden back-to-back emission in PDI, shows up in the experiments as a minimum of intensity rather than a strict zero. A closer inspection can be made by taking cuts at fixed θ_b or fixed θ_c. For the present discussion, we restrict these cuts to the regions of maximum intensity and hence better statistics. At θ fixed at 41 and 55°, see Fig. 2 of [7], there is good or very good *relative* agreement between experiments and the two first order theories shown there. At θ fixed at 263 and 277°, Fig. 3 of [7], the agreement is still acceptable but it has degraded. Moreover, the experiments are measured on an absolute scale, and both theoretical models *do not* produce the correct magnitude as they had to be rescaled by different factors, 0.7 and 3.2.

We think that these deviations are a manifestation of non-first order effects. The best proof for it was supplied by Berakdar [16] who developed a new approach based on Green functions using an iterative procedure where the interaction potential of a system of N particles is expressed in terms of the potentials of N-1 interacting particles. This approach is applied in Fig.3 of [16] to our data at 5 keV yielding 2 important observations: (i) the FBA results are insensitive to the charge state of the projectile. Hence, the difference between electron and positron impact results clearly shows the importance of non-FB effects, even at 5 keV. (ii) there is a factor ~3 difference in magnitude between experiments and FB-CCC results, while a considerably better agreement is achieved with Berakdar's incremental approach, in shape *and* in magnitude.

At this point, we conclude that we observe for DI large non FB effects at 5 keV, substantially larger than the 5-10% seen in (e,2e) SI, and the ~30% seen in (e,2e) ionization + excitation. These effects mostly affect the magnitude, while the shape of the cross section is less affected.

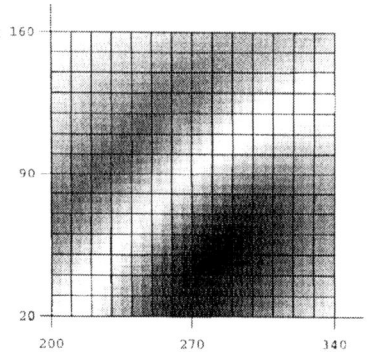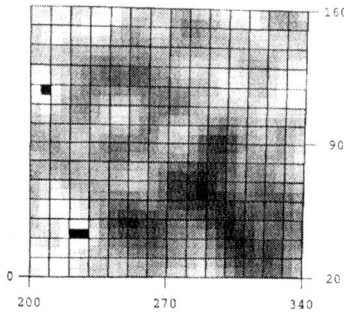

FIGURE 1. Contour plots of the (e,3e) cross section vs ejection angles θ_b (vertical axis) and θ_c (horizontal axis). Left : FB-CCC. Right : experiments. E_0=601eV, E_b=E_c=11eV, θ_a=+1.5°.

At 1.1 keV impact energy, it is seen from the contour plots of Fig. 6 of [11] that the intensity distribution has again the characteristic mussel-shell type distribution. The forward structure is still present in both experimental and CCC theoretical distributions, but the backward lobe has not survived: some intensity destruction has occured there. By taking cuts at fixed θ_b or θ_c [8] in the region of high intensity, large discrepancies are observed with the first-order models: large shifts of the lobes, e.g. see Fig. 1(q-r) of [8], but also a failure of the theories to consistently reproduce the *relative* scale of the experimental data. These large discrepancies are definitely appreciably larger than what was observed at 5 keV.

This observation is confirmed and amplified at lower impact energy, 600 eV. The experimental and theoretical results are shown as contour plots in Fig. 1. Being optimistic, one can say that theory and experiments do bear some resemblance. However, the experiments show significant differences from theory. (i) the back-to-back nodal line is only vaguely present. (ii) the intensity corresponding to the backward structure is almost completely destroyed, as it was already observed at 1.1 keV [8]. (iii) the mussel-shell intensity distribution has changed considerably: the experimental distribution is stretched along a line which is perpendicular to the one for the CCC distribution. And (iv) most important, it shows the appearance of two islands of intensity, or two wings.

We believe this redistribution of intensity and the appearance of these islands to be a pure manifestation of non-first order effects, through destructive and constructive interferences. Such interference effects were also noticed by Berakdar [16] who noted "the presence of additional subsidiary peaks, ... which are absent in the FBA (CCC) results. ... The origin of these peaks are interference effects between the various terms in the sum (8) when evaluating the cross sections."

These non-first order effects are obviously much larger here for (e,3e) processes than they are for (e,2e) at the same impact energy. To visualize them in more detail, we consider again in Fig. 2 cuts at fixed θ_b or θ_c in the region of high intensity where the statistics are best. The distribution is no longer single-lobed, but it has additional side lobes or wings. Moreover, the main lobe exhibits a large shift in its angular position and is much narrower, as seen e.g. at 280 and 290°.

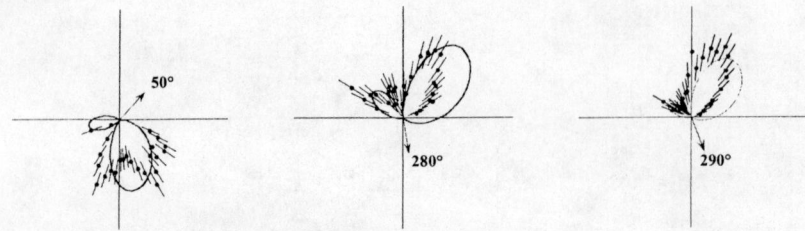

FIGURE 2. (e,3e) angular distributions with one ejection angle fixed at 50, 280 or 290°. E_0=601eV, E_b =E_c =11eV, θ_a=+1.5°. Full line: FB-CCC. Full circles: experiments.

(e,3-1e) Double Ionization

Finally, we present results of our latest DI experiment on He, at 640eV incident energy. In this (e,3-1e) experiment the slowest ejected electron remains undetected. Its energy is known and is fixed at 10 eV, whereas integration is performed over its direction. The scattered projectile is detected at θ_a = +1.5° with energy E_a = 500 eV in coincidence with the fast ejected electron, say b, with E_b = 51 eV. The idea behind these experiments is the following: in an earlier paper [17], we presented (e,3-1e) measurements on Ar at 5 keV incident energy, where the 2 ejected electrons were detected in coincidence, ignoring the scattered one. The conclusion from this work was that at 5 keV the DI process can mostly be attributed to a shake off (SO) process where the fast electron is ejected along the momentum transfer direction in an e-e binary collision. Subsequent relaxation of the target releases the slow ejected electron with a more or less isotropic distribution.

Though similar, our new measurements are different in that they emphasize non first-order mechanisms in the following way: (i) by the choice of a simpler target (He), (ii) by the lower impact energy, 0.6 keV, (iii) by the choice of detecting in coincidence the electron pair (scattered - fast ejected) instead of (slow ejected - fast ejected). This way, the momentum transfer is fixed, while it was not in the Ar case. So that a first order DI process (SO or TS1) should yield a peaked distribution along **K**, in "a kind of binary and recoil (e,2e) distribution". Deviations from this picture would be clear signature of non-first order processes, in particular the TS2 mechanism.

The experimental results are compared in Fig. 3 to theoretical calculations from the "usual" 1st-order models, the 3C model by Dal Cappello and the C4FS model by Berakdar. Theories show "a kind of (e,2e) distribution" where the cross section is symmetric about ±**K** directions. In contrast, the experimental data show (i) no symmetry about **K**, and (ii) large shifts in the positions of the maxima and minima. These observations are clear evidence for the presence of strong non-first order effects, and hence a strong contribution of the TS2 mechanism in the DI process.

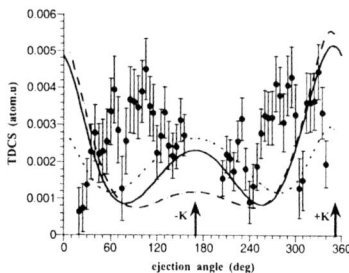

FIGURE 3. (e,3-1e) angular distribution at E_0=640eV, E_b=51eV, E_c (undetected) =10eV, E_a=500eV, θ_a=1.5°. Full line: FB-3C, dashed and dotted lines: C4FS, with different initial state wavefunctions.

Though these data as they stand are sufficient to support the given conclusion, it would be interesting to strengthen it by performing the complementary (e,3-1e) experiments corresponding to the same kinematics, i.e. by detecting the pair (scattered 500eV - slow ejected 10eV), or the pair (fast ejected 51eV - slow ejected 10eV). One could also vary the kinematics, by varying the impact energy or the E-sharing or the momentum transfer.

ACKNOWLEDGMENTS

The authors are grateful to S. Vuidot, S. Roussin and T. Mathieu for their help in the data analysis on the computer.

REFERENCES

1. Lahmam-Bennani, A. *et al*, *Phys. Rev. Lett.* **63**, 1582 (1989).
2. Lahmam-Bennani, A. *et al*, *J. Phys. B* **25**, 2873 (1992) ; El Marji, B. *et al*, *J. Phys. B* **30**, 3677 (1997); Schröter, C. . *et al*, *J. Phys. B* **31**, 131 (1998).
3. El Marji, B. *et al*, *Phys. Rev. Lett.* **83**, 1574 (1999).
4. Taouil, I. *et al Phys. Rev. Lett.* **81**, 4600 (1998).
5. Dörn, A. *et al*, *Phys. Rev. Lett.* **82**, 2496 (1999).
6. Lahmam-Bennani, A. *et al*, *Phys. Rev.* A**59**, 3548 (1999).
7. Kheifets, A. *et al*, *J. Phys. B* **32**, 5047 (1999).
8. Lahmam-Bennani, A. *et al*, *J. Phys. B* **34**, 3073 (2001).
9. Dörn, A. *et al*, *Phys. Rev. Lett.* **86**, 3755 (2001); *J. Phys. B submitted* (2001).
10. Duguet, A. *et al*, *Rev. Sci. Instrum.* **69**, 3524 (1998).
11. Lahmam-Bennani, A. *et al*, "Is the Optical Limit Approached in Coplanar (e,3e) Experiments on He at High and Intermediate Energies?", in *Many Particle Spectroscopy of Atoms and Molecules, Clusters and Surfaces*, ed by J. Berakdar and J. Kirshner, New York: Plenum, Dordrecht: Kluwer, 2001.
12. Bray, I. and Fursa, D.V., *Phys. Rev. A* **54**, 2991-3004 (1996).
13. Ehrhardt, H. *et al*, *J. Phys. B* **20**, L193 (1987).
14. Dupré, C. *et al*, *J. Phys. B* **25**, 259 (1992); Avaldi, L. *et al*, *J. Phys. B* **31**, 2981 (1998).
15. Marchalant, P. *et al*, *J. Phys. B* **32**, L705 (1999); Fang, Y. *et al*, *J. Phys. B* **34**, L19 (2001).
16. Berakdar, J., *Phys. Rev. Lett.* (2000) **85**.
17. El Marji, B. *et al*, *J. Phys. B* **29**, L157 (1996).

Distinguishing between Target Structure and Ionization Mechanisms in (e,3e) Experiments

M. A. Coplan*, J. W. Cooper*, J. H. Moore[†], J. P. Doering[¶],
and R. W. van Boeyen[¶]

*Institute for Physical Science and Technology, University of Maryland,
College Park, MD 20742, USA

[†]Department of Chemistry and Biochemistry, University of Maryland,
College Park, MD 20742, USA

[¶]Department of Chemistry, The Johns Hopkins University, Baltimore, MD 21218, USA

Abstract. Electron impact double ionization with full determination of the kinematics of the collision, (e,3e), can, in principle, provide direct information on the correlated motion of the ejected electrons at the instant of ejection, provided that the mechanism of ejection is known. The symmetries of the observed double ionization cross sections in a variety of kinematic regimes can be used to experimentally investigate ejection mechanisms. The sensitivities of various experimental geometries and kinematic regimes to different ionization mechanisms are discussed in the context of the double ionization of the 3s electrons of magnesium. The practical implications for extracting information about electron correlation are examined.

INTRODUCTION

Electron impact ionization studies serve many different purposes. Our interest is principally to use them to learn more about the electronic structure of atoms and molecules. Single ionization or (e,2e) experiments have been very useful for the study of the momentum distributions of the valence and inner shell electrons in a large number of atomic and molecular targets from atomic hydrogen [1] to complex polyatomic molecules [2]. For studies of two-electron properties, such as correlation, double ionization, (e,3e) experiments, are most appropriate. For the (e,2e) work, a great deal of effort has been spent in establishing the ejection mechanism and then devising appropriate experimental geometries to constrain the cross section measurements to kinematic regimes where the mechanism is well established [3]. With double ionization, the situation becomes considerably more complex. In this paper we discuss double ionization mechanisms and describe our current experiment for extracting electron correlation information from (e,3e) cross section measurements.

CP604, *Correlations, Polarization, and Ionization in Atomic Systems*
edited by D. H. Madison and M. Schulz

DOUBLE IONIZATION MECHANISMS

In order to use electron impact double ionization for studying electronic structure, it is necessary that the ejection of the two electrons occur in a direct way over a time that is short compared to an electron period. Only in this way do the trajectories of the electrons after ejection reflect their motion within the target. Auger and shake-off are indirect double ionization mechanisms that compete with direct mechanisms, but can be distinguished from them by the angular distributions of the ejected electrons [4].

The simplest formulation of direct double ionization is based on a hard sphere collision model where an incident electron strikes a target consisting of two electrons bound to an ion core. As a result of the collision, the incident electron is scattered and two electrons are ejected from the target. This can be represented by the momentum conservation equation

$$\mathbf{p}_0 + \mathbf{q}_1 + \mathbf{q}_2 + \mathbf{q}_{ion_i} = \mathbf{p}_s + \mathbf{p}_1 + \mathbf{p}_2 + \mathbf{q}_{ion_f}, \tag{1}$$

where $\mathbf{p}_0$, $\mathbf{p}_s$, $\mathbf{p}_1$, and $\mathbf{p}_2$ are the momenta of the incident, scattered, and ejected electrons respectively, $\mathbf{q}_1$ and $\mathbf{q}_2$ are the momenta of the electrons in the target at the instant of the ejections, and $\mathbf{q}_{ion_i}$ and $\mathbf{q}_{ion_f}$ are the momenta of the ion core at and after the instant of the collision. In an impulsive collision where no momentum is transferred directly to the core $\mathbf{q}_{ion_i} = \mathbf{q}_{ion_f}$ so that

$$\mathbf{q}_1 + \mathbf{q}_2 = \mathbf{p}_1 + \mathbf{p}_2 - \mathbf{K}, \tag{2}$$

where $\mathbf{p}_0 - \mathbf{p}_s = \mathbf{K}$, the momentum transfer vector. As a result, measurement of $\mathbf{p}_0$, $\mathbf{p}_s$, $\mathbf{p}_1$, and $\mathbf{p}_2$ for each collision is sufficient to give $\mathbf{q}_1 + \mathbf{q}_2$, the net momentum of the two electrons in the target. By repeating the measurement many times it is possible to determine the two-electron net momentum distribution. For the purpose of studying the correlated motion of the two target electrons, it is also important to know the distribution of the relative motion, $\mathbf{q}_1 - \mathbf{q}_2$, of the two electrons [5]. This can only be accomplished if the details of the ionization mechanism are known.

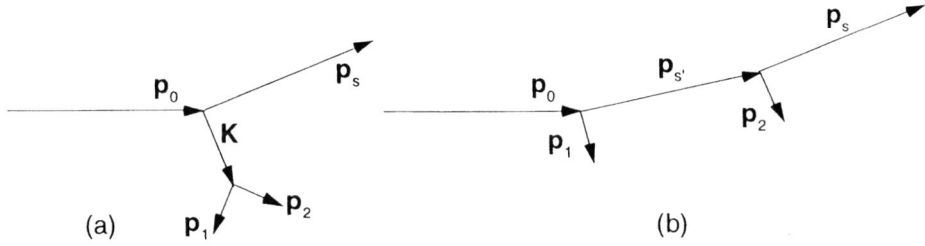

Figure 1. Double ionization by the (a) TS1 and (b) TS2 mechanisms.

TS1 and TS2 are two direct double ionization mechanisms that have been proposed [6]. For the TS1 mechanism, the incident electron collides with one electron in the target and is scattered. The struck electron collides with a second target electron and they both leave the target with momenta p_1 and p_2. If we require that the two ejected electrons have the same energy, and that $q_1 + q_2 = 0$, the electrons will emerge from the collision at nearly 90° to each other with the vector $p_1 + p_2$ along K. This is illustrated in Fig. 1(a).

For the TS2 mechanism, the incident electron first strikes one electron knocking it from the target and then collides with the second electron knocking it out in turn, Fig. 1(b). If we require that the ejected electrons have the same energy, the directions at which the TS2 ejected electrons emerge from the target are tightly constrained.

Figures 2(a) and 2(b) show in perspective the kinematics of the TS1 and TS2 mechanisms scaled for a 1000 eV incident electron, a 838 eV scattered electron at 22° to the incident electron direction, and two 70 eV ejected electrons. These are the conditions for the ionization of the two 3s electrons of magnesium. The momentum transfer vector, K, lies in the plane of p_0 and p_s. In Fig. 2(a) the base of the 90° cone represents the locus of the ends of all possible p_1 and p_2 vectors. The plane defined by p_1 and p_2 can assume all possible orientations about K. Detectors placed at positions equivalent to the location of the base will intercept the two ejected electrons. In Fig. 2(b) we show the first collision of the TS2 mechanism in which the incident electron

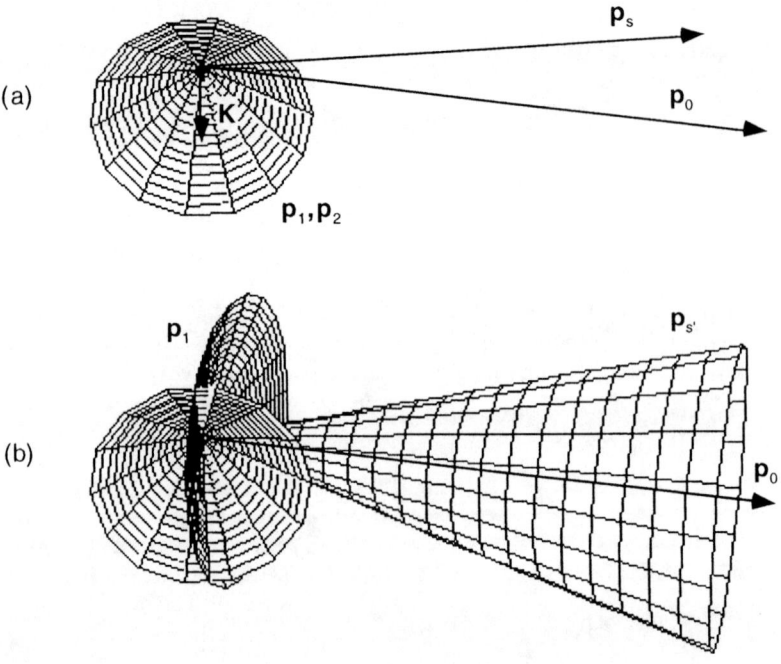

Figure 2. Kinematics for (a) TS1 and (b) the first collision of TS2.

with momentum $\mathbf{p}_0$ ejects the first electron from the target with momentum $\mathbf{p}_1$. The momentum of the ejected electron is $\mathbf{p}_1$ and the base of the $\mathbf{p}_s$ cone is the locus of the ends of all possible scattered electron momentum vectors. The base of the $\mathbf{p}_1$ cone is the locus of the ends of all possible ejected electron momentum vectors, however if $\mathbf{p}_1$ is to be detected at 45° to $\mathbf{K}$ as for the TS1 mechanism, only two $\mathbf{p}_1$ vectors are allowed. Their positions are at the intersection of the $\mathbf{p}_1$ cone with the 90° cone with axis along $\mathbf{K}$. The second ejected electron momentum vector, $\mathbf{p}_2$, is not shown in Fig. 2(b), but, as for the TS1 case, it is at 90° to $\mathbf{p}_1$ and in a plane containing $\mathbf{p}_1$ and $\mathbf{K}$. This plane is very nearly perpendicular to the scattering plane.

EXPERIMENT

We have constructed an experiment to take advantage of the differences between the TS1 and TS2 mechanisms in order to distinguish between them. This is shown in Fig. 3. The experiment consists of a scattered electron detector/analyzer that defines

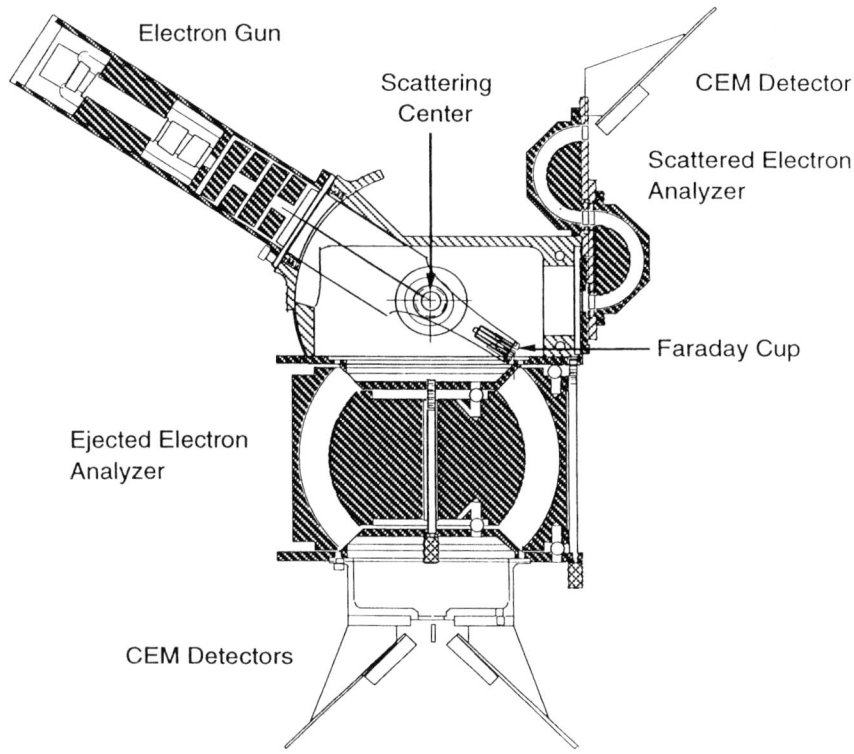

Figure 3. Cross-section of the double ionization spectrometer in the scattering plane showing the electron gun, ejected and scattered electron analyzers, detectors, Faraday cup, and scattering center. The geometry is for 1000 eV incident electrons on a magnesium target and two 70 eV 3s ejected electrons.

the scattering plane and a multiple-detector/analyzer centered on the momentum transfer direction and positioned to accept electrons of equal energy emerging from the scattering center at 45° to the momentum transfer direction. For a TS1 mechanism, coincidences will be observed between all oppositely placed detectors. For a TS2 mechanism, the coincident detection of ejected electrons will only occur in a plane nearly perpendicular to the scattering plane.

The above analysis is based on a number of simplifying assumptions:

1. $\mathbf{q}_1 = \mathbf{q}_2 = 0$
2. Zero electron binding energies
3. No interactions of the incident electron with the core or the scattered and ejected electrons with each other and the residual ion
4. $\mathbf{q}_{ion_i} = \mathbf{q}_{ion_f} = 0$

If the knocked-out electrons have non-zero values of $\mathbf{q}_1$ and $\mathbf{q}_2$ at the instant of ejection, they will be registered by detector pairs other than those that are opposite each other. In fact, it is the distribution of coincidence events among all the detector pairs that provide the information about two electron momentum distributions. For binding energies much less than the incident, scattered and knocked-out electron kinetic energies, the changes in kinematic geometries differ from the zero-binding energy kinematics by a few degrees. These differences can be calculated precisely from the energy level structure of the target.

If, in the course of the double ionization collision, sufficient momentum is transferred to the ion core to raise it to an excited state, the incident electron energy minus the binding energies of the two electrons will no longer be equal to the kinetic energies of the scattered and ejected electrons. Energy analysis of the incident, scattered and knocked-out electrons allows rejection of the $\mathbf{q}_{ion_i} \neq \mathbf{q}_{ion_f}$ events that result in core excitation.

Because of the long-range nature of the coulomb potential, the trajectories of the electrons participating in the double ionization will have their trajectories distorted from those of the simple direct knock-out model. These distortions can be reduced, as has been the case in (e,2e) single electron momentum density experiments, by working at high incident, scattered and knocked-out electron energies. In this case, *high* will have to be determined experimentally, the criterion being momentum densities that are energy independent [3].

No matter what experimental geometry is chosen, it is unlikely that all observed double ionization events will be the result of a single mechanism. There are a number of ways of distinguishing among the different possible mechanisms within the context of our experiment. The $\mathbf{p}_1$, $\mathbf{p}_2$ analyzer along the momentum transfer direction can discriminate between TS1 and TS2 ejection through angular correlation analysis as has been shown. Another way of separating TS1 and TS2 events is by the energy dependences of the cross sections. For TS1, we expect the double ionization cross section to vary as, $1/E_0^2$ where E_0 is the energy of the incident electron. For TS2 the variation is as $1/E_0^4$ for fixed ejected electron energies. There are also a number of geometric elements in a double ionization experiment [7, 8, 9, 10] that can be used to distinguish among double ionization events. The variation in cross section with the angle between p_1 and p_2, the angle between $p_1 + p_2$ and p_0, and the angle between the p_1, p_2 and p_0, p_s planes can all be used for this purpose. Finally, there is the extraction of $|q_1 - q_2|$ distributions from the experimental data. For the TS2 mechanism $|q_1|$ and $|q_2|$ for each of the knocked-out electrons can be separately determined from measurements of azimuthal angles with respect to the $q_1 = q_2 = 0$ plane, while for the TS1 mechanism only $|q_1 + q_2|$ will be able to be determined.

CONCLUSIONS

An electron impact double ionization spectrometer has been constructed for measuring two-electron momentum distributions of the 3s electrons in magnesium. The geometry of the spectrometer has been chosen in a way that fixes the collision mechanism and therefore separates the physics of the ionizing collision from the structure of the target.

ACKNOWLEDGEMENT

Research supported by NSF grant PHY-99-87870.

REFERENCES

1. McCarthy, I. E., Weigold, E., Zhang, X., and Zheng, Y., *J. Phys. B: At. Mol. Opt. Phys.* **22**, 931-938 (1989).
2. Meville, J. J., Zheng, Y., Hollenbone, B. P., Cann, N. M., Brion, C. E., Kim, C.-K., Wolf, S, *Can. J. Phys.* **74**, 773 (1997).
3. Pinkás, A. A., Coplan, M. A., Moore, J. H., Jones, S., Madison, D. H., Rasch, J., Whelan, C. T., Allan, R. J., and Walters, H. R. J., "Triple Differential Cross Sections for Electron Impact Ionization of Helium, Neon, and Argon from 0.1 to 1 keV, Theory and Experiment Compared," in *New Directions in Atomic Physics*, edited by C. T. Whelan, et al., Kluwer/Plenum, New York, 2001, pp. 319-332.
4. Carlson, T. A. and Krause, M. O., *Phys. Rev.* **140**, 1057 (1965).
5. Wang, J. and Smith, V. H., *J. Chem. Phys.* **99**, 9745 (1993); Banyard, K. E. and Mobbs, R. J., *J. Chem. Phys.* **88**, 3788 (1988); Mobbs, R. J. and Banyard, K. E., *ibid* **78**, 6106 (1983).
6. Lahmam-Bennani, A., Ehrhardt, H., Dupré, C., and Duguet, A., *J. Phys. B: At. Mol. Opt. Phys.* **24**, 3645-3653 (1991).
7. Treiman, S. B. and Yang, C. N., *Phys. Rev. Letters* **8**, 140 (1962).
8. McGuire, J. H., *Adv. At. Mol. Opt. Phys.* **29**, 217 (1991).
9. Yu. F., Pavlitchenkov, A. V., Levin, V. G., and Neudatchin, V., *J. Phys. B* **11**, 3587 (1978).
10. Levin, V. G., Neudatchin, V., Pavlitchenkov, A. V., and Smirnov, *J. Phys. B* **17**, 1525 (1984).

Ionization of Helium by an Intense High Frequency and Ultrashort Laser Field: a Progress Report

B. Piraux*, G. Lagmago-Kamta+ and J. Bauer*

**Laboratoire de Physique Atomique et Moléculaire,*
Université Catholique de Louvain, 2, Chemin du Cyclotron,
B1348 Louvain-la-Neuve, Belgium
+*Department of Physics and Astronomy, The University of Nebraska,*
116, Brace Laboratory, Lincoln, NE 68588-0111, USA

Abstract. The interaction of helium with a very intense ultrashort laser pulse (whose photon energy exceeds the double ionization potential) is studied by solving the corresponding time-dependent Schrödinger equation in its full dimensionality. At low intensity, our results for the total ionization yield agree very well with those obtained by using accurate one-photon total ionization rates. A momentum space analysis of the ionizing two-electron wavepacket allows us to extract qualitative but also quantitative information about various single and double ionization processes.

INTRODUCTION

The self-amplified spontaneous emission at VUV wavelengths in a free electron laser (FEL) has been recently observed at the DESY Laboratory (Hamburg, Germany) [1]. This breakthrough opens the route to an experimental study of the interaction of matter with a very intense electromagnetic field of much higher frequency than can presently be achieved. In the short term *i.e.* within 2 or 3 years, the FEL at DESY is expected to deliver ultrashort pulses (in the femtosecond domain) whose wavelength ranges from 6.4nm to 20nm and with a peak intensity of about $5x10^{17}$Watt/cm². In this new regime of frequency and intensity, the basic atomic processes are not well understood and experiments are not easy to design as long as quantitative atomic data are not available. These are the principal reasons that motivate the present work on the interaction of helium with strong high frequency fields.

CP604, *Correlations, Polarization, and Ionization in Atomic Systems*
edited by D. H. Madison and M. Schulz

At field intensities of the order of 5×10^{17}Watt/cm^2, the strength of the atom-field interactions may be of the same order or even exceed the strength of the electron-electron interactions. This means that our problem becomes essentially a time-dependent three-body problem. We are therefore inevitably led to solve numerically the corresponding time-dependent Schrödinger equation (TDSE). In the case of a two-electron system, this represents a very challenging task although important progress has been achieved in the last two years [2]. In the present contribution, we describe our numerical approach to solve the TDSE. We analyze the total wavefunction of the system after its interaction with the pulse both in position and momentum space and extract qualitative and quantitative information about various ionization processes. In the low intensity regime, our results for the total ionization agree very well with those obtained by averaging on the intensity profile of the pulse, the one-photon total ionization rates of Pont and Shakeshaft [3].

Atomic Structure Calculations

Our numerical approach is of configuration interaction (CI) type. It is based on an expansion of the total wavefunction in terms of an antisymmetrized product of Coulomb-Sturmian functions of the electron radial coordinates, and spherical harmonics coupled for a total angular momentum L. The key features of the present method are the use of different numbers of Coulomb-Sturmian functions for each electron, the inclusion of many different pairs of Sturmian non-linear parameters in the expansion as well as complex scaling.

A Coulomb-Sturmian function for a given angular momentum l is given by:

$$S_{l,n}^{\kappa}(r) = N_{n,l}^{\kappa} r^{l+1} e^{-\kappa r} L_{n-l-1}^{2l+1}(2\kappa r), \qquad (1)$$

where $N_{n,l}^{\kappa}$ is a normalization constant and L_{n-l-1}^{2l+1} denotes a Laguerre polynomial. κ is the Sturmian non-linear parameter and the index n takes the values $l+1$, $l+2$, $l+3$, ... The Coulomb-Sturmian functions form a complete basis set and they are exact solutions of the Schrödinger equation for a single electron in the field of a nucleus of charge Z. If $\kappa = Z/n$, the Coulomb-Sturmian function coincides with the corresponding hydrogenic wavefunction of principal quantum number n. It is therefore clear that both the non-linear parameter κ and the index n (which fixes the number of nodes) control the radial spread of the function. The introduction of several sets of Sturmian non-linear parameters allows to span a larger region of space and permits a simultaneous description of many eigenstates. We demonstrated recently [4], that our method provides very accurate energies for the high lying singly and doubly excited states even for large total angular momenta while keeping the size of the basis reasonable. In the case of the doubly excited states, complex scaling has been introduced to have direct access to their energy width.

The present approach is probably the most efficient one for *ab-initio* calculations dealing with asymmetrically excited states since its convergence and accuracy *increases* with the degree of excitation of one of the two electrons.

As for all CI approach, convergence is slow for the ground state due to a cusp in the two-electron wavefunction, which is a consequence of the singularity of the Coulomb repulsion operator when both electrons 'occupy the same position'. However, by choosing various sets of Sturmian non-linear parameters, we obtain a reasonable and sufficient accuracy (5 digits) with a relatively small basis size. Finally, let us mention that within the present approach, the calculation of the atomic wavefunction in the momentum space is straightforward. This requires to perform the Fourier-Bessel transform of the Coulomb-Sturmian functions which gives Gegenbauer polynomials.

Time Propagation

We adopt the velocity form for the interaction Hamiltonian. We propagate the full wavefunction in the atomic basis (in which the atomic Hamiltonian is diagonal) for the following reasons:(i) the system of coupled first order differential equations to solve, for the amplitude of each atomic state, is a non-stiff problem; (ii) before the time-propagation, the accuracy of the eigenstate energies can be easily assessed; (iii) during the time propagation, the evolution of the population of a given atomic state can be traced (or even removed) from the wavefunction. In addition, we use the interaction picture which permits to extract from the wavefunction, the fast oscillations due to the field free atomic energies. The time propagation is performed by means of an explicit embedded Runge-Kutta method [5] which allows an automatic control of the time step during the propagation.

The accuracy and the convergence of the results depend crucially on the density of states in all continua taken into account. It is important to make sure that within the frequency bandwidth of the pulse, the number of continuum states present and well reproduced by our basis is sufficiently large. Otherwise, those states may be 'viewed' by the system as pseudo bound states which may lead to entirely spurious dynamical effects. Within the present approach, the choice of appropriate Sturmian non-linear parameters is important in this context.

In all calculations presented here, we used a basis that takes into account 5 total angular momenta and all together, about 13000 atomic states.

Results and Discussion

We consider a sine square pulse whose full duration is 10 optical cycles. The photon energy is equal to 3.271 a.u. which drives the system 10 eV above the double ionization threshold. In Fig.1, we show our results for the total ionization yield P_{ion} as a function of the peak field intensity.

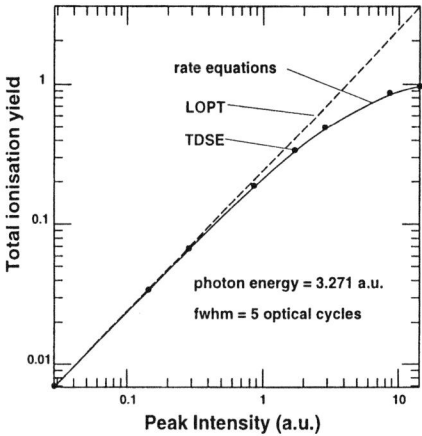

FIGURE 1. Total ionization yield as a function of the peak field intensity in a.u. for helium exposed to a sine square pulse whose total duration is 10 optical cycles and photon energy, 3.271 a.u.. TDSE results are compared to those obtained with lowest order perturbation theory and the rate equations.

The ionization yield is given by 1 minus the total population left in all states below the double ionization threshold at the end of the interaction. Our results are compared to those obtained with the lowest order of perturbation theory (LOPT) and with what is referred to as the rate equation given by:

$$P_{ion} = 1 - exp\left(\int_{-\infty}^{\infty} \Gamma(I(t))dt\right);\qquad(2)$$

where $I(t)$ is the instantaneous field intensity and Γ the one-photon total ionization rate of Pont and Shakeshaft [3]. The agreement between our results and those obtained by the rate equation is excellent. We performed similar calculations at a lower photon energy (3 a.u.): in that case, saturation occurs at lower intensity as expected and multiphoton absorption becomes significant at high intensity so that Eq. 2 is not valid any more. In Fig. 2, we show for a photon energy of 3 a.u., a radial contour plot of the angles-integrated two-electron probability density both in the position and momentum space at the end of the interaction. The peak intensity is 1.4 a.u. or 5×10^{16} Watt/cm^2. It is important to stress that all negative energy components of the total wavefunction have been substracted so that we are actually dealing here with the *ionized* two-electron wavepacket. In the position space, the probability density exhibits a series of maxima along the axes which are the signature of single ionization which is the dominant process here. The maximum around $(r_1, r_2) = (4,4)$ is attributed to direct (non-sequential) double

111

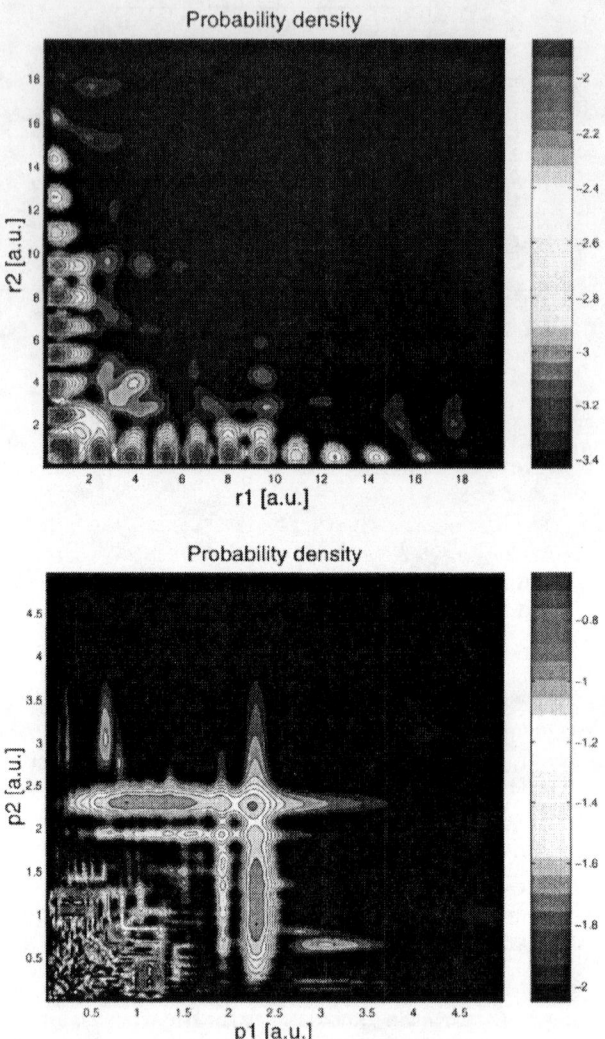

FIGURE 2. Radial contour plot of the angles-integrated two-electron probability density in the position space (above) and momentum space (below) at the end of the interaction of He with a 10 optical cycle pulse whose frequency is 3 a.u. and peak intensity 5×10^{16} Watt/cm^2. Logarithmic scale is used.

ionization. The elongated shape of the maxima along each axis indicates also that sequential double ionization is becoming important. At higher intensity or for longer pulses, we checked that this is the dominant double ionization process. The contour plot of the probability density in the momentum space is very instructive.

The long stripes at $p_1 = 2.3$ a.u. (or $p_2 = 2.3$ a.u.) correspond to single ionization with He$^+$ left in its ground state. We have checked by making a cut through the stripe and parallel to the vertical axis, that the distribution in p_2 coincides within a factor with the momentum distribution of the ground state of He$^+$.

More intriguing is the second stripe that occurs around $p_1 = 1.9$ a.u. (or $p_2 = 1.9$ a.u.). This structure is in fact present for a wide intensity interval ranging from very low values, well in the perturbative regime till higher values where saturation occurs. The structure stays present if we double the pulse duration. In addition, simulations based on a different approach show also this structure for a wide range of photon energies ranging from the second single ionization potential till above the double ionization potential [6]. A more deep analysis is necessary to understand the physical origin of this effect. The shorter stripes around $p_1 = 1.2$ a.u. and $p_1 = 1.1$ a.u. correspond to single ionization with He$^+$ left in the n=2 and n=3 levels respectively. By projecting the total wavefunction on a product of a bound state (1s, 2s and 2p) and a continuum state (radial Coulomb function) of He$^+$, the ratio of the probability for He$^+$ to stay in the n=2 levels to the probability to stay in its ground state is consistent, at low field intensity, with the calculations of Burke et al [7]. Note that the position of various maxima (position of the stripes) do not coincide exactly with the expected value. This is attributed to the ponderomotive and AC Stark shifts which represent about 3 percent of the photon energy in the present case. On the other hand, the width of the stripes and in particular the one corresponding to He$^+$ left in its ground state is consistent with the frequency bandwidth of the pulse which in the present case is equal to 0.46 a.u..

We tried in this short contribution to give a brief account of our method and to show its potentiality. The basis used to develop the helium wavefunction is particularly adapted to asymmetrically excited states which makes this basis specially adapted to the study of the dynamics of two-electron Rydberg wavepackets. We presented here results about the ionization of helium by a strong high frequency laser pulse. They demonstrated the existence of new unexpected features. This requires a deeper and more quantitative analysis which we have planned. In addition, there is no particular obstacle to extend our calculations to longer and more realistic pulse durations, higher intensities and other frequencies.

Acknowledgments

J.B. who is on leave of absence of the Department of Nuclear Physics and Radiological Safety of the University of Lodz, Poland, thanks the 'Fonds National de la Recherche Scientifique de la Communauté Francaise de Belgique' for his post doctoral position at the Université Catholique de Louvain, Louvain-la-Neuve, Belgium.

113

References

1. Andruszkow J., Aune B., Ayvazyan V. *et al, Phys. Rev. Lett.* **85**, 3825-3829 (2000).
2. Parker J.S., Moore L.R., Meharg K.J., Dundas D.D., and Taylor K., *J. Phys. B: At. Mol. Opt. Phys.* **34**, L69-L78 (2001).
3. Pont M., Shakeshaft R., *J. Phys. B: At. Mol. Opt. Phys.***28**, L571-L577 (1995).
4. Lagmago-Kamta G., Piraux B. and Scrinzi A., *Phys. Rev. A***63**, 040502-4(R) (2001).
5. Hairer E. NØrsett S.P. and Wanner G.,*Solving Ordinary Differential Equations II, Nonstiff Problems*, Springer-Verlag, Berlin, 1987, pp. 167-175.
6. Bachau H., private communication.
7. Jacobs V.L. and Burke P. G., *J. Phys. B: At. Mol. Phys.***5**, L67-L70 (1972).

Role of the He bound state wavefunction in the interpretation of coincidence experiments

L.U.Ancarani[1] and Yu.V.Popov[2]

[1] *L.P.M.C., Institut de Physique, Technopôle 2000, 57078 Metz, France*
[2] *Institute of Nuclear Physics, Moscow State University, Moscow 119899 Russia*

Abstract. In the present contribution we focus on the importance of the initial bound state wavefunction, and in particular on the essential role played by the electron-electron corelation, in the interpretation of several coincidence experiments on helium. Theoretical issues are briefly presented in order to illustrate that wavefunctions used in practical atomic calculations are far from the formal solutions, and questions are also raised with respect to the interpretation of recent transfer ionization measurements.

Simple bound state wavefunctions, used with success for calculations of elastic and inelastic scattering, are not necessarily satisfactory to describe the details of many-electrons ionization processes. Many efforts were concentrated to get the proper final state wavefunctions close to the formal solution of the Schrödinger equation [1-3]; however, the probing functions used up to now for the bound state are far from the formal solution. In this contribution we shall concentrate on the role played by the initial state wavefunction, starting from a few examples where the helium wavefunction maybe responsible for the lack of agreement between theory and experiment. We then discuss some theoretical issues about the formal solution, and finally present some preliminary interpretation of recent proton-helium transfer ionization measurements.

Several coincidence experiments on helium remain unexplained by available collision models which describe the continuum wavefunctions in various degree of sophistication. However, (e,2e) and especially (e,3e) ionization measurements in which the two electrons are active, have demonstrated the essential role played by the electron-electron correlation in the target ground state and that simple helium wavefunctions are not good enough. Also, the correct asymptotic behaviour of the wavefunction may lead to some qualitative interpretations which cannot be obtained with "traditional" wavefunctions.

A first example comes from the high energy dipolar (e,2e) regime where one observes in the triple differential cross section a persistent discrepancy in the vicinity of the recoil peak between the experimental data and the results of several calculations and references therein) with simple or complicated final state wavefunctions and "traditional" helium ground state wavefunctions [4]. It was shown [5] that it is the asymptotic behavior of both wavefunctions (bound and scattered) which is mainly responsible for the ratio of the binary to recoil peak heights. For the ionization-excitation process, in which the singly charged helium is left in the excited $n = 2$ state, discrepancies also exist [4]; it seems, however, that these should be attributed to second Born terms rather than to the helium wavefunction [6]. Another example comes from the surprising results on helium obtained in a series of (e,3e) experiments at high impact energies and small momentum transfers [7]. Further recent measurements in similar kinematics [8], on the other hand, show good agreement with theoretical calculations using conventional helium wavefunctions. Finally, in proton-helium transfer ionization process, unexpected discrete structure in the three-body final state momentum distributions were observed [9]. These and other unexplained results may be possibly related to the incorrect asymptotic behaviour of traditional trial wavefunctions and/or to unusual correlations in the helium ground state.

It is important to underline that no probing helium wavefunction used up to now in practical calculations (generally of the Hartree-Fock or Hylleraas type) has the correct behavior both at small and large hyperradius ρ even if very accurate energies can be obtained. The fact is that one has to compromise between building a wavefunction as close as possible to the formal solution on one hand, and practical for numerical calculations on the other.

In [10] we have outlined some deficiencies of widely used wavefunctions for the helium bound state are outlined, namely: the minimization procedure (Ritz variational method) to get the helium energy spectrum is not unique and conclusive on the quality of the proposed wavefunction [11]; the "local" energy $E(\vec{r}_1, \vec{r}_2) = \frac{\hat{H}\Phi}{\Phi}$ is never a constant value over the whole space; the eigenfunctions cannot be power series of the variables r_1, r_2 and r_{12} (the two radial coordinates and the electron-electron distance) and it has been suggested that a formal solution might contain logarithms [12-13]; the Hylleraas approximate basis is not a formal solution of the Schrödinger equation [12]; the cusp conditions are not often satisfied; finally, the asymptotic behavior was not well established until recently [10].

The fact that many issues about the helium ground state wavefunction are still unclarified has motivated us to study the Schrödinger equation for helium S-bound states, as far as possible without any preliminary approximations and simplifications [10]. With an original transformation we have obtained a system of channeling equations which have confirmed some known results and found some new ones. On one hand, a recurrancy system which gives automatically the proper behavior near the triple collisions point (the hyperradius $\rho \to 0$) has been derived. It was found that the helium bound state wavefunction includes logarithmic terms which are a result of the electron-electron interaction. This result supports the *ab initio* assumption of Bartlett [12] and Fock [13]. On the other hand, the study at large hyperradii leads to the following results:

1) The existence of two physical channels is identified: they coincide with those in nuclear physics if the interaction potential is zero, and determine the exact global asymptotics of the helium bound wavefunctions. The binary channel visually coincides with the traditional binary global asymptotics of the convergent wavefunction of fully stripped helium (see [14], or BBK-function [15], for example). It is interesting to note that, for this channel, the plane angle θ_{12} between electrons is a constant of motion, and this suggests that an asymptotic angular quantization may exist [16].

2) The rigorous solutions of the Hylleraas equation indicate that there is no place for the Hylleraas type exponent $e^{\delta r_{12}}$.

3) It is shown that the Sturmian basis is not adequate for describing the formal solution of helium bound states. Even the well-known binary channel wavefunction cannot be represented in the form $f(r_1)f(r_2)g(r_{12})$ or as a finite sum of such products.

These old and new results indicate that no trial helium wavefunction used in practical calculations respects at the same time all these demanding formal properties.

We now turn to another recent source of information on the helium ground state wavefunction which comes from experiments [9] in simple capture (SC) $p + He \rightarrow H^0 + He^+$ and transfer ionization (TI) $p + He \rightarrow H^0 + He^{2+} + e^-$ processes with fast projectile protons. Using the COLTRIMS technique, the momenta of fragments have been measured in coincidence for the first time, and this with high resolution, so that the experiment was kinematically complete. For TI, the momenta of He^{2+} and the polar θ_p and azimuthal φ_p scattering angles of the outgoing H atom were measured. The obtained differential cross sections $(d\sigma/d\theta_p)$ distinctly show a very large peak at extremely small hydrogen scattering angles θ_p (about 0 - 0.4 mrad) that cannot be attributed to the traditional electron capture channels (electron-electron Thomas, nuclear-electron Thomas). No theoretical model can presently describe such behaviour. A first interpretation of the data [17] seem to indicate that a new transfer ionization channel, strongly connected to the electron-electron correlation in the helium system, may be responsible for these new observations. Starting from the Born series approach it was found in a theoretical analysis [18] that, within the Plane Wave Born Approximation for proton-atom collision, strong similarities in the formulation appear between (e,2e) reactions and SC on one hand, and (e,3e) reactions and TI on the other. Contrary to (e,3e), in TI the velocity of the fast electron (captured by the proton) cannot be exactly measured. Transfer ionization processes, however, allow one to reach kinematical domains so far unreachable in (e,3e) spectroscopy and therefore provide invaluable information. Similarly to electron momentum spectroscopy, the differential cross section is directly related to the Coulomb-Fourier transform of the initial helium wavefunction [18-19] and the influence of electron-electron correlations could be tested.

117

The results of calculations [18] using uncorrelated and several correlated wavefunctions clearly show that the correlation in the initial state plays a very important role for TI but not for SC. Agreement between theory and experiment is improved with more correlated wavefunctions, particularly for TI where the two electrons are active in the process. Although some discrepancies with the measurements are present the model [18] is able to reproduce the presence and some of the features of the large peaks observed [9] and so far unexplained.

In an attempt to qualitatively interpret their measurements within the shake-off model, a quite unusual behaviour at asymptotic distances is proposed for the helium ground state wavefunction [17]. This makes sense since the proton passes at a rather large distances from the helium nucleus $(2 - 3$ a.u.$)$ and thus tests that part of the target. In this respect, we expect that the transfer ionization process will allow us to gain direct and invaluable information on the correlation which asymptotically governs the helium ground state.

The electron-electron correlation in the helium ground state plays an essential role in several ionization processes, in particular when the two electrons are active. In this contribution we have raised a few questions on the use in practical calculations of non formal helium wavefunctions and on the relevance that this may have in explaining some discrepancies between theoretical and experimental cross sections.

ACKNOWLEDGEMENTS

We thank C. Dal Cappello, H. Schmidt-Böcking, L. Schmidt, V. Bylik, J.H. McGuire, O. Chuluunbaatar, S.I. Vinitsky, H. J. Lüdde, E. Engel, C.T. Whelan and Y. Grandati for showing interest in our work. This publication is partially supported by the Russian Ministry for Industry, Science and Technologies (contract 108-39(00)-II).

REFERENCES

1. J.Berakdar, Phys. Rev. A **53**, 2314 (1996); **54**, 1480 (1996); **55**, 1994 (1997).
2. G. Gasaneo et al., Phys. Rev. A **55**, 2809 (1997).
3. Y.E. Kim and A.L. Zubarev, Phys. Rev. A **56**, 521 (1997).
4. C. Dupré et al., J. Phys. B: At. Mol. Phys. **25**, 259 (1992); P.J. Marchalant, C.T. Whelan and H.R.J. Walters, J. Phys. B: At. Mol. Phys. **31**, 1141 (1998).
5. Yu.V. Popov et al., Few Body Systems suppl. **10**, 235 (1999).
6. P. Marchalant et al., J. Phys. B: At. Mol. Phys. **33**, L749 (2000).
7. A. Kheifets et al., J. Phys. B: At. Mol. Phys. **32**, 5047 (1999).
8. A. Dorn et al., Phys. Rev. Lett. **86** 3755 (2001).
9. V. Mergel, Ph.D. thesis, Univ. Frankfurt, Shaker Verlag 1996; V. Mergel et al., Phys. Rev. Lett. **79** 387 (1997); V. Mergel et al., Phys. Rev. Lett. **86** 2257 (2001).

10. Yu.V. Popov and L.U. Ancarani, Phys. Rev. A **62** 42702 (2000).

11. H. Conroy, J. Chem. Phys. **41**, 1327 (1964)

12. J.H. Bartlett, J.J. Gibbons and C.G. Dunn, Phys. Rev. **47**, 679 (1935); J.H. Bartlett, Phys. Rev. **51**, 661 (1937)

13. V.A. Fock, Izvest. Akad. Nauk SSSR (Rus.) **18**, 161 (1954); Kgl. Norske Videnskabers Selskabs Forh. **31**, 138 (1958)

14. S.P.Merkuriev, Ann. Phys. (NY) **130**, 395 (1980).

15. M.Brauner, J.Briggs and H.Klar, J. Phys. B: At. Mol. Phys. **22**, 2265 (1989).

16. Yu.V.Popov and L.U.Ancarani, contribution to this volume.

17. H. Schmidt-Böcking *et al*, private communication.

18. Yu.V. Popov *et al*, submitted (2001).

19. V.G. Neudatchin, Yu.V. Popov and Yu.F. Smirnov, Physics Uspekhi **42**, 1017 (1999).

20. Yu.V. Popov, C. Dal Cappello and K. Kuzakov. J. Phys. B **29**, 5901 (1996).

Experimental investigation of the asymptotic momentum wave function of the He ground state

H. Schmidt-Böcking[a], V. Mergel[a], R. Dörner[a], O. Jagutzki[a], L. Schmidt[a],
T. Weber[a], C. L. Cocke[b], H. J. Lüdde[c], E. Weigold[d], Yu. V. Popov[e],
H. Cederquist[f], H. T. Schmidt[f], R. Schuch[f], J. Berakdar[g]

a. Institut für Kernphysik, Universität Frankfurt, August-Euler-Str. 6, 60486 Frankfurt, FRG
b Department of Physics, Kansas State University, Manhattan, Kansas 66506, USA
c. Institut für Theoretische Physik, Universität Frankfurt, Robert-Mayer-Str. 6, 60487 Frankfurt, FRG
d. Res. School of Phys. Sc. and Eng., Austr. Nat. Uni., Canberra 0200, Australia,
e. Nuclear Physics Institute, Moscow State University, Moscow 119899 Russia.
f. Department of Physics, Stockholm University, S 10405 Stockholm, S
g. MPI für Mikrostrukturphysik, Weinberg 2, 06120 Halle, FRG

Abstract. The correlated Kinematical Transfer Ionization channel cKTI in fast 4-body (p + He $\Rightarrow$ H° + He2+ + e) processes has been used to probe the highly correlated contributions to the asymptotic part of the He ground state wave function. In this reaction one electron with controlled initial momentum (2.5 to 7.5 a.u.) in the He ground state is kinematically captured by the proton at large impact parameters. The measured 3-particle final-state momentum distributions show features of a well-structured three-particle momentum wave function. We conclude that cKTI must almost exclusively proceed via the highly correlated and, virtually excited, non-s^2 contributions in the He ground state.

INTRODUCTION

Long-range correlation remains one of the most fundamental puzzles in the quantum mechanical world. Such correlations play a significant role in nature: they underlie much of the details in chemical bonding and prominent solid-state phenomena such as collective excitations (plasmons) and superconductivity. Furthermore, long-range correlation forces play a fundamental role in living matter, Bose-Einstein-Condensates (BEC) and halo nuclei [1]. Correlation in the asymptotic part of the atomic wave function plays a crucial role when atoms approach each other and may strongly influence the behavior of the atom in its chemical environment. Experimentally very little is known about this asymptotic part of the wave function, since direct measurements are rather difficult.

In atomic physics, the 3-particle wave function of ground state He provides the exemplary study of correlation in few-body systems [2]. It is one of the chemically least active atoms in nature and the non-s^2 part of the ground-state wave function is very small and is therefore difficult to probe with standard techniques such as spectroscopy. We will show that the high momentum components in the asymptotic part of the He ground state wave function contain significant non-s^2 contributions, both

CP604, Correlations, Polarization, and Ionization in Atomic Systems
edited by D. H. Madison and M. Schulz
© 2002 American Institute of Physics 0-7354-0048-2/02/$19.00

electrons necessarily composing an entangled $^1S_0$ state. In a Multi-Configuration Interaction (MCI) approach [3,4] these weak contributions are described by off-diagonal matrix elements (also called pseudo, off-shell, or virtually excited states) and together they are responsible for less than 10^{-6} of the total electronic density of the $^1S_0$ ground state. Nevertheless, these small fractions are of fundamental importance for the interaction of He with its environment [5] and [6]. It is to be stressed that while these non-s^2 asymptotic components are of less importance when it comes to the ground-state energy, the response of the ground-state He atom in certain reactions, as the one described below, proceeds only through the pathways of virtual states.

In this paper it is shown that the very weak non-s^2 components are essential for the correlated Kinematical Transfer Ionization (cKTI) process. This observation is unraveled by a novel multi-fragment coincidence technique. In the cKTI one considers the reaction $p+He => H^\circ + He^{2+} + e$. The projectile proton is very fast with respect to the mean velocity of the bound electrons. One of the electrons of He is resonantly captured by the fast proton at relatively large distance from the He nucleus [7,8]. From the simultaneously measured transverse nuclear momentum exchange we have qualitative information on the nuclear impact parameter and select for our analysis only the very small H° deflection angles, i.e. distant collisions. Furthermore the initial velocity of the captured electron always exceeds the mean K shell velocity in He. We can conclude that the electron is captured in momentum space from the asymptotic part of the He ground state and as we discuss below very unlikely from regions close to the He nucleus.

In all cKTI processes, the He^{1+} recoil instantaneously fragments and the second electron is emitted into the H° scattering plane with a well-defined momentum vector. From the momentum vectors of the scattered H° and the recoiling He^{2+} target nucleus, we directly determine the correlated momentum wave function of both electrons in the asymptotic part of the initial ground state of He [8]. This method allows the measurement of the local parts of the asymptotic wave function (on the level of one part in 10^8 - 10^{10} of the total wave function) not observable by other techniques such as energy resolved spectroscopy of the ground state.

In the study presented here the cKTI transfer ionization channel [7,8] is chosen. Here, a fast proton captures at distant collisions (H° deflection angles below $5 \cdot 10^{-4}$ rad) one He electron (named number 1) nearly exclusively to the H° 1s state. The second electron (2) is simultaneously left in the continuum of He^+. In this channel the proton transfers a virtual photon and thus energy, but only a negligible amount of momentum to the He system. The perturbation of the correlated momentum wave function of He by the proton is rather small and can be neglected in comparison with the large initial momentum vectors of the two electrons. Since the force of the fast departing neutral H° on the remaining He^{1+} decreases rapidly, final-state interactions between the scattered projectile and the continuum fragments is negligibly small in this process.

The cKTI transfer ionization channel competes with other TI reaction channels [9, 10,11,12,13,14,15] where each channel leads to a characteristic location in the H° and He^{2+} final-state momentum phase space.

EXPERIMENT

To experimentally distinguish the different channels the recoil momenta and projectile momentum transfer have to be measured in coincidence with extremely high resolution (better than one part in 10^5 with respect to the projectile momentum). That is, the recoil ion longitudinal and transverse energy distribution have to be detected with 100 µeV resolution compared to, e.g., the 1 MeV proton impact energy E_p, to separate the cKTI channel from the other competing processes. There are 9 degrees of freedom in the final state and, thus momentum and energy conservation requires that 5 final state momentum components (3 of the He^{2+} recoil and the 2 $H°$ transverse momentum components) are measured in coincidence. Further, very high detection efficiencies are needed as the cKTI cross section is typically only of the order of barns. Using the high momentum resolution and multi-coincidence efficiency of cold target recoil ion momentum spectroscopy (COLTRIMS) [16] the complete final-state momentum distributions for fast (400 to 1400keV) p + He => $H°$ + He^{2+} + e transfer ionization processes (TI) were systematically measured at the 2.5 MV van de Graaf accelerator of the Institut für Kernphysik of the Universität Frankfurt. Details of the $H°$ and He^{2+} coincidence experiment are given in reference [9]. The momentum resolution obtained is estimated to be < 0.3 a.u.

RESULTS AND DISCUSSION

For the cKTI process we can deduce from the measured final-state momentum distributions the initial-state momentum vector k_{e1} of electron 1. In Fig. 1 the pure

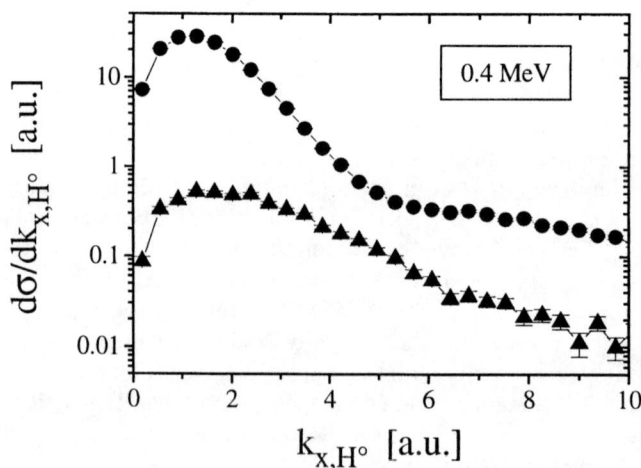

FIGURE 1. Singly differential pure capture (•) and TI (▲) cross sections as function of the $H°$ transversal momentum $k_{x,H°}$ for 0.4 MeV proton impact energy.

single-electron capture and TI cross sections are shown as functions of the $H°$ transversal momentum $k_{x,H°} = m_p v_p \Theta_{H°}$ at 0.4 MeV. At very small $H°$ deflection angles $\Theta_{H°}$ (region of peak structure) the proton is deflected only by the initial transverse momentum of the captured electron. Therefore the $H°$ transverse momentum $k_{x,H°}$ is identical to the electron 1 transverse momentum in the initial state (measured with respect to the incoming projectile direction). Since for cKTI the initial-state velocity of electron 1 and that of the projectile must match (Brinkman-Cramer type capture), the complete initial-state momentum vector $\mathbf{k}_{e1} = (k_x, k_y, k_z)$ of electron 1 (where k_x is the transverse momentum component, k_y is the component perpendicular to the $H°$ scattering plane and for each event is always set to zero, k_z is the longitudinal component) can be deduced from the measured $H°$ transverse momentum with

$$\mathbf{k}_{e1} = (k_{x,H°}, 0, \sqrt{(m_e v_p)^2 - (k_{x,H°})^2}\,). \qquad (1)$$

The final-state momentum vector of electron 2 is deduced from the measured recoil momentum vector and the $H°$ transverse momentum vector.

The surprising features of these measured correlated momentum patterns (see Fig. 2) are [7]:

1. The cKTI occurs only if the mean momenta of all 4 particles (the two electrons, the α nucleus, and the proton) are located in one plane.
2. The cKTI is most probable when electron (1), the recoiling He^{2+}, and electron (2) always share comparable momenta. In contrast, the cKTI process is unlikely if the momentum of these particles is peaked at zero.
3. Electron (2) is predominantly emitted into the backward direction.

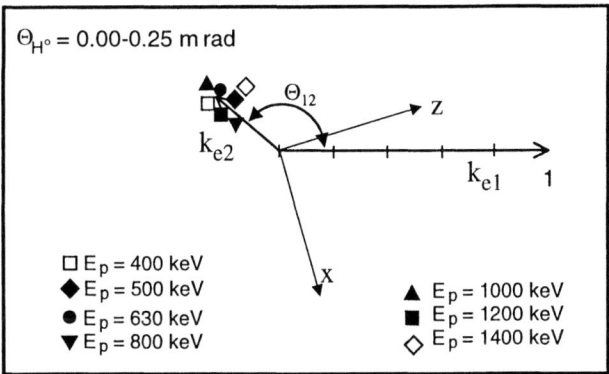

FIGURE 2. Initial state momentum relation between the two electron momenta $\mathbf{k}_{e1}$ and $\mathbf{k}_{e2}$ in He ground state derived from the data (see text) for different proton impact energies E_p. Momenta are scaled to $\mathbf{k}_{e1} = 1$.

In [8] it is shown in detail that the present data are in full agreement with earlier, but less differential measurements [11] and that the experimental findings cannot be explained by any of the previously known TI channels or reaction mechanisms. Further, the observed final-state momentum pattern cannot be created by any uncorrelated scattering of three Coulomb particles fulfilling the energy, momentum and angular momentum conservation laws [8].

Concerning observation 1: As suggested in [8,9], 4-particle momenta in one plane are expected if in the cKTI process angular momentum is transferred from the initial He to the scattered H° (i.e. the cKTI process only occurs when the proton impact vector is parallel to the 3-particle momentum plane). Furthermore, as discussed in [8,9], at these high velocities the proton neither excites the electrons into a He and projectile p state (transfers to the electrons angular momentum) nor does it exchange angular momentum in the elastic collision with the He^{2+} nucleus. Thus, the angular momentum of the electrons in the final-state must reflect properties inherent in the initial He ground state as highly correlated, higher angular momentum components of both electrons in the $^1S_0$ ground state. According to Multi Configuration Interaction (MCI) calculations the global He ground state wave function contains about 1% non-s^2 contributions (mainly p^2 contributions) and each electron can indeed provide such angular momentum. Since for all cKTI events, He^{2+} and electron (2) mean momentum vectors are in the H° scattering plane, the cKTI process must proceed via the capture of non-s^2 electrons in the asymptotic region and the observed momentum patterns must reflect the initial angular momentum correlation of the non-s^2 contributions in the He ground state asymptotic wave function.

Concerning observation (2) and (3) above: the question is how can one explain the high shake-off energy of electron (2), sometimes above 200eV? Since electron (1) is captured from a non-s^2 component of the initial state electron 2 can never have zero velocity. Furthermore since the capture of electron 1 proceeds always via virtual states with high initial momentum components, the momentum conservation favors high momentum components of electron 2. In addition since electron 1 is mostly captured when it has high forward momentum the backward emission of electron 2 at very small H° deflection angles is favorable due to momentum conservation and the observed correlation structure.

To learn more about the structure of the asymptotic correlated momentum wave function, we need to unveil the different data to look for any possible scaling behavior in the final-state momentum patterns. We therefore calculate from the final state H° momentum the momentum of electron 1 in the initial state with equation 1. The initial state momentum relation between the two electrons in the He asymptotic wave function derived from the data is presented in Fig. 2, where the measured momentum relations are shown for all investigated impact energies E_p at two very small H° scattering angles. While the momentum vector of the initial state $\mathbf{k}_{e1}$ is always plotted to the right and normalized to 1, the momentum vector $\mathbf{k}_{e2}$ is scaled in its length too for all ten different energies of the projectile E_p. The sign of the angle between the electrons Θ_{12} is defined by the H° deflection, – sign being opposite to the H° deflection. The z direction is defined by the incoming projectile and the x direction by the H° transverse deflection. For all systems investigated the relative angle between electron 1 and 2 appears nearly constant with $\Theta_{12} = -140°$ +/- 15°. Also the ratios of the

124

magnitudes of the momenta are constant within the experimental uncertainty of about 20%. It is striking to see that the cKTI process always yields a discrete momentum pattern.

The asymmetry of the momentum pattern in the final state (seen from above the plane in a clockwise direction) is striking. We find that electron 2 is always emitted opposite to the proton. This directional asymmetry points to the important role of high angular momentum components of the He wave function. Classically we can consider that electron (1) is captured by the proton when it is generally closer to the proton than to the α nucleus and r_{el} (the electron (1) position vector relative to the α nucleus) always points in the positive x direction. Furthermore the captured electron (1) has to move forward, i.e. in the positive z direction. This means that the orientation of the electron angular momentum, i.e. its m-value, always points upward with respect to the H° scattering plane. Since electron (2) must have exactly the opposite angular momentum, its angular momentum is also oriented with m= -1 (for p^2 components).

So far all theoretical attempts to explain these structures have failed completely. The theoretical studies of Popov et al. [18] demonstrate that correlated helium wave functions better reproduce the single differential cross sections both for TI and SC reactions, especially for large projectile energies, but many details (absolute values and angular position of maxima) were not well described. This can be understood in the following way. The authors used trial variational helium wave functions, which concentrated on the space domain out to 0.6 a.u. and did not include contributions of highly excited (continuum) states, which we observe. For projectile energies $E_p > 1$ MeV the impact parameter becomes comparable to the mean helium radius, and variational wave functions give better and better results. But these functions are not able to describe properly the asymptotic space domain important at smaller projectile energies. Recently Popov and Ancarani showed [17] that the asymptotic behavior of all "traditional" helium ground state wave functions is not correct, and the mutual angle of the bound helium electrons in the space asymptotic domain is an integral of motion (i.e. it is fixed).

These virtual highly excited continuum non-s^2 components have similarities with halo states in nuclei (e.g. ^{6}He or ^{11}Li, with a core and two nucleons rotating at large radii outside the classical nuclear potential), which always fragment into the core plus two free nucleons (Borromean states) [1], when one of the two nucleons is knocked off.

We conclude that the kinematical structures observed by Mergel et al. [7] can be interpreted by the kinematics of a cKTI process proceeding via generalized shake-off processes from non-s^2 contributions in the He ground state wave function. Furthermore we have shown that the non-s^2 contributions in the He groundstate wave function contain interesting properties with respect to the hidden world of correlation. The very fast electrons captured far from the He nucleus are possibly those with the strongest dynamical e-e-correlations. As shown in [8] the relative contribution of non-s^2 components increases with increasing momentum since the cKTI process becomes relatively more important with increasing proton energy.

ACKNOWLEDGEMENT

We acknowledge the numerous fruitful disputs and discussions with our colleges and friends. U. Ancarani, H. Belcik, S. Berry, J. Briggs, S. Brodzki, R. Dreizler, J. Feagin, B. Fricke, S. Fritsche, Ch. Greene, W. Greiner, J. McGuire, S. Hagmann, M. Horbatsch, E. Horsdal, P. Hoyer, D. Ionescu, D. Madison, J. Macek, S. Manson, R. Moshammer, C.D. Lin, R. Olson, J. Reading, R. Rivarola, J.M. Rost, V. Schmidt, J. Ullrich and many other collegues. V.M. acknowledges the support by the Studienstiftung des Deutschen Volkes. This work was supported by the DFG, the BMBF, GSI-Darmstadt, Graduiertenprogramm des Landes Hessen, Humboldt-Foundation, and Roentdek GmbH.

Electronic address: schmidtB@hsb.uni-frankfurt.de

REFERENCES

1. Zhukov et al. Physics Reports 231, 151 (1993)
2. Tanner G. et al., Rev. Mod. Phys. Vol. 72, No2, 497 (2000)
3. T. Kinoshita, Phys. Rev. $\underline{115}$, 366 (1959)
4. G. W. F. Drake, in Long Range Casimir Forces: Theory and Recent Experiments on Atomic S.D Systems, edited by F.S. Levin and D. A. Micha (Plenum, New York, 1994) p.107
5. V. M. Efimov, Comments Nucl. Part. Phys. $\underline{19}$, 271 (1990)
6. J. Macek, Z. Phys. D $\underline{3}$, 31 (1986)
7. Mergel et al. Phys.Rev.Lett. $\underline{86}$, 2257 (2001)
8. H. Schmidt-Böcking et al., J. Phys. B to be submitted (2001)
9. V. Mergel , PhD thesis, Shaker Verlag, ISBN 3-8265-2067-X (1996)
10. J.S. Briggs and K. Taulbjerg, J. Phys. B: At. Mol. Opt. Phys. 12 2565 (1979)
11. J.P. Giese and E. Horsdal, Phys. Rev. Lett. 60, 2018 (1988)
12. J. Palinkas et al., Phys. Rev. Lett. 63, 2464-67 (1989)
13. T. Ishihara and J. H. McGuire, Phys. Rev. A 38, 3310-3318 (1988)
14. J.H. McGuire et al., Phys. Rev. Lett. 62, 2933-36 (1989)
15. J. H. McGuire et al., J. Phys. B 28, 913 (1995)
16. R. Dörner et al., Physics Reports 330, 96-192 (2000)
17. Yu.V. Popov and L.U. Ancarani, PRA, 62, 42702 (2000)
18. Yu.V. Popov et al. to be published

Triply Differential Single Ionization Cross Sections in Fast Ion-Atom Collisions: Large versus Small Perturbation

M. Schulz[1], R. Moshammer[2], A.N. Perumal[2], D.H. Madison[1],
R.E. Olson[1], S. Jones[1], M. Foster[1], and J. Ullrich[2]

[1]*Physics Department and Laboratory for Atomic, Molecular, and Optical Research,
University of Missouri-Rolla, Rolla, MO 65409, USA*

[2]*Max-Planck-Institut für Kernphysik, Saupfercheckweg 1, 69117 Heidelberg, Germany*

Abstract. We compare experimental triply differential cross sections for single ionization of He for small (100 MeV/amu C^{6+} projectiles) and large perturbation (3.6 MeV/amu Au^{53+} projectiles). For electrons emitted into the scattering plane at the small perturbation, the data are well described by our calculations. For the large perturbation as well as outside the scattering plane for the small perturbation, in contrast, serious discrepancies especially with the continuum distorted wave – eikonal initial state (CDW-EIS) model are found.

One of the fundamental interests underlying essentially all research in atomic collision physics is the many-body problem. Our understanding of this topic is still rather incomplete especially for the intermediate particle numbers (between 3 and approximately 10) one typically encounters in atomic collisions. From a theoretical point of view the basic problem is that the Schrödinger equation is not solvable in closed form for more than two mutually interacting particles. It is therefore crucially important to have detailed experimental data in order to guide theoretical modeling efforts.

Kinematically complete experiments on ionization processes by electron impact have been performed for more than three decades [1-5] since the pioneering work by Ehrhardt et al. [6]. The vast majority of these studies were restricted to electrons emitted into the scattering plane defined by the initial projectile momentum vector $\mathbf{p_o}$ and the momentum transfer vector $\mathbf{q} = \mathbf{p_o} - \mathbf{p_f}$, where $\mathbf{p_f}$ is the scattered projectile momentum. For large projectile electron energies, i.e. for small perturbation Q/v (where Q and v are the projectile charge and velocity), the data are in general well reproduced by 1st order treatments [e.g. 7]. For large perturbation, in contrast, the description of the data remains challenging. Only recently significant theoretical progress was made with the development of asymptotically correct three particle

CP604, *Correlations, Polarization, and Ionization in Atomic Systems*
edited by D. H. Madison and M. Schulz
© 2002 American Institute of Physics 0-7354-0048-2/02/$19.00

continuum wavefunctions by Brauner, Briggs, and Klar (BBK) [8,9] and the convergent close coupling approach (CCC) [10]. For the e + H collision system, even mathematically complete solutions of the three-body Coulomb problem were reported [11-13].

For heavy ion impact the first kinematically complete experiment on single ionization was performed 7 years ago [14] and fully differential cross sections were only reported this year [15]. For such projectiles the continuum distorted wave – eikonal initial state approach (CDW-EIS) [16] usually works very well in describing total cross sections as well as differential ionized electron spectra for a wide range of perturbations. CDW-EIS was therefore generally viewed as the state-of-the-art theory for single ionization by heavy ion impact. However, recently Moshammer et al. [17] reported a complete break-down of CDW-EIS in describing doubly differential single ionization cross sections as a function of the electron energy and the projectile scattering angle for large perturbations.

The current status of research that emerges from single ionization studies both for electron and heavy ion impact shows that our understanding of the many-body problem is particularly incomplete under conditions where higher order contributions are important or even dominant (i.e. for large perturbations). In the present work we therefore pursued two approaches to investigate the role of such higher order contributions: First, we studied single ionization for a very small perturbation of $Q/v=0.1$ (100 MeV/amu C^{6+} + He) for a specific scattering geometry, where higher order effects are expected to be important although the total cross section is dominated by first order contributions. Second, we studied a collision system with very large perturbation of 4.4 (3.6 MeV Au^{53+} + He) for which already the total cross section is dominated by higher order processes.

The experiments were performed at GSI in Darmstadt (Au^{53+} projectiles) and GANIL in Caen (C^{6+} projectiles). The projectile beams were intersected with a cold (<1K) He beam from a supersonic gas jet. The projectiles which did not change their charge state were selected by a switching magnet and detected by a scintillator. The recoil ions and the ionized electrons were extracted in the longitudinal direction (defined by the initial projectile direction) by a weak electric field and detected by two position-sensitive channel plate detectors. Both detectors were set in coincidence with the projectile detector. A uniform magnetic field confined the transverse motion of the electrons so that all electrons with a transverse momentum of less than 3.5 a.u. (2 a.u. for C^{6+} projectiles) were guided onto the detector. The transverse momenta of the electrons and the recoil ions were calculated from their position on the respective detector and their longitudinal momentum components were determined from the time of flight. The momentum vector of the scattered projectile was deduced from momentum conservation.

From the data we obtained triply differential cross sections (TDCS) at fixed electron energies and fixed magnitudes of the momentum transfer vector for electrons ejected into the scattering plane as a function of the polar electron emission angle. These cross sections are shown for 100 MeV/amu C^{6+} + He for an electron energy of E_e = 6.5 ± 3.5 eV and a momentum transfer of q = 0.88 ± 0.11 a.u. in Fig.1. The so-called binary peak (near 90°) and recoil peak (near 270°), well known from electron

128

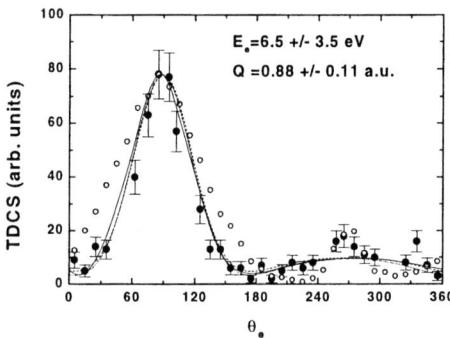

Figure 1. TDCS for electrons emitted into the scattering plane for 100 MeV/amu C^{6+} + He. Dashed curve: 1st Born; solid curve: BBK; open circles: CTMC

impact studies, can be seen. As expected, the location of these peaks agrees with the direction of **q** and –**q**, respectively. The dashed curve in Fig 1 represents a 1st Born approximation [18] and the solid curve a calculation based on the BBK approach [19]. The 1st Born approximation is in good agreement with both the experimental data and the BBK calculation. This confirms the expectation that for small perturbations the cross sections in the scattering plane are dominated by first order contributions.

One might expect that higher order contributions are more important in the plane perpendicular to the scattering plane and containing the initial projectile beam axis. An electron initially at rest can only be ejected in the direction of the momentum transfer vector (i.e. into the scattering plane) by a first order process because of momentum conservation. Therefore, an electron can only be ejected into the perpendicular plane either due to its momentum distribution in the initial bound state or by a higher order process, which therefore should be more important than in the scattering plane.

The TDCS for the perpendicular plane integrated over electron energies from 3 to 50 eV and over momentum transfers from 0.66 to 2 a.u. are shown in Fig. 2. The

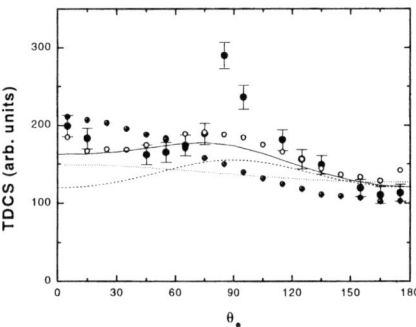

Figure 2. TDCS for electrons emitted into the perpendicular plane for 100 MeV/amu C^{6+} + He. Dashed curve: 1st Born; dotted curve: CDW-EIS; solid curve: BBK; open circles: CTMC; open circles with crosses: CTMC without nuclear-nuclear interaction

129

angular axis only covers the range from 0 to 180° because in the perpendicular plane the cross sections necessarily exhibit mirror symmetry with respect to 180°. Two prominent features are observed in the data: first, there is a peak structure near 90°, although it is not as pronounced as in the scattering plane. Second, in the forward direction (0°) the cross sections are almost twice as large as in the backward direction (180°).

The 1st Born calculation (dashed curve in Fig. 2), which works very well in describing the data in the scattering plane, essentially collapses in the perpendicular plane. It neither reproduces the peak near 90° nor the forward-backward asymmetry in the data. This shows that indeed in the perpendicular plane higher order contributions are significant. Somewhat improved agreement with the data is obtained with the BBK (solid curve), the CDW-EIS (dotted curve) and the classical trajectory Monte Carlo (CTMC) calculations (open circles). All of these higher order calculations predict a forward-backward asymmetry, although in the BBK and even more so in the CDW-EIS approach it is not as strong as in the data. Only the CTMC model predicts a maximum near 90°, however it is not as pronounced as in the data.

In order to study the origin of the observed features in the TDCS in the perpendicular plane we have performed test calculations where the projectile-target nucleus interaction was removed from the BBK and the CTMC calculations (the CDW-EIS does not include this interaction to begin with). In the case of the BBK approach the result is essentially identical to the 1st Born approximation, i.e. the forward-backward asymmetry has disappeared. For the CTMC model, the corresponding result is shown as the open circles with crosses. Here, the peak near 90° has disappeared, but the forward-backward asymmetry still remains.

The comparison between these test calculations and with the original results shows that the role of the projectile-target nucleus interaction is very different in the BBK and the CTMC model. In the BBK method it leads to the forward-backward asymmetry. In the CTMC approach, on the other hand, it is clearly not the interaction between the two nuclei leading to this asymmetry since it is still present in the test calculation. Here, this interaction leads to the peak structure near 90°. The only difference between the 1st Born approximation and the CDW-EIS method is that the latter includes the post-collision interaction (PCI) between the outgoing projectile and the ionized electron. Therefore, in the CDW-EIS the forward-backward asymmetry can be associated with the PCI. This comparison between the various higher order treatments shows that these models do not even agree on which mechanism leads to what features in the cross sections for the perpendicular plane.

For the Au^{53+} projectiles (large perturbation) the situation is fundamentally different from the one for the C^{6+} projectiles (small perturbation) in that for the former even the total cross section is no longer dominated by first order contributions. For large perturbations we therefore expect pronounced higher order effects already in the scattering plane.

In Fig. 3 we show the TDCS for the Au^{53+} projectiles in the scattering plane for an electron energy of 55 ± 5 eV and a momentum transfers of 0.65 ± 0.15. Here, we also observe two peaks which are, however, significantly shifted compared to the direction of $\mathbf{q}$ and $-\mathbf{q}$. One of these peaks asymptotically approaches $\mathbf{q}$ for large magnitudes of $\mathbf{q}$

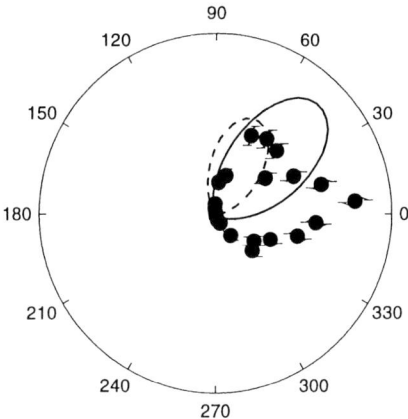

Figure 3. TDCS for electrons emitted into the scattering plane for 3.6 MeV/amu Au^{53+} + He. Dashed curve: 1st Born; solid curve: CDW-EIS multiplied by 35

and we therefore associate it with the binary peak. For the second peak the shift is much larger (more than 100°) and it stays nearly fixed at 0° for all electron energy/momentum transfer combinations studied. By elimination we associate this peak with the recoil peak. Such a shift, although significantly smaller, is a phenomenon well-known from electron impact studies and it is mainly due to the PCI. Our data show that the PCI affects the recoil peak particularly strongly, but its impact on the binary peak is comparatively weak. We also note that for small momentum transfers the recoil peak is the dominating feature in the data.

The dashed and solid curves in Fig. 3 show 1st Born and CDW-EIS calculations. Not surprisingly, the 1st Born approximation is in poor agreement with the data. The recoil peak is very small and the shift of the binary peak is not reproduced by this calculation. The CDW-EIS model qualitatively reproduces the shift of the binary peak. However, in the magnitude this calculation drastically differs from the data in some cases by more than an order of magnitude. Even more surprising is the complete absence of the recoil peak in the CDW-EIS calculation, which is the dominant contribution in the data for small momentum transfers. It is remarkable that a model which utterly fails to reproduce the TDCS can, for exactly the same collision system, very well describe doubly differential ionized electron spectra [20].

In summary, we have performed kinematically complete experiments and presented triply differential single ionization cross sections for fast heavy ion impact at very small and very large perturbation. While the data for small perturbation in the scattering plane are well described by a first order treatment, both first order and higher order models are in poor agreement with the data in the perpendicular plane. Here, various higher order approaches do not even agree on the mechanisms leading to the characteristic features in the cross sections. For the large perturbation pronounced higher order effects are observed already in the scattering plane. Here the CDW-EIS approach, which so far was considered the state-of-the-art theory for ionization by ion impact, fails to describe the data qualitatively and quantitatively. We conclude that our understanding of higher order contributions to single ionization is far from being complete.

One of us (M.S.) is grateful for the hospitality of the Max Planck Institut für Kernphysik. This work was supported by GSI, NSF, DOE and the Deutsche Forschungsgemeinschaft within the Leibniz-program. Fruitful discussions with Drs. B. Najjari and A.B. Voitkiv are acknowledged.

REFERENCES

1. Ehrhardt, H. , Jung, K., Knoth, G., and Schlemmer, P., Z. Phys. D1, 3 (1986) and references therein
2. Stefani, G., Avaldi, L., and Camilloni, R., J. Phys. B13, L673 J. Phys. B23, L227 (1990)
3. Lahmann-Bennani, A., J. Phys. B24, 2401 (1991) and references therein
4. Dorn, A., Elliot, A., Guo, X., Hurn, J., Lower, J., Mazevet, S., McCarthy, I.E., Shen, Y., and Weigold, E., J. Phys. B30, 4097 (1997)
5. Dorn, A., Kheifets, A., Schröter, C.D., Najjari, B., Höhr, C., Moshammer, R., and Ullrich, J., Phys. Rev. Lett. 86, 3755 (2001)
6. Ehrhardt, H., Schulz, M., Tekaat, T., and Willmann, K., Phys. Rev. Lett. 22, 89 (1969)
7. Marchalant, P., Whelan, C.T., and Walters, H.R.J., J. Phys. B31, 1141 (1998)
8. Brauner, M., Briggs, J.S., and Klar, H., J. Phys. B22, 2265 (1989)
9. Jones, S. and Madison, D.H., Phys. Rev. Lett. 81, 2886 (1998)
10. Bray, I. and Stelbovics, A.T., Phys. Rev. Lett. 69, 53 (1992)
11. Pindzola, M.S. and Robicheaux, F., Phys. Rev. A54, 2142 (1996)
12. Rescigno, T.N., Baertschy, M., Issacs, W.A., and McCurdy, C.W., Science 286, 2474 (1999)
13. Buffington, G.D., Madison, D.H., Peacher, J.L., and Schultz, D.R., J. Phys. B32, 2991 (1999)
14. Moshammer, R., Ullrich, J., Unverzagt, M., Schmitt, W., Jardin, P., Olson, R.E., Mann, R., Dörner, R., Mergel, V., Buck, U., and Schmidt-Böcking, H., Phys. Rev. Lett. 73, 3371 (1994)
15. Schulz, M., Moshammer, R., Madison, D.H., Olson, R.E., Marchalant, P., Whelan, C.T., Walters, H.R.J., Jones, S., Foster, M., Kollmus, H., Cassimi, A., and Ullrich, J., J. Phys. B34, L305 (2001)
16. Crothers, D.S.F. and McCartney, M., Comp. Phys. Comm. 72, 288 (1992)
17. Moshammer, R., Perumal, A.N., Schulz, M., Kollmus, H., Mann, R., Hagmann, S., and Ullrich, J., submitted to Phys. Rev. Lett.
18. Madison, D.H., Phys. Rev. A8, 2449 (1973)
19. Jones, S., and Madison, D.H., Phys. Rev. Lett. 81, 2886 (1998)
20. Moshammer, R., Fainstein, P.D., Schulz, M., Schmitt, W., Kollmus, H., Mann, R., Hagmann, S., and Ullrich, J., Phys. Rev. Lett. 83, 4721 (1999)

DIFFERENTIAL ENERGY LOSS MEASUREMENTS FOR SINGLE AND MULTIPLE IONIZATION BY POSITRONS AND ELECTRONS

R.D. DuBois, Kh. Khayyat, C. Lloyd and C. Doudna

University of Missouri-Rolla, Rolla, MO, USA

Abstract. Single and multiple ionization yields, as a function of projectile energy loss and for forward scattering angles, have been measured for 750 eV positron and 800 eV electron impact on argon. Energy losses extend between zero and approximately 60% of the initial kinetic energy. For positron impact, the percentage of double ionization is found to increase rapidly for the first 80 eV of energy loss, then more slowly until the L-shell threshold. For still higher energy losses the percentage of double ionization saturates at approximately 15%. This is in excellent agreement with photo double ionization measurements. For electron impact, considerably higher amounts of multiple ionization are observed for conditions where the projectile energy loss is large.

This paper reports recent progress on positron impact differential ionization studies being performed at the University of Missouri-Rolla. Goals of these studies are to supplement existing information about inelastic collisions between particles and antiparticles and ultimately to provide complete kinematic information about such interactions. To date emphasis has been on three areas, namely a) differential information about multiple, as well as single, ionization processes, b) comparing data for positron and electron impact, and c) studying inelastic processes over a large range of energy transfer, in particular for large energy transfers.

These particular areas were chosen because the amount of differential ionization information for antiparticle impact is rather limited (see refs. 1-7) and no differential data for multiple electron removal exists. Second, by performing differential studies for both positron and electron impact, we hope to identify charge dependent effects on the interaction dynamics. With regard to this, multiple ionization data may be especially useful since total cross section measurements [8] have shown factors of two differences in the double ionization cross sections at higher impact energies whereas the single ionization cross sections are nearly identical. Finally, large energy transfers have been investigated. By large energy transfer we mean interactions where the

CP604, *Correlations, Polarization, and Ionization in Atomic Systems*
edited by D. H. Madison and M. Schulz
© 2002 American Institute of Physics 0-7354-0048-2/02/$19.00

projectile loses 50% or more of its initial kinetic energy. Such processes have never been previously investigated. Hence, nothing is known about the differential cross sections in this region. One reason is because this region cannot be explored using electron impact simply because the scattered projectile cannot be distinguished from the ionized electron. This means that since large energy losses are much less likely to occur than small energy losses one always observes features associated with energy transfers less than 50%.

This paper addresses all of the areas mentioned above. Specifically, single and multiple electron removal from argon has been measured for 750 eV positron and for 800 eV electron impact. In both cases, the single and multiple ionization was measured as a function of projectile energy loss and for projectiles scattered into a cone of angles centered about zero degrees. From these data and from comparisons between the positron and electron impact data, new information and insights into lepton impact ionization of atoms is obtained.

EXPERIMENTAL METHOD

In brief, the experimental method consisted of injecting a beam of positrons (or electrons) into a scattering chamber where they intersected a jet of argon atoms. Projectiles scattered into horizontal and vertical angles between roughly ±17° entered a specially designed electrostatic energy analyzer and were directed on a position sensitive detector. The spectrometer focused different energy particles onto different points along one axis of this detector. These scattered projectiles were measured in coincidence with target ions that were extracted from the interaction region using a weak electric field and then counted using a second position sensitive detector.

From time differences between a recoil ion signal and a scattered projectile signal, a time-of-flight spectrum for recoil ions was generated. By integrating intensities for the Ar^{q+} peaks in the time-of-flight spectrum, information about the single and multiple ionization yields. Data were collected and stored event by event. Then, by placing software restrictions on the projectile detector positions, information about the single and multiple ionization yields were obtained. Detector positions are analogous to selecting specific post collision energies for the scattered projectiles; thus we obtain information about the ionization yields as a function of projectile energy loss. Due to weak positron beam intensities and hence low statistics, data were summed for a small range of energy losses. Finally, these data were corrected for recoil ion detection efficiencies using information taken from DuBois [9] Last, by adjusting the voltages placed on the spectrometer, different energy loss regions could be investigated. As a result, energy losses between 0 and approximately 80% were studied.

For these studies the positron beam was produced using a radioactive ^{22}Na source and a biased tungsten moderator. The beam transport system consisted of electrostatic lenses and deflectors; no magnetic guidance was used. The electron beam was produced by simply reversing all potentials and extracting secondary electrons produced at the moderator. In both cases, energy calibrations of the projectile spectrometer and position sensitive detector were performed by biasing the moderator to particular values and noting where the primary beam landed on the detector. However, because of the large diameter of the primary beam, the large range of scattering angles accepted by the spectrometer, and lensing effects in the spectrometer associated with attempts to smooth the spectrometer field just before the projectile detector, an extended rather than a point beam was observed. Thus, the energy scales may have uncertainties of 10 eV or larger. This certainly introduces problems near the various threshold reasons, but the interest of the present studies was on overall behaviors and relative importance of multiple ionization as a function of projectile energy loss rather than on precisely determined threshold energies and behaviors.

RESULTS

As an example of what is observed as a function of energy loss, time of flight spectra for 750 eV positron impact on argon are shown in Figure 1. Three different energy losses are shown. The largest peak in each spectrum has been arbitrarily normalized to unity in order to illustrate how the relative amounts of multiple and single ionization change with energy loss. Please note that in these spectra the relative peak intensities have not been corrected for recoil ion detection efficiencies. Doing so would decrease the multiple ionization peak intensities on the order of 30-40% with respect to the single ionization intensity. Also seen in the spectra are small contributions of molecular oxygen and nitrogen ions which come from ionization of the background gases in the scattering chamber. The spectra, from bottom to top in the figure, indicate increasing degrees of "noise" due to lower statistics. This is because the relative cross sections decrease several orders in magnitude from the lowest to highest energy losses shown.

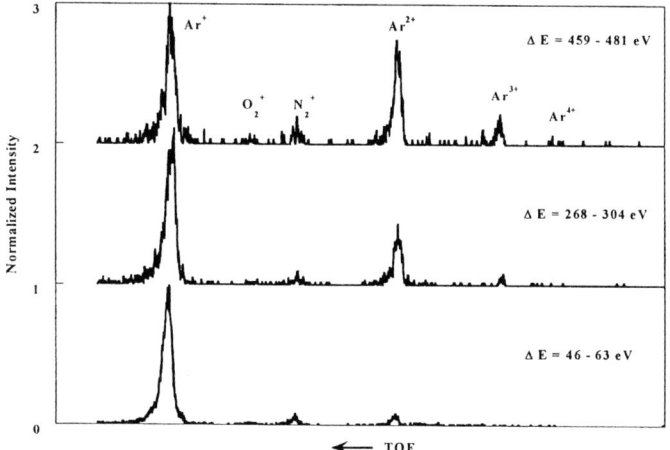

Figure 1. Time of flight spectra for 750 eV positron impact on argon. The spectra are measured for different energy losses, ΔE, and for all projectile scattering angles between 0 and $\pm 17°$.

These spectra immediately provide several important pieces of information. First, for the smallest energy loss, bottom spectrum, the dominant process is seen to be single electron removal. With increasing energy loss the relative amount of double and triple ionization dramatically increases. The reader is reminded that the smallest energy loss shown is sufficient to remove one or two electrons from the M shell of argon. The middle spectrum is for energy losses sufficient to remove several M-shell electrons but also is capable of removing an L-shell electron. Subsequent Auger processes can then lead to double, triple, or higher degrees of target ionization. Note that a small Ar^{3+} peak is visible in the middle spectrum.

The upper spectrum is taken for conditions where the projectile has lost approximately 60% of its initial kinetic energy. Here the amount of double ionization has again increased, triple ionization is now on the few percent level, and a hint of quadruple ionization can be seen.

Since one of our interests is to investigate charge effects on ionization dynamics, we also performed similar measurements for 800 eV electron impact. A small portion of these data are shown in Figure 2. For the example shown here, the energy loss is sufficient to remove M or L-shell electrons, as was the case for the top spectrum in the previous figure. However, as clearly seen, the degree of multiple ionization is significantly larger than for positron impact. This is consistent with the fact that at these impact energies integrated double ionization cross sections for electron impact are larger than those measured for positron impact. However, we point out that the present work samples but a small portion of the overall phase space and that more experimental information is needed in order to correlate these differences in the differential double ionization yields with those observed in total cross section measurements.

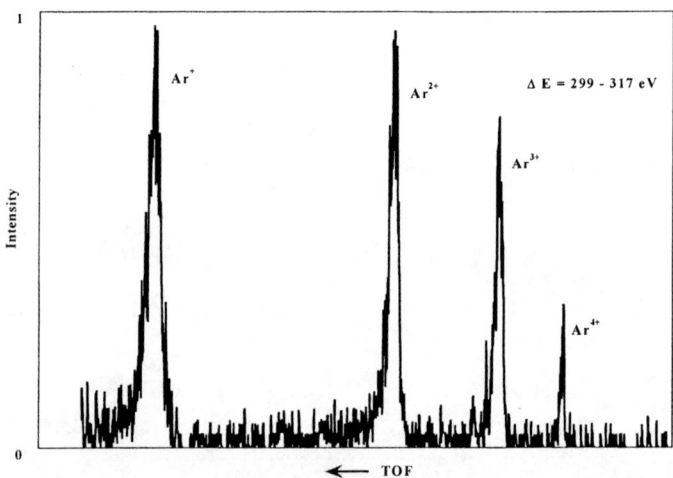

Figure 2. Time of flight spectrum for 800 eV electron impact on argon. Conditions are as for figure 1.

Using spectra like those shown in Figure 1, the relative amounts of single and double ionization were determined as a function of projectile energy loss, i.e., $f(2+) = I(2+)/[I(1+) + I(2+) + I(3+) \ldots]$. Here $f(2+)$ is the fraction of double ionization and $I(q+)$, etc., is the intensity of the $q+$ ionization peak divided by its relative detection efficiency. The results are shown in Figure 3. The arrows between 0 and 50 eV energy loss near the bottom axis indicate the energy thresholds for removing one or two (shorter, longer arrows) M-shell electrons and the arrow near 250 eV indicates the threshold for removing an L-shell electron. These data are for 750 eV positron impact on argon and for positrons scattered within a cone of half angle approximately 17 degrees centered on the forward direction.

As seen, the differential double ionization increases exponentially for energy losses up to approximately 80 eV and then at a slower rate until the L-shell opens. At that point, the percentage of double ionization remains relatively constant at approximately 15%. Please note that for experimental reasons discussed earlier, the energy resolution was rather broad and the calibration could have uncertainties on the order of 10 eV. This is the reason that some single and double ionization intensities are observed below the thresholds for single and double M-shell ionization.

Also shown in the figure is a horizontal dashed line, labeled hv, which indicates the photo double ionization values of Schmidt et al. [10] As seen, our values at higher energy losses are in excellent agreement with these photoionization values. Also for reference, on the right hand axis of the figure we indicate the overall ratio of double to single ionization, designated by σ_{tot}, as measured by Bluhme et al. [8]

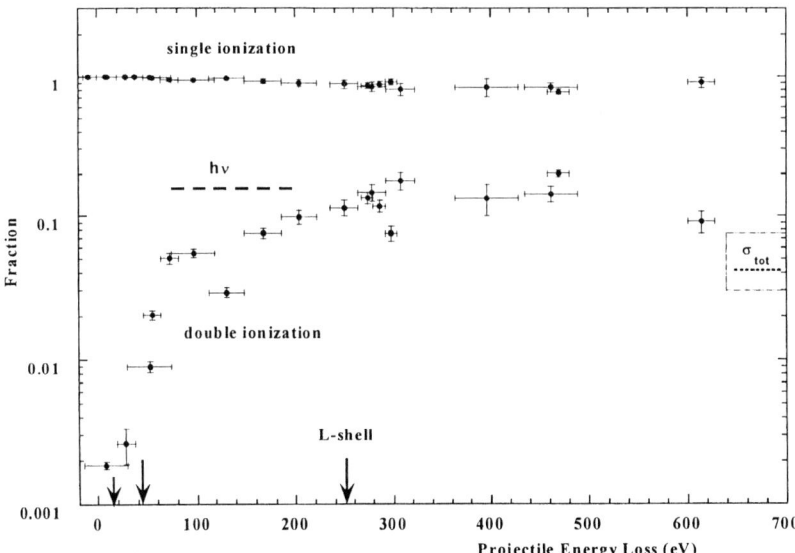

Figure 3. Fractions of single and double ionization of argon versus projectile energy loss for 750 eV positron impact. Arrows indicate thresholds for various processes and dashed lines show total double ionization fractions measured for photoionization or positron impact, designated by hv and σ_{tot} respectively. See text for details.

SUMMARY

Differential energy loss data have been presented for positron and electron impact ionization of argon. In both cases, multiple as well as single ionization are observed and the relative amount of multiple ionization is found to increase with increasing energy loss. This is not unexpected. However, these data provide some interesting insights into the electron removal processes. For example, for positron impact no triple ionization is observed unless the energy loss is sufficient to open the L-shell, although no obvious change is observed in the amount of double ionization as the L-shell is opened. Another important piece of information from these data is that for large energy losses, the positron impact data are in excellent agreement with photo double ionization values. Lastly, although not pointed out above, by measuring the multiple ionization as a function of energy loss, it is possible to separate double, or multiple, outer shell processes from inner shell processes followed by Auger cascades. This is quite important for testing multiple ionization models.

This work supported by the National Science Foundation, Grant No. PHY9732150. R.D.D. acknowledges support by the Deutsche Forschungsgemeinschaft during preparation of this manuscript.

REFERENCES

1. J. Moxom, G. Laricchia, M. Charlton, G.O. Jones and Á. Kövér, J. Phys. B **25**, L613 (1992).
2. Á. Kövér, G. Laricchia, and M. Charlton, J. Phys. B **26**, L575 (1993) and J Phys. B **27**, 2409 (1994).
3. A. Schmitt et al., Phys Rev A **49**, R5 (1994).
4. Á. Kövér, R.M. Finch, G. Laricchia, and M. Charlton, J. Phys. B **30**, L507 (1997).
5. Á. Kövér and G. Laricchia, Phys. Rev. Lett. **80**, 5309 (1998).
6. Kh. Khayyat et al., J Phys B **32**, L73 (1999).
7. Á. Kövér, K. Paludan and G. Laricchia, J Phy B **34**, L219 (2001).
8. H. Bluhme, H. Knudsen, J.P. Merrison and K.A. Nielsen, J. Phys B **32**, 5835 (1999).
9. R. D. DuBois, Phys. Rev. A **36**, 2585-2593 (1987).
10. V. Schmidt, N. Sandner, H. Kuntzemüller, P. Dhez, F. Wuilleumier and E. Källne, Phys Rev A **13**, 1748 (1976).

Complete experiments for electron excitation of pseudo-two-electron atoms

Albert Crowe, Danica Cvejanovic and Derek Brown

Department of Physics, University of Newcastle upon Tyne, NE1 7RU, UK

Abstract. The current status of coherence experiments for the pseudo-two-electron atoms, Mg, Ca, Sr, and Ba is summarised. Comparisons are made between experiment and theory for the ns^2 ^{1}S - $nsnp$ ^{1}P excitation in each atom. The behaviour of the coherence parameters across the different atoms is also compared.

INTRODUCTION

Major advances have been made in understanding single atomic electron excitation processes in simple atoms through experimental studies using electron-photon correlation and superelastic scattering techniques, complemented by theoretical advances using the Convergent Close Coupling (CCC) and R-matrix with pseudo-states (RMPS) methods. This is particularly true of atomic hydrogen [1] and the pseudo-one-electron systems, lithium [2], sodium [3], potassium [4], rubidium [5] and the simplest two electron system, helium [6]. Unlike helium, however, the available experimental tests for the one-electron systems have been largely confined to simple S - P transitions, a consequence of the sole use of superelastic techniques for the alkalis. While of major importance, there is clear evidence that they provide a less severe test of theory than optically forbidden transitions [7].

For heavy atoms, current theoretical models provide an inadequate description of low-intermediate energy electron impact excitation. The recent work of Guo et al [8] on differential cross section ratios for the individual levels of the $4p^5 5s$ configuration of Kr shows a good example of this.

THE TWO ELECTRON ATOMS

The pseudo-two-electron atoms provide a good platform to systematically introduce and understand each of the increasing complexities in turn as the principal quantum number is increased in these group IIA elements with ns^2 outer shell configurations. The basic characteristics of each atom together with the current status of coherence studies are summarised below. With the exceptions of helium and barium, all experimental studies have been confined to excitation from the ground state to the $nsnp$ ^{1}P$_1$ states.

CP604, *Correlations, Polarization, and Ionization in Atomic Systems*
edited by D. H. Madison and M. Schulz

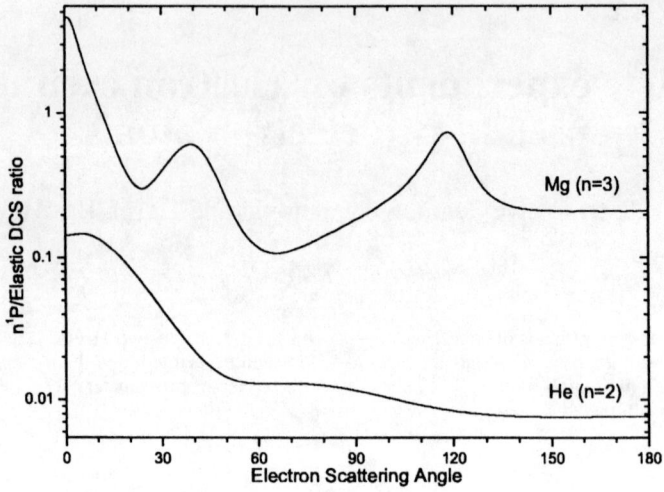

FIGURE 1. n^1P/elastic differential cross section ratios at 40 eV for He ($n = 2$) and Mg ($n = 3$). The curves are CCC calculations ([10] for He and [11] for Mg).

Helium ($1s^2$)

This simplest two electron system, with the lower lying ($L \leq 2$) excited states obeying LS coupling, has been extensively studied with the experimental data reproduced in detail by CCC and RMPS theories [9].

Beryllium ($1s^2 2s^2$)

In principle Be enables the influence of the core electrons to be studied, but no experiments have been carried out due to its extreme toxicity.

Magnesium ($1s^2 2s^2 2p^6 3s^2$)

Given the toxicity of beryllium, magnesium provides the first opportunity to study the influence of the core electrons in a LS-coupled system. The strong coupling of the low-lying P states to the ground state, together with electron correlation effects may also play major roles in the excitation process.

The distinction between excitation of the 2^1P state in helium (excitation energy 21.2 eV) and the 3^1P state of magnesium (excitation energy 4.35 eV) can be seen by comparing their differential cross sections with the corresponding elastic cross sections. Figure 1 shows the situation at an electron energy of 40 eV.

The only published experimental coherence parameters in magnesium are those of Brunger et al [12] for the 3^1P state at 20 and 40 eV and scattering angles in the range 5 - 20°, obtained using the electron-polarized photon correlation method. Figures 2(a), (b) show their measured Stokes parameters P_1 and P_2 at 40 eV, together with new data from this laboratory [13] at scattering angles out to 120°. The overall agreement between experiment and the most recent calculations [11, 14, 15] is good, especially for P_2. Interestingly, the relativistic distorted wave (RDW) approach of Kaur et al [14] performs well at this energy, with the exception of P_1 in the angular range 25 - 55°. In this context, it must be remembered that an incident energy of 40 eV corresponds to nine times the threshold energy compared to less than twice for helium (2^1P). It may be, as with helium [7], that the ^1S - $^{1,3}D$ excitation will provide a sterner test of theory [16].

Calcium (... $3s^23p^64s^2$)

In calcium an additional feature is the empty 3d shell in the ground state.

The 4^1P state has been studied using both the superelastic scattering [17, 18] (25.7, 45 eV, 3 - 100°) and electron-polarized photon correlation [19, 20] (45, 60, 100 eV, 10 - 45°) methods. In figures 2(c), (d), the measured Stokes parameters P_1 and P_2 at 45 eV [20] are again compared with the RDW theory of Srivastava et al [21]. Excellent agreement is obtained.

Strontium (... $3s^23p^63d^{10}4s^24p^65s^2$)

Now the n = 3 shell is filled, but the 4d shell is empty. Excitation of the 5^1P state has been studied experimentally by Hamdy et al [22] at 45 eV and scattering angles out to 113° and by Beyer et al [23] at 30.3 and 58.4 eV using the scattered electron-polarized photon correlation method. Figures 2(e), (f) show that the data at 45 eV are well predicted by the RDW theory of Srivastava et al [21].

Barium (... $4s^24p^64d^{10}5s^25p^66s^2$)

This heaviest alkaline earth has been studied experimentally using the superelastic scattering method. The studies by Zetner et al are more comprehensive than for the lighter two-electron atoms and involve both the $6s^2\ ^1S_0$ - $6s6p^1P_1$ [24, 25, 26, 27] and the $6s5d^1D_2$ - $6s6p^1P_1$ [28, 29, 30] excitation processes.

Ba ($6s^2\ ^1S_0$ - $6s6p\ ^1P_1$)

The experiments of Zetner et al cover the energy range 5 - 80 eV. Their P_1, P_2 data at 36.67 eV [25] are shown in figures 2(g),(h). Again there is good agreement with the RDW calculations [21] at 40 eV.

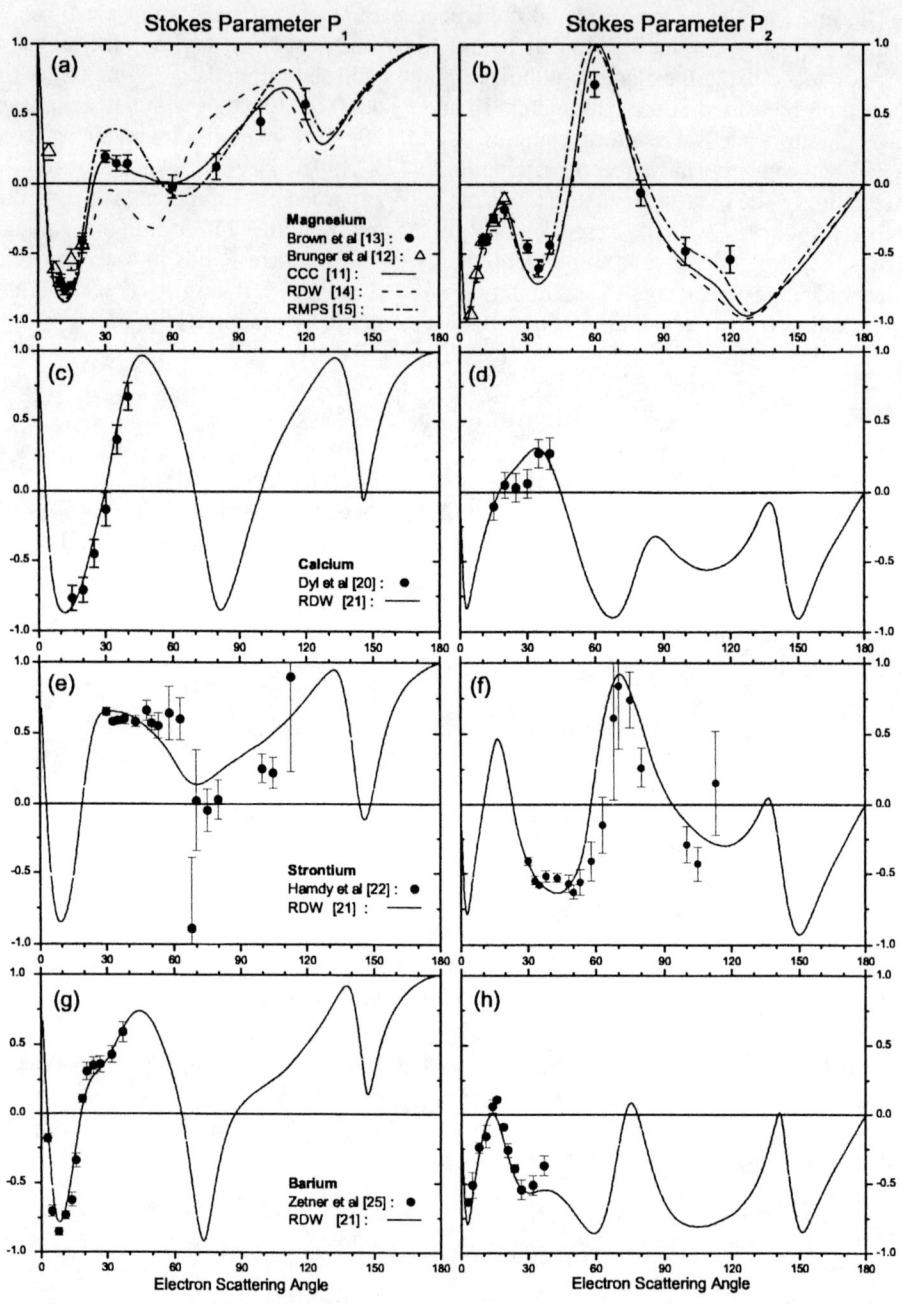

FIGURE 2. Stokes parameters P_1 and P_2 for the n^1P states of Mg (40 eV), Ca (45 eV), Sr (45 eV) and Ba (40 eV).

Ba ($6s5d\ ^1D_2$ - $6s6p\ ^1P_1$)

The pioneering experiments of Zetner's group [28, 29, 30] involve collisional de-excitation from the laser excited $^1P_1$ state to the lower lying $5d\ ^1D_2$ state. Four coherence parameters have been determined at 10, 20 and 40 eV. In this case the agreement between experiment and CCC predictions is less good quantitatively.

DISCUSSION AND CONCLUSIONS

It is clear that much can be learned about the dynamics of excitation in increasingly heavy atoms from a systematic study of the pseudo-two-electron atoms. The great majority of studies so far have involved the ns^2 - $nsnp\ ^1P$ processes. Experimentally, more data are available for barium than the lighter atoms, although they are restricted to scattering angles below 60°. The RDW is the only theory available for all of Mg, Ca, Sr, Ba and in each case excellent agreement is seen with experiment (figure 2). This agreement with perturbative theoretical approaches is in contrast to the situation for helium and is an outcome which requires further attention.

Figure 2 enables a comparison to be made between the Stokes parameters P_1 and P_2 for the different atoms. In general, a complex situation is observed. In particular, there are major differences in the behaviour of both P_1 and P_2 for the two lightest elements, Mg and Ca, while the Ca and Ba parameters are most alike. A co-ordinated experimental and theoretical approach is needed to promote an understanding of these observations and to test the calculations at incident energies much closer to the excitation thresholds. Similarly, observations on other excited states are required to test new calculations on these states [16, 30].

ACKNOWLEDGMENTS

We are grateful to Klaus Bartschat, Igor Bray, Dmitry Fursa, Rajesh Srivastava and Al Stauffer for details of their calculations prior to publication.

REFERENCES

1. Yalim, H., Cvejanovic, D., and Crowe, A., *J Phys B*, **32**, 3437 (1999).
2. Karaganov, V., Bray, I., and Teubner, P. J. O., *Phys Rev A*, **59**, 4407 (1999).
3. Shurgalin, M., Murray, A. J., MacGillivray, W. R., Standage, M. C., Madison, D. H., Winkler, K. D., and Bray, I., *J Phys B*, **32**, 2439 (1999).
4. Stockman, K. A., Karaganov, V., Bray, I., and Teubner, P. J. O., *J Phys B*, **34**, 1105 (2001).
5. Hall, B. V., Shurgalin, M., Murray, A. J., MacGillivray, W. R., and Standage, M. C., *Aust J Phys*, **52**, 515 (1999).
6. Igual-Ruiz, N., Donnelly, B. P., McLaughlin, D. T., Cvejanovic, D., Crowe, A., Fursa, D., Bartschat, K., and Bray, I., *J. Phys B*, **34**, 2289 (2001).

7. Bartschat, K., "Complete Experiments in Electron-Atom Collisions - Benchmarks for Atomic Collision Theory", in *Complete scattering experiments*, edited by U. Becker and A. Crowe, Kluwer/Plenum, New York, 2001, p. 61.
8. Guo, X., Mathews, D. F., Mikaelian, G., Khakoo, M. A., Crowe, A., Kanik, I., Trajmar, S., Zeman, V., Bartschat, K., and Fontes, C. J., *J Phys B*, **33**, 1895 (2000).
9. Cvejanovic, D., and Crowe, A., *(in these proceedings)* (2001).
10. Fursa, D. V., and Bray, I., *J Phys B*, **30**, 757 (1997).
11. Fursa, D. V., and Bray, I., *Phys Rev A*, **63**, 032708 (2001).
12. Brunger, M. J., Riley, J. L., Sholten, R. E., and Teubner, P. J. O., *J Phys B*, **22**, 1431 (1989).
13. Brown, D. O., Cvejanovic, D., and Crowe, A., *J Phys B*, **34**, (to be published) (2001).
14. Kaur, S., Srivastava, R., McEachran, R. P., and Stauffer, A. D., *J Phys B*, **30**, 1027 (1997).
15. Bartschat, K., *(private communication)*, (2001).
16. Srivastava, R., McEachran, R. P., and Stauffer, A. D., *J Phys B*, **34**, 2071 (2001).
17. Law, M. R., and Teubner, P. J. O., *J Phys B*, **28**, 2257 (1995).
18. Teubner, P. J. O., Karaganov, V., R, L. M., and Farrell, P. M., *Can J Phys*, **74**, 984 (1996).
19. Chwirot, S., Dziczek, D., Srivastava, R., D, D., and Dygdala, R. S., *J Phys B*, **29**, 5919 (1996).
20. Dyl, D., Dziczek, D., Piivinski, M., Gradziel, M., Srivastava, R., S, D. R., and Chwirot, S., *J Phys B*, **32**, 837 (1999).
21. Srivastava, R., Zuo, T., P, M. R., and Stauffer, A. D., *J Phys B*, **25**, 3709 (1992).
22. Hamdy, H., Beyer, H. J., and Kleinpoppen, H., *J Phys B*, **26**, 4237 (1993).
23. Beyer, H. J., Hamdy, H., Zohny, E. I. M., Mahmoud, K. R., El-Fayoumi, M. A. K., and Kleinpoppen, H., *Z Phys D*, **30**, 91 (1994).
24. Zetner, P. W., Li, Y., and Trajmar, S., *J Phys B*, **25**, 3187 (1992).
25. Zetner, P. W., Li, Y., and Trajmar, S., *Phys Rev A*, **48**, 495 (1993).
26. Li, Y., and Zetner, P. W., *Phys Rev A*, **49**, 950 (1994).
27. Johnson, P. V., Spanu, C., Y, L., and Zetner, P. W., *J Phys B*, **32**, 3437 (1999).
28. Li, Y., and Zetner, P. W., *J Phys B*, **28**, 5151 (1995).
29. Li, Y., and Zetner, P. W., *J Phys B*, **29**, 1803 (1996).
30. Johnson, P. V., Eves, B., Zetner, P. W., Fursa, D., and Bray, I., *Phys Rev A*, **59**, 439 (1999).

Electron-photon correlations in electron-impact excitation of alkaline-earth atoms

D. V. Fursa[*] and I. Bray[†]

[*]Electronic Structure of Materials Centre, The Flinders University of South Australia, G.P.O. Box 2100, Adelaide 5001, Australia
[†]Centre for Atomic, Molecular and Surface Physics, School of Mathematical and Physical Sciences, Murdoch University, Perth 6150, Australia

Abstract. We present a study of electron-photon correlations for the $ns^{2\,1}S$-$nsnp\,^1P^o$ transitions in Mg, Ca, Sr, and Ba atoms. Effects of inter-channel coupling and approximations in the target wave functions were investigated.

INTRODUCTION

Modern electron-atom scattering theories, such as Convergent Close-Coupling (CCC) method [1] and R-Matrix with Pseudo States (RMPS) method [2], have proved to be very successful in describing electron scattering from hydrogen[1], alkali atoms [3] and helium [4, 5]. The one-electron approximation for target wave functions was mostly adequate in those studies. Electron scattering from alkaline-earth atoms brings the next level of complexity due to substantially more complex target wave functions. Here, two-electron excitations in the target wave functions become progressively more important as we go from light to heavier alkaline-earth atoms. In addition, we observe a breakdown of the nonrelativsitic approximation for heavy alkaline-earth atoms. The ability of electron-atom scattering theories to adequately account for these new features is yet to be systematically demonstrated. Study of electron-photon correlations is known to offer a very sensitive test of various aspects of the scattering process. Here we look at electron-photon coherence parameters for the $ns^{2\,1}S$-$nsnp\,^1P^o$ transitions in Mg, Ca, Sr, and Ba for which a great deal of theoretical and experimental results are available.

THEORETICAL METHOD

We calculate wave functions of the alkaline-earth atoms in a model of two active electrons above an inert Hartree-Fock core. For each of the alkaline-earth atoms the set of one electron orbitals is obtained by diagonalizing the corresponding one-electron positive ion (for example Mg^+) in a Sturmian (Laguerre) basis. Standard configuration-interaction (CI) calculations are then performed in the space of two valence electrons and energy levels and wave functions of the target atom are determined. One and two-electron polarization potential is added to account for polarizability of the core [6]. The set of closed-coupling (CC) equations is obtained by expanding the total wave function

CP604, Correlations, Polarization, and Ionization in Atomic Systems
edited by D. H. Madison and M. Schulz
© 2002 American Institute of Physics 0-7354-0048-2/02/$19.00

TABLE 1. Characteristics of the ground state and first excited $^1P^o$ state of alkaline-earth atoms and helium.

	Excitation energy	Oscillator strength		Spectroscopic factors	
		CI	FC	^1S	^1P
He	21.1	0.276	0.282	0.995	0.999
Mg	4.3	1.72	1.89	0.934	0.942
Ca	2.9	1.73	2.17	0.933	0.828
Sr	2.7	1.80	2.21	0.942	0.838
Ba	2.2	1.69	2.39	0.931	0.706

in the target state basis. These equations are formulated and solved in momentum space. The use of a Sturmian basis allow us to perform a discretization of the target continuum and thereby model coupling to the ionization channels in the scattering calculations.

In Table 1 we present some important properties of the first excited $^1P^o$ states of alkaline-earth atoms and compare them with corresponding values in helium. The excitation energies of $nsnp$ $^1P^o$ levels in alkaline-earth atom are about five to six times smaller than in helium. The oscillator strengths are in the range of 1.7 - 1.8 a.u. which is much larger than in helium (0.276 a.u.). This suggests that the $^1P^o$ level is coupled much stronger to the ground state in the alkaline-earth atoms than in helium. In a close-coupling calculation this would manifest itself in a very fast rate of convergence for this alkaline-earth atom transition.

The choice of two-electron configurations is of great importance for an accurate calculation of target wave functions. In helium, we have used the Frozen-Core model where one of the electrons fixed in the He^+ $1s$ orbital. The success of the FC model in the e-He scattering calculations was largely due to the rather small probability of two-electron excitations in helium which are more than ten times smaller than transitions with excitations of only one electron [7]. In the alkaline-earth atoms the two-electron excitations play a much more prominent role. There are two-electron excited states in the discrete spectrum of theses atoms, while there are none in helium. The accuracy of the FC model for the alkaline-earth atoms as opposed to helium can be assessed by comparing the values of oscillator strengths for these transitions calculated in the FC model and in the CI model, see Table 1. For helium there is very little difference while for the alkaline-earth atoms we observe substantial and increasing difference as we progress from the lighter Mg atom to the heavier Ba atom. The relative weight of the FC configurations in the ground state and $^1P^o$ wave functions is given by the spectroscopic factor of the FC configurations [6]. They are presented in Table 1. For helium ground and 2^1P^o states the FC configurations are by far the most important. The ground states of the alkaline-earth atoms have relatively large and very similar FC spectroscopic factors. For the $^1P^o$ states it diminishes from Mg to Ba to a relatively small value (0.706) confirming that two-electron excitations are more important in heavy atoms.

RESULTS

We present results of our calculations of parameters γ and $L_\perp$ in Figs. 1 and 2 for Mg (40 eV), Ca (45eV), Sr (30.3 eV), and Ba (20 eV). The results of the RDWA calculations [8, 9] and experimental data indicate that relativistic effects are not important for these transitions, since both find that P^+ is approximately unity.

We find very good agreement between all theoretical results and experiment for the γ parameter. For the $L_\perp$ parameter the agreement between the experiment, present CCC, RDWA [8, 9], and DWA [16, 14, 17], and calculations are very good in the forward scattering region for Mg, Ca, and Sr but not for Ba where large differences between theoretical results are apparent. At intermediate scattering angles the CCC method is in a good agreement with experiment for Mg, Ca, and Ba but for Sr we found a large discrepancy at around 50° with the DWA and RDWA results supporting the experiment. This discrepancy is rather unexpected and prompted us to look at the influence of channel coupling effects and the accuracy of the target wave functions on the $L_\perp$ parameter.

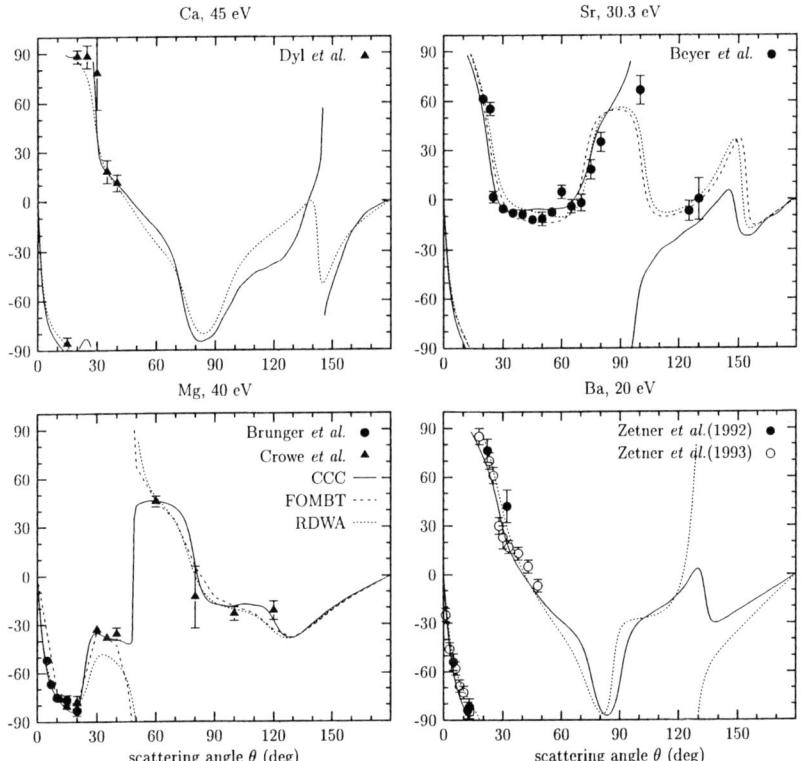

FIGURE 1. The γ parameter for excitation of the $(3s3p)\,^1P^o$ state of Mg at 40 eV, $(4s4p)\,^1P^o$ state of Ca at 45 eV, $(5s5p)\,^1P^o$ state of Sr at 30.3 eV, $(6s6p)\,^1P^o$ state of Ba at 20 eV incident electron energies. The measurements are due to Brunger *et al.*[10] (Mg), Crowe *et al.*[11] (Mg), Law and Teubner [12] (Ca), Dyl *et al.*[13] (Ca) Beyer *et al.*[14] (Sr), Li and Zetner [15] (Ba). The theory is as described in text.

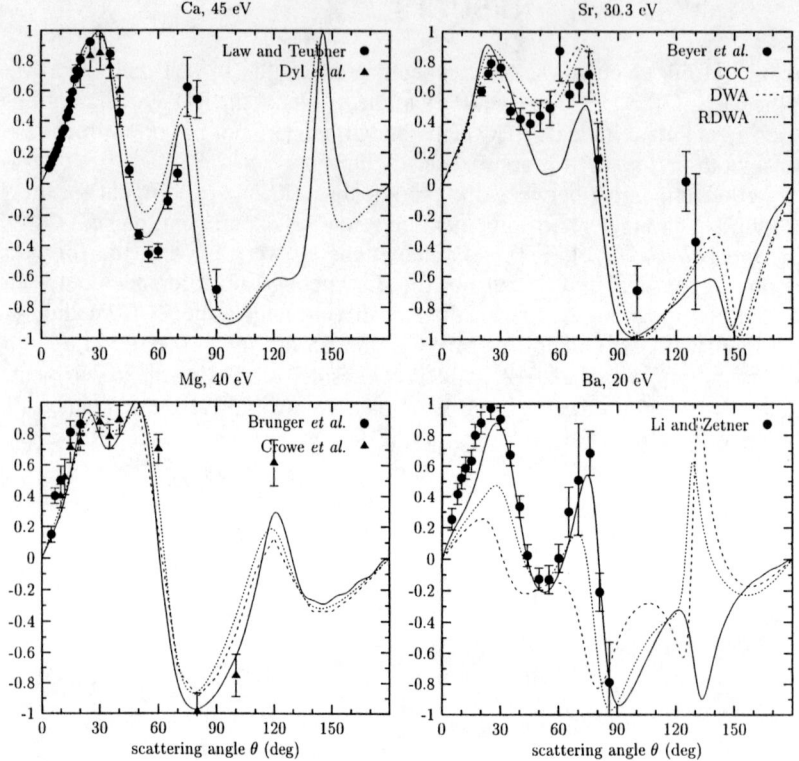

FIGURE 2. Same as in Fig. 2 but for the $L_\perp$ parameter.

In Fig. 3 we present results of 2-state CC calculations performed using the same CI wave function model as in the CCC calculations. The difference between the CCC and CC(2) calculations provides an estimate of importance of inter-channel coupling. We generally observe little difference between CCC and CC(2) results for all four alkaline-earth atoms. This is consistent with the noted very large values of oscillator strengths for these transitions.

We have also performed 2-state CC calculations using the FC model (CC(2)-FC) for the description of the ground and ^1P states (see Fig. 3). The difference between the two CC(2) calculations provides us with an estimate of importance of electron correlations effects in the target wave functions. We find little difference between the results of the two models, though the differences are larger for heavier atoms (Sr and Ba) which is expected as the FC model becomes less accurate for the heavier atoms, see Table 1.

The observed lack of sensitivity of the $L_\perp$ parameter to the details of the inter-channel coupling and the accuracy of the target wave functions can be partially explained by noting that incident electron energies of 40 eV in Mg, 45 eV in Ca, 30.3 eV in Sr, and 20 eV in Ba when measured in the threshold units (see Table 1) are in the range of 10 to 16. In helium the 10 times of the 2^1P^o excitation threshold corresponds to 200 eV incident electron energy which is considered to be a large energy where good agreement between experiment and various theoretical methods was found [4].

It is interesting to note that there is a striking similarity between helium 200 eV $L_\perp$ parameter and corresponding $L_\perp$ values in the alkaline-earth atoms. In Fig. 3 we presented in addition the e-He 20 eV CCC results [4]. For forward scattering, up to 40°, there is virtually no differences between the $L_\perp$ parameter in the helium and alkaline-earth atoms. Comparing helium and alkaline-earth atoms $L_\perp$ values at intermediate scattering angles we see that an additional maximum starts to develop at about 50° in Mg and becomes very pronounced in Ca, Sr, and Ba at 75°. In the backward angles one more maximum develops in Mg and Ca which becomes inverted in Sr and Ba. The nature of these structures in the $L_\perp$ parameter is not understood and requires further investigation.

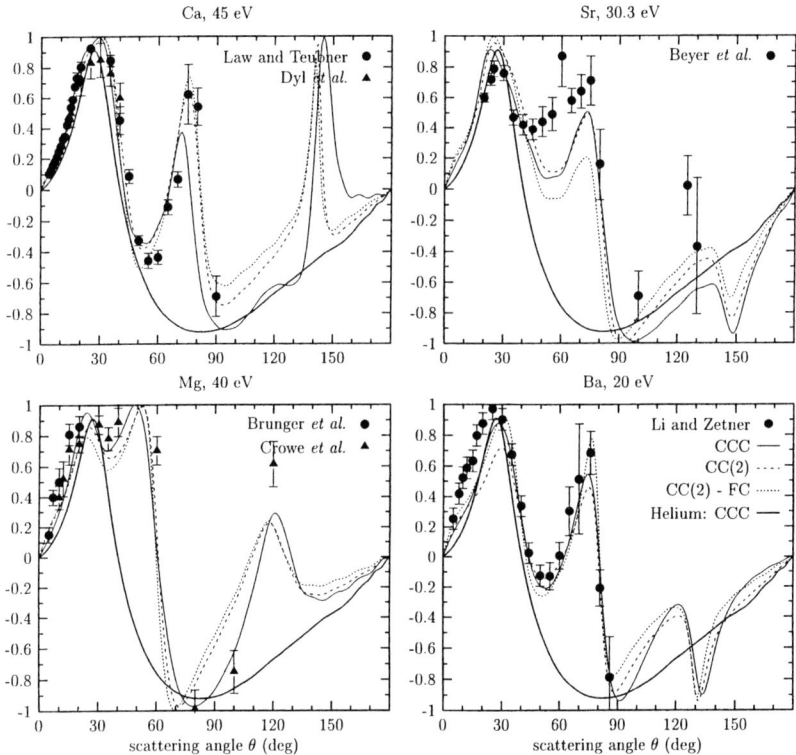

FIGURE 3. Same as in Fig. 2. In addition the results of the 2-state CC calculations in the CI and frozen-core models are presented. The CCC calculation is for helium $2\,^1P$ excitation by 200 eV electrons.

CONCLUSIONS

We have found that the EICP parameters for the $ns^2\,^1S$-$nsnp\,^1P^o$ transitions in Mg, Ca, Sr, and Ba are largely insensitive to the electron correlations in the target wave functions and the details of channel coupling in the scatetring calculations, at least at the considered (relatively large) incident electron energies. We observed a striking similarity between the $L_\perp$ parameter for helium and for alkaline-earth atoms. The "non-helium" features of the $L_\perp$ parameter in the alkaline-earth atoms, in our opinion, deserve a careful study. Clearly, in order to test modern theoretical methods the experimental acitivity should be directed to lower energies and large scattering angles. Note, however, that other transitions involving target states which are strongly affected by electron correlations and break-down of the nonrelativistic approximation in the target wave functions might provide a better testing ground for the theoretical methods.

ACKNOWLEDGMENTS

We are grateful to Albert Crow, Peter Zetner, Al Stauffer, Rajesh Srivastava and George Csanak for providing the data. Support of the Australian Research Council and the Flinders University of South Australia is acknowledged.

REFERENCES

1. Bray, I., and Stelbovics, A. T., *Phys. Rev. A*, **46**, 6995–7011 (1992).
2. Bartschat, K., Hudson, E. T., Scott, M. P., Burke, P. G., and Burke, V. M., *J. Phys. B*, **29**, 115–123 (1996).
3. Bray, I., *Phys. Rev. A*, **49**, 1066–1082 (1994).
4. Fursa, D. V., and Bray, I., *Phys. Rev. A*, **52**, 1279–1298 (1995).
5. Bartschat, K., *J. Phys. B*, **31**, L469–L476 (1998).
6. Fursa, D. V., and Bray, I., *J. Phys. B*, **30**, 5895–5913 (1997).
7. Füling, S., Bruch, R., Rauscher, E. A., Neill, P., Träbert, E., Heckmann, P. H., and McGuire, J. H., *Phys. Rev. Lett.*, **68**, 3152–3155 (1992).
8. Srivastava, R., Zuo, T., McEachran, R. P., and Stauffer, A. D., *J. Phys. B*, **25**, 3709–3720 (1992).
9. Kaur, S., Srivastava, R., McEachran, R. P., and Stauffer, A. D., *J. Phys. B*, **30**, 1027–1042 (1997).
10. Brunger, M. J., Riley, J. L., Scholten, R. E., and Teubner, P. J. O., *J. Phys. B*, **22**, 1431–1442 (1989).
11. Crowe, A., Cvejanovic, D., and Brown, D., *private communications* (????).
12. Law, M. R., and Teubner, P. J. O., *J. Phys. B*, **28**, 2257–2267 (1995).
13. Dyl, D., Dziczek, D., Piwinski, M., Gradziel, M., Srivastava, R., Dygdala, R. S., and Chwirot, S., *J. Phys. B*, **32**, 837–844 (1999).
14. Beyer, H. J., Hamdy, H., Zohny, E. I. M., Mahmoud, K. R., El-Fayoumi, M. A. K., Kleinpoppen, H., Abdallah Jr., J., Clark, R. E. H., and Csanak, G., *zpd*, **30**, 91–97 (1994).
15. Li, Y., and Zetner, P. W., *Phys. Rev. A*, **49**, 950–955 (1994).
16. Meneses, G. D., Pagan, C. B., and Machado, L. E., *Phys. Rev. A*, **41**, 4740–4750 (1990).
17. Clark, R. E., Abdallah Jr., J., Csanak, G., and Kramer, S. P., *Phys. Rev. A*, **40**, 2935–2949 (1989).

Electron excitation of the D states of Mg: Calculation of the Stokes parameters

R. Srivastava*, R. P. McEachran† and A. D. Stauffer**

*University of Roorkee, Roorkee-247667, India
†Australian National University, Canberra, ACT 0200, Australia
**York University, Toronto, Canada M3J 1P3

Abstract. We have used the RDW approximation to study the excitation of the 3^3D and 4^3D states of magnesium by electron impact at 20 and 40 eV. We have investigated in detail the differential cross sections and Stokes parameters for the fine-structure transitions from the triplet states and find that the results are very different from the average values predicted by non-relativistic calculations.

INTRODUCTION

Until recently, most studies on electron impact coincidence parameters, both experimental and theoretical, have considered the case of S to P transitions in an atomic system. There is now growing interest in determining these parameters for S to D transitions. This case is more challenging since it requires five independent quantities for a 'complete experiment'.

In this paper we investigate the effect of the fine-structure of the excited nD states of magnesium with $n = 3$ and 4 on the behaviour of the differential cross sections and Stokes parameters following electron excitation. We carry out our calculations in the Relativistic Distorted-Wave Approximation (RDW) using j-j coupled Dirac-Fock wave functions which provide us with distinct bound state wave functions for the fine-structure states of the atom.

THEORETICAL BACKGROUND

Details of the RDW method for magnesium have been given by [1] and further results for this system are presented in [2].

The ground $^1S_0$ state of this atom has the configuration $1s^2 2s^2 2\bar{p}^2 2p^4 3s^2$ in the relativistic j-j coupling notation. The excited D states are represented by an outer shell with configuration either $3s n\bar{d}$ or $3s nd$, $n = 3$ and 4, and give rise to singlet D states with total angular momentum $J = 2$ as well as triplet D states with $J = 1$, 2 and 3. It is these triplet fine-structure states that are the focus of our interest here. We have used the GRASP92 program of [3] to calculate configuration interaction wave functions for these states based on the configurations detailed above. We denote these wave functions as Single Configuration Ground State (SCGS) since the ground state is represented by

CP604, *Correlations, Polarization, and Ionization in Atomic Systems*
edited by D. H. Madison and M. Schulz
© 2002 American Institute of Physics 0-7354-0048-2/02/$19.00

a single configuration. We have also included the configurations $2\bar{p}^2$, $2\bar{p}2p$ and $2p^2$ in a further calculation which we denote as Multi-Configuration Ground State (MCGS) since these additional configurations contribute to the ground as well as excited states. This latter calculation yields results for the energy levels of the excited states which are in much better agreement with experiment than the former. In particular, the singlet and triplet energy levels are inverted in the SCGS calculation.

We have used the formulae in [4] to calculate the differential cross sections as well as the Stokes parameters P_1, P_2, P_3 and P_4. This information is almost sufficient to completely characterize the excitation process. Only a phase factor is undetermined. We have determined the Stokes parameters for all possible radiative decays $^3D_J \rightarrow {}^3P_{J'}$ where $J = 1$, 2 or 3 and $J' = 0$, 1 or 2. Since the selection rule $|J - J'| \leq 1$ must be satisfied, six such fine-structure transitions are possible.

If the fine-structure is not resolved then the measurements include radiation from all the fine-structure transitions. Theoretically, this corresponds to a calculation in LS coupling which ignores the spin-orbit interaction terms. In our case, we can average our results over the various transitions weighted with the corresponding transition probability.

Bartschat and Madison [5] have analyzed the Stokes parameters in the case of forward (or backward) scattering. Under the assumption of LS coupling and the conservation of orbital and spin angular momentum, they were able to derive a unique value for the Stokes parameters independent of the values of the scattering amplitudes. In particular, P_1 and P_4 must have a value of 0.3174 for ^3D states. In our calculations, orbital and spin angular momentum are not separately conserved so that no such unique limit exists. In fact, we find that spin-orbit coupling produces significant spin-flip in the excitation of some of these fine-structure levels.

RESULTS

All the results shown here are calculated using our MCGS wave functions.

In figure 1 we show the ratios $r_J = \sigma_J/\sigma_1$ where σ_J is the differential cross sections for exciting the fine-structure levels of the 3^3D and 4^3D states with angular momentum J at both 20 and 40 eV. The results for the sum of these cross sections have been presented in [2]. We see that these ratios are approximately constant over much of the angular range and have values which are in the neighbourhood of the ratios of the statistical weights $2J + 1$. The largest deviation from the constant value is for r_2 at 40 eV for both the 3^3D and 4^3D states at forward scattering angles. Measurements of these ratios would be of interest to verify the deviation from their statistical weights.

The results we have obtained for the average value of the Stokes parameters P_1 and P_4 in the forward direction vary somewhat with the transition and the energy but all fall in the range $0.29 - 0.31$ which is quite close to the LS value of 0.3174. This indicates that the LS results are reasonably consistent with the more elaborate relativistic treatment of these processes. However, we find that the values of the Stokes parameters for the individual fine-structure transitions vary considerably from the average values. This is shown in figures 2 and 3 where we display the Stokes parameters for the transitions from

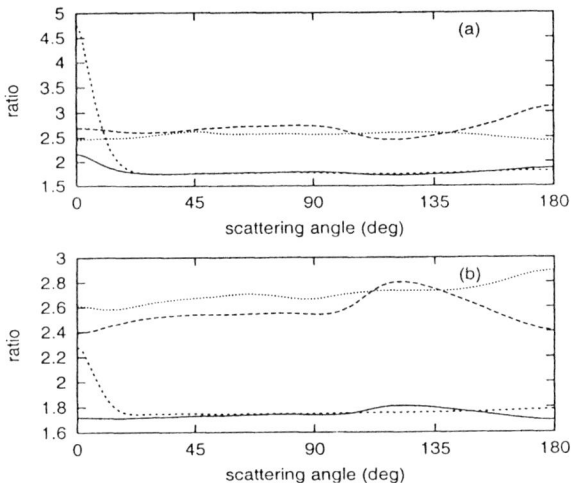

FIGURE 1. Cross section ratios, (a) 3^3D, (b) 4^3D. 20 eV: _____, r_2; _ _ _ _, r_3. 40 eV: _ _ _ _ _, r_2; ......, r_3.

the various fine-structure levels for incident electrons with energies of 20 and 40 eV. We remark that the parameters for the 4^3D state have almost identical values and are not shown separately. We also note from the figure that the shape of the curves for the various transitions are very similar, differing only in magnitude. However, the variation in magnitude (and sign) among the various parameters for the transitions is substantial. In particular, since the angular momentum $L_\perp$ perpendicular to the collision frame has the opposite sign to P_3, it will have negative values for small scattering angles for a number of the fine-structure transitions in contradiction to the usual propensity rules. This is markedly so at the higher energy of 40 eV.

CONCLUSIONS

We have shown that the magnitude of the Stokes parameters for the various fine-structure transitions differs considerably from the average value given by a non-relativistic approximation using *LS* coupling. This indicates that the spin-orbit interaction is substantial for the triplet D states of magnesium. This is already obvious from the fine-structure splitting of the energy levels of this state. We have provided additional evidence of the importance of taking this interaction into account in electron impact excitation.

It would be extremely interesting to verify this conclusion by experimental measurement. This could be done in a time-reversed experiment where the individual fine-structure levels could be optically pumped from the lower lying P levels and then de-

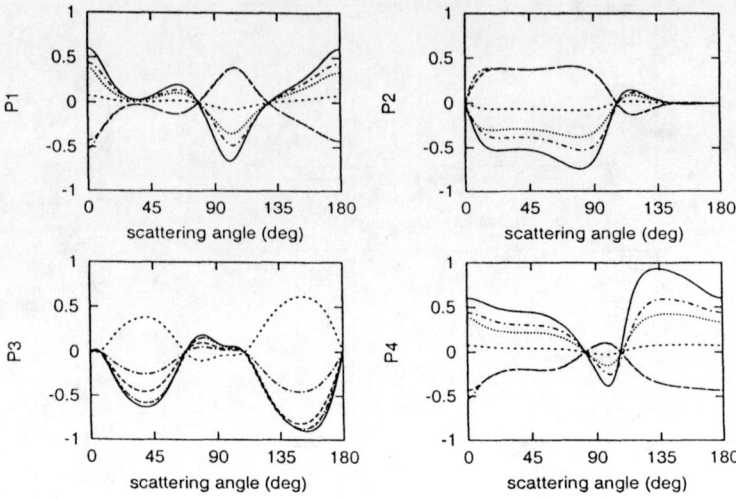

FIGURE 2. Stokes parameters for the 3^3D states at 20 eV for the transitions J to J'.
————, 1 – 0; _ _ _ _, 1 – 1; ....., 1 – 2; ......, 2 – 1; _·_·_, 2 – 2; _··_··, 3 – 2.

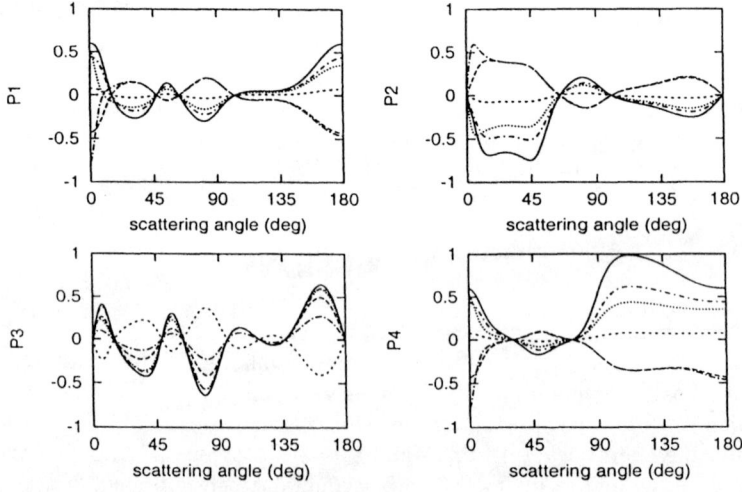

FIGURE 3. Stokes parameters for the 3^3D states at 40 eV. The legend is the same as for figure 2.

excited to the ground state by electron impact. However, the fact that the scattering cross section is small in this case (it is basically an exchange transition) and that the atom has to be excited to the P state before optically pumping to the D state would make this measurement very difficult.

ACKNOWLEDGMENTS

RS would like to thank the Department of Atomic Energy (DAE), Govt. of India, for financial support. RPM and ADS are grateful for a grant from the Natural Science and Engineering Research Council of Canada.

REFERENCES

1. S. Kaur, R. Srivastava, R. P. McEachran and A. D. Stauffer, J. Phys. B **30**, 1027 (1997)
2. R. Srivastava, R. P. McEachran and A. D. Stauffer, J. Phys. B **34**, 2071 (2001)
3. F. A. Parpia, C. Froese Fischer, C. and I. P. Grant, Comput. Phys. Commun. **94**, 249 (1996)
4. K. Bartschat, K. Blum, G. F. Hanne, and J. Kessler, J. Phys. B **14**, 3761 (1981)
5. K. Bartschat and D. H. Madison, J. Phys. B **21**, 153 (1988)

Excitation of the 3D states of helium by electron impact

Danica Cvejanović* and Albert Crowe[†]

*Schuster Laboratory, The University of Manchester, M13 9PL, UK
[†]Physics Department, University of Newcastle, NE1 7RU, UK

Abstract. Electron impact excitation of the 3D states in helium has been studied over a large range of scattering angles and incident electron energies starting from near the excitation threshold. Experimental coherence parameters are reproduced by both the Convergent close coupling and R-matrix with pseudo states models. In excitation of both the 3^1D and the 3^3D states negative values of $L_\perp$ are observed under very different excitation conditions, contradicting a proposed propensity rule.

INTRODUCTION

The study of coherent excitation of atoms with angular momentum $L \geq 2$ by electron impact has presented substantial new challenges for both experiment and theory, even in a simple atom like helium. After two decades of study of electron impact excitation of the 3D states in helium, existing experimental and theoretical data can be regarded now as giving an accurate picture, even though the majority of experimental results have not been obtained from complete experiments. An outline of existing experimental data on the 3^1D state has been given recently by Cvejanović et al [1] and for the 3^3D state [2, 3, 4, 5] and references therein. A new generation of theories, Convergent Close Coupling (CCC) [6, 7] and R-matrix with pseudo states (RMPS) [8] have dramatically improved our ability to model electron impact excitation in helium, reproducing coherence parameters for excitation of the 3D states right down to the excitation threshold. These calculations now cover the same set of kinematic parameters as existing experiments and show excellent agreement with the experimental data and between themselves.

In the polarization correlation method, used in the present study and in the majority of experiments on excitation of the D states, a subset of identically prepared states is observed. These correspond to a well defined momentum transfer, while the existence of a plane of symmetry, defined by the incident and scattered electron momenta, has a consequence that only the $M = 0, \pm 2$ magnetic sublevels are observed. The states excited by electrons of energy E, scattered at an angle θ, can be described by the wave function

$$|\Psi> = a_{+2}(E,\theta)|2+2> + a_0(E,\theta)|20> + a_{-2}(E,\theta)|2-2>, \qquad (1)$$

where $a_{0,\pm2}$ correspond to complex scattering amplitudes,

$$a_{+2} = \alpha_{+2}e^{i\Phi_{+2}}; \quad a_0 = \alpha_0 e^{i\Phi_0}; \quad a_{-2} = \alpha_{-2}e^{i\Phi_{-2}}. \qquad (2)$$

CP604, Correlations, Polarization, and Ionization in Atomic Systems
edited by D. H. Madison and M. Schulz
© 2002 American Institute of Physics 0-7354-0048-2/02/$19.00

A complete experiment needs to determine the magnitudes and relative phases of the scattering amplitudes (2).

An alternative, and physically more transparent description, is based on the charge cloud parameters,

$$(P_\ell, \gamma, L_\perp, \rho_{00}). \tag{3}$$

These parameters (3) are directly related to the scattering amplitudes (2). However, they do not provide a complete description in the case of D states. The polarization pattern of the D state decay, characterized by the Stokes parameters from which the parameters (3) are derived, and which are measured in $(e^-; \gamma^{3D-2P})$ coincidence experiments, is not uniquely related to the wavefunction (1). A detailed analysis relevant for coherent excitation of the D states and the new parameters,

$$(\tilde{L}_\perp^\pm, \tilde{\gamma}^\pm), \tag{4}$$

leading to a complete determination of function (1) has been published by Andersen and Bartschat [9, 10]. A complete experiment on D state excitation has to be based on the detection of triple coincidences, $(e^-; \gamma^{3D-2P}; \gamma^{2P-1S})$, [11].

EXPERIMENTAL

The experimental setup and details relevant to the measurement of Stokes parameters in the case of the 3D states have been described in detail by Fursa et al [3] and references therein. Briefly, electrons produced by an oxide cathode are focussed into the interaction region where they are crossed with the atomic beam. Electrons scattered at a pre-selected angle are energy analyzed and those with energy loss corresponding to excitation of the 3D states are detected by a channel electron multiplier in coincidence with polarized 3D-2P decay photons. Photons are detected in two directions: perpendicular to the scattering plane for Stokes parameters P_1, P_2 and P_3, and in the scattering plane, but at the right angles to the incident electron beam, for P_4.

RESULTS AND DISCUSSION

The new parameters $(\tilde{L}_\perp^\pm, \tilde{\gamma}^\pm)$, specific to D state excitation [9, 10], relate to the two decay routes, rather than directly to the transferred angular momentum $L_\perp$ and the charge cloud alignment angle γ. For decay perpendicular to the scattering plane, the route identified by + corresponds to $3D(M = +2) \rightarrow 2P(M = +1) \rightarrow 1S(M = 0)$ and the route − to $3D(M = -2) \rightarrow 2P(M = -1) \rightarrow 1S(M = 0)$. The $\pm$ sign relates to the helicity of P-S decay photons, identifying uniquely the transition through the $P(M = +1)$ or $P(M = -1)$ magnetic substates. While some of the attractive transparency in physical meaning of the parameter set (3) is lost in the new parameters, they permit a full description of excitation of the D states. Simple relationships exist between $\tilde{L}_\perp^\pm$ and $L_\perp$ and consequently the magnitudes of the scattering amplitudes (2),

$$\tilde{L}_\perp^\pm = \pm 2\frac{\alpha_{\pm2}^2 - \alpha_0^2/6}{\alpha_{\pm2}^2 + \alpha_0^2/2}. \tag{5}$$

The parameters $\tilde{\gamma}^\pm$ are related directly to the relative phases $\beta_{\pm2}$ [10],

$$\beta_{\pm2} = \pm(\Phi_{\pm2} - \Phi_0), \tag{6}$$

with

$$\beta_{\pm2} = -2\tilde{\gamma}^\pm \pm \pi. \tag{7}$$

It is the parameters $\tilde{\gamma}^\pm$ which are not uniquely determined from measurements of the Stokes parameters in $(e^-;\gamma^{3D-2P})$ experiments, resulting in an uncertainty in the sign, but not the size of the relative amplitudes. Two methods have been recently applied to resolve this ambiguity. Experimentally the sign can be obtained only from a triple coincidence experiment, $(e^-;\gamma^{3D-2P};\gamma^{2P-1S})$, permitting observation of a coherent sum of the two $\pm$ decay channels. A pioneering triple coincidence experiment has been performed for an incident electron energy of 60 eV and scattering angle of 40° [11, 12]. However, due to the low counting rates triple coincidence experiments are very difficult and are not realistic over a wide range of energies and angles. In Newcastle an alternative approach is used. This is based on the use of a reliable theory to distinguish between the two solutions for $\tilde{\gamma}^\pm$ in $(e^-;\gamma^{3D-2P})$ experiments and consequently establish the correct overall phase difference and remove the ambiguity. The procedure has been described in detail by Fursa et al [3].

The complexity in excitation of the 3D states of helium is manifest in the difficulties which theoretical models had in reproducing experimental data prior to development and application of the CCC and RMPS methods. One of the main goals of recent studies has been to establish the validity of these new theoretical models, by comparing experimental data with theoretical predictions over a wide range of energies from near threshold to medium energies and over a large span of scattering angles. Recent studies also include coherence parameters for the 3^3D state at 23.45 eV where excitation is dominated by negative ion resonances.

In figure 1 coherence parameters for excitation of the 3^1D state measured in Newcastle at an electron impact energy of 26.5 eV [1, 13] are shown. Parameters from both the incomplete set (3) and the new complete set (4) are included, $\tilde{\gamma}^+$ and $\tilde{\gamma}^-$ being obtained by comparison with CCC calculations. The experimental data are compared with CCC [4], RMPS [8] and 29-state R-matrix (R29) [14] results. In general good agreement is observed with both the CCC and RMPS theories. The agreement in the case of CCC theory gives an indication of the validity of the frozen core model at impact energies close to the excitation threshold. The largest differences between the CCC and RMPS models are observed at scattering angles above 120° where there are no experimental data.

In excitation of the 3^1D state an interesting situation, different from observations in other excited states of helium including its triplet counterpart, is observed at small scattering angles and electron impact energies above 40 eV. This is specially pronounced in the 60 eV data [1] which are compared with results at 29.6 eV in figure 2. The

158

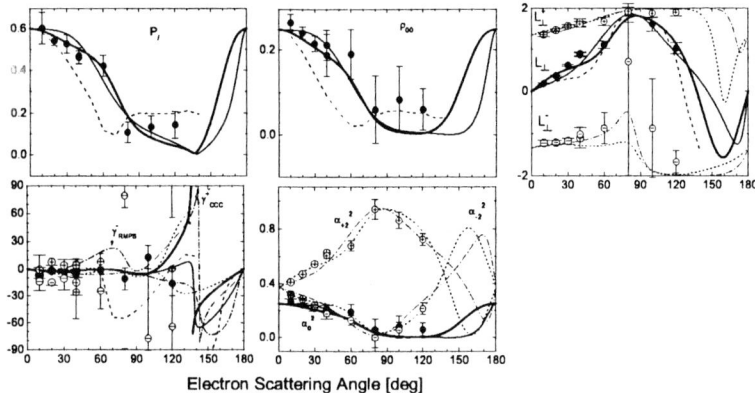

FIGURE 1. Coherence parameters for the 3^1D state at 26.5 eV. Experiments from Newcastle [1, 13]: Incomplete set, •; Complete set, Open symbols with sign inside: +, for $\tilde{L}^+, \tilde{\gamma}^+$ and –, for $\tilde{L}^-, \tilde{\gamma}^-$. The same convention using $\pm$ signs is used for the ± 2 scattering amplitudes $\alpha_{+2}^2, \alpha_{-2}^2$, while • represents α_0. Theories: CCC [4], complete set and α_0, broad full line; incomplete set and $\alpha_{\pm 2}^2$, dotted line; RMPS [8], complete set and α_0, thin line; incomplete set and $\alpha_{\pm 2}^2$, chained line; R29 [14], dashed line.

characteristic angular behaviour at 60 eV is supported by both the CCC [15] calculations and previous experiments [11, 16, 17, 18, 19]. At lower impact energies the Stokes parameters and correspondingly the charge cloud parameters show a slow and smooth, almost linear, angular variation up to a scattering angle of 40°. This picture is strongly in contrast with the observation at 60 eV (figure 2) where the parameters exhibit a rapid variation with scattering angle.

The behaviour of $L_\perp$ is especially interesting as angular momentum transferred in electron impact excitation of atoms has attracted considerable theoretical attention [20, 21, 22, 23] and references therein. Positive values of $L_\perp$ have been observed with overwhelming regularity at small scattering angles , leading to formulation of the Andersen - Hertel propensity rule [21]. In contrast to this, slightly negative values are observed in excitation of the 3^1D state at 40 eV, clearly negative values at 45 eV and progressively more negative values at 60 eV [1] and 80 eV [16]. The behaviour of $L_\perp$ in excitation of the 3^3D state is very different and in agreement with the propensity rule. The 3^1D results are clearly demonstrated by all available experimental data shown in figure 2 and by the CCC calculations , but a more detailed insight into the mechanism responsible is still not clear.

The behaviour of $L_\perp$ for the 3^1D state at 60 eV is mirrored in the behaviour of the other charge cloud parameters and by α_M^2. The angular behaviour of α_M^2 at low impact energies is again similar to that in other states and characterized by a slow variation, with excitation of the $M = +2$ magnetic sublevel being the increasingly dominant process. By

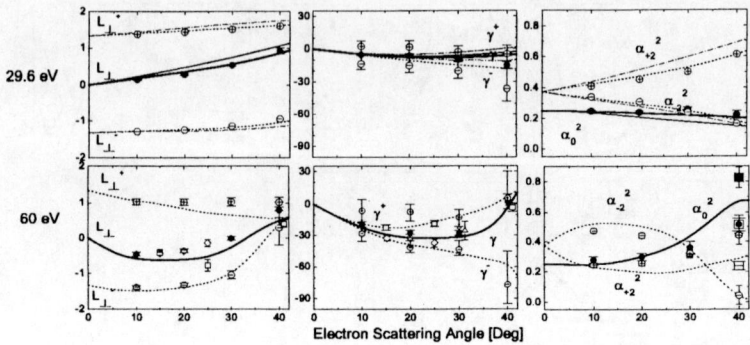

FIGURE 2. Coherence parameters for the 3^1D state at small scattering angles and energies of 29.6 and 60 eV. Experiments from Newcastle [1, 24] and theories, CCC [3] and RMPS [8] same as figure 1. Previous experiments: Square, reference [16, 11]; Diamond, reference [17, 18]; Triangle, reference [19].

contrast, at 60 eV the $M = -2$ sublevel dominates the excitation process at the lowest scattering angles, reaching a maximum around 20°, after which it falls off rapidly to zero at 40°. The probability for excitation of the $M = 0$ sublevel, i.e. α_0^2, shows a strong increase from zero scattering angle and dominates the excitation at 40°. The α_{+2}^2 shows the least variation in this angular range. Consequently, for atoms excited by 60 eV electrons scattered at 40°, decay to the ground state through the $^1P(M = -1)$ channel arises almost entirely from excitation of the $3^1D(M = 0)$ state, a unique situation in excitation of the D states. In terms of the excited state charge cloud, the situation evolves rapidly from one where the charge cloud is predominantly in the scattering plane to one where it is perpendicular to the scattering plane.

As a consequence of the dominant excitation of the $M = 0$ magnetic sublevel at 60 eV and 40°, the parameter $\tilde{L}_\perp^-$ has a positive value, very similar to $L_\perp$ and $\tilde{L}_\perp^+$, while $\tilde{\gamma}^-$ and $\tilde{\gamma}^+$ have very different values with $\tilde{\gamma}^+$ being similar to γ. All three parameters, $\gamma, \tilde{\gamma}^+, \tilde{\gamma}^-$ exhibit strong variation in the angular range 10-40°.

Excitation influenced by negative ion resonances has been investigated both experimentally and theoretically for excitation of the 3^3D state [5]. Negative ion resonances have been observed experimentally [25] at 23.45 eV and confirmed by RMPS and CCC theories [25]. Both the CCC and RMPS calculations indicate rapid variation of the coherence parameters in the resonance region. The experimental energy resolution in the incident beam, which is typically around 0.5 eV, prevents a direct comparison with the theoretical calculations. However, when averaged over the incident electron profile the theoretical results show very good agreement with experiment. An interesting feature observed in this resonance region is yet another example of clearly negative values of $L_\perp$. This departure from the general propensity rule may be a consequence of the temporary attachment of an electron in the negative ion but in this case it may be a more general near-threshold phenomena. Slightly negative $L_\perp$ values are predicted by both the CCC and RMPS calculations in the near-threshold excitation of the 2^1P state [5].

CONCLUSION

Recent experimental studies at Newcastle have been directed towards a complete mapping of data for excitation of both the 3^1D and the 3^3D states from near the excitation threshold to medium energies and for a large span of scattering angles. Theoretical CCC and RMPS models are remarkably good in predicting coherence parameters for all the kinematic situations studied experimentally. Negative $L_\perp$ values at small scattering angles observed in 60 eV excitation of the 3^1D state and in the near-threshold region for the 3^3D excitation deserve further theoretical consideration.

REFERENCES

1. Cvejanović, D., McLaughlin, D.T., and Crowe, A., *J. Phys. B: At. Mol. Opt. Phys.*, **33**, 3013 (2000).
2. Crowe A., Donnelly, B.P., McLaughlin, D.T., Bray I., and Fursa D.V., *J. Phys. B: At. Mol. Opt. Phys.*, **27**, L795 (1994).
3. Fursa D.V., Bray I., Donnelly, B.P., McLaughlin, D.T., and Crowe A., *J. Phys. B: At. Mol. Opt. Phys.*, **30**, 3459, (1997).
4. Fursa D.V., Bray I., Donnelly, B.P., McLaughlin, D.T., and Crowe A., *Phys. Rev. A*, **56**, 4606, (1997).
5. Crowe A., Cvejanović, D., McLaughlin, D.T., Donnelly, B.P., Fursa, D., Bray, I. and Bartschat K., *J. Phys. B: At. Mol. Opt. Phys.*, **33** 2571, (2000).
6. Fursa D.V., and Bray I., *Phys. Rev. A*, **52**, 1279, (1995).
7. Fursa D.V., and Bray I., *J. Phys. B: At. Mol. Opt. Phys.*, **30**, 757, (1997).
8. Bartschat K., *J. Phys. B: At. Mol. Phys.*, **32**, L355, (1999).
9. Andersen N., and Bartschat K., *Adv. At. Mol. Phys.*, **36**, 1, (1996).
10. Andersen N., and Bartschat K., *J. Phys. B: At. Mol. Opt. Phys.*, **30**, 5071, (1997).
11. Mikosza, A.G., Williams, J.F., and Wang J.B., *Phys. Rev. Lett.*, **79**, 3375, (1997).
12. Mikosza, A.G., *The Physics of Electronic and Atomic Collisions, AIP Conference Proceedings 500*, Ed. Y Itikawa et al, New York: AIP, p297, (2000).
13. McLaughlin, D.T., Donnelly, B.P., and Crowe A., *Phys. Rev. A*, **49**, 2545, (1994).
14. Ratnavelu, K., Fon, W.C., and Berrington K.A., *J. Phys. B: At. Mol. Opt. Phys.*, **28**, L429, (1995).
15. Fursa D.V. and Bray I., *Private communication*, (1999).
16. Mikosza, A.G., Hippler R., Wang, J.B., and Williams J.F., *Z. Phys. D*, **30** 129, (1994).
17. Batelaan, H., van Eck, J., and Heideman, H.G.M., *J. Phys. B: At. Mol. Opt. Phys.*, **21**, L741, (1988).
18. Batelaan, H., van Eck, J., and Heideman H.G.M., *J. Phys. B: At. Mol. Opt. Phys.*, **24**, 5151, (1991).
19. Beijers, J.P.M., Doornenbal, S.J., van Eck, J., and Heideman H.G.M., *J. Phys. B: At. Mol. Phys.*, **20**, 6617, (1987).
20. Madison D.H. and Winters K.H., *Phys. Rev. Lett.*, **47**, 1885, (1981).
21. Andersen N., and Hertel I.V., *Comments At. Mol. Phys.*, **19**, 1, (1986).
22. G Csanak and D C Cartwright *J. Phys. B: At. Mol. Phys.*, **20**, L603, (1987).
23. Bartschat, K., Andersen, N., and Loveall D., *Phy. Rev. Lett.*, **83**, 5254, (1999).
24. Donnelly, B.P., McLaughlin, D.T., and Crowe A., *J.Phys. B. At. Mol. Opt. Phys.*, **27**, 319, (1994).
25. Cvejanović, D., Clague, K., Fursa, D., Bartschat, K., Bray, I., and Crowe A., *J. Phys. B: At. Mol. Opt. Phys.*, **33**, 2265, (2000).

Electron Scattering from the $2p^5 3s$ Configuration of Neon – Differential Cross Section Ratio Tests of Target Wave Functions

M. A. Khakoo, J. Wrkich and M. Larsen

Physics Department, California State University,
Fullerton, CA 92834, USA

Abstract. Differential cross section ratios for the electron impact excitation of the $2p^5 3s$ levels of Neon have been measured at electron impact energies of 20eV, 25eV, 30eV, 40eV, 50eV and 100eV. Comparison of these ratios with available experimental and theoretical models is made.

INTRODUCTION

Electron impact excitation of the ground state configuration (np^6) of a rare gas to its first excited state configuration ($np^5 n+1s$) has been shown to be a special system where ratios of differential cross sections (DCSs) provide valuable insights on relativistic interactions in the target structure and scattering dynamics[1]. We have recently made a range of accurate absolute DCS measurements and DCS ratios for the electron impact excitation of the $2p^6 \rightarrow 2p^5 3s$ configurations of Ne at incident electron energies (E_0) of 20eV, 25eV, 30eV, 40eV, 50eV and 100eV and for scattering angles (θ) from $1°$ to $130°$. We define two of these DCS ratios for Ne as

$$r = \frac{DCS(3s[3/2]_2)}{DCS(3s'[1/2]_0)} \quad \text{and} \quad r' = \frac{DCS(3s[3/2]_1)}{DCS(3s'[1/2]_1)} \tag{1}$$

where the excited levels of Ne can be expressed in the intermediate coupling [2] in terms of LS-coupling components, in ascending order of excitation energy, as

$$|3s[3/2]_2\rangle = |3^3P_2\rangle,$$

$$|3s[3/2]_1\rangle = \beta|3^1P_1\rangle - \alpha|3^3P_1\rangle,$$

$$|3s'[1/2]_0\rangle = |3^3P_0\rangle,$$

and

$$|3s'[1/2]_1\rangle = \alpha|3^1P_1\rangle + \beta|3^3P_1\rangle. \tag{2}$$

The ratio r is that of the pure LS-coupled metastable levels. In the limiting case of near-degeneracy and absence of spin-orbit effects in the excitation process, r tends to

CP604, *Correlations, Polarization, and Ionization in Atomic Systems*
edited by D. H. Madison and M. Schulz
© 2002 American Institute of Physics 0-7354-0048-2/02/$19.00

the ratio of the statistical weights of the J=2:J=0 levels viz. 5:1. Deviation of r from the value of 5 indicates non-degeneracy-induced effects, spin-orbit interactions or second-order effects in the scattering process. On the other hand, in the optical limit (high E_0 and small θ), application of dipole selection rules show, within the above single-configuration coupling scheme, the limit for r':

$$\text{Dipole Lim r'} = \beta^2/\alpha^2 \quad . \tag{3}$$

Deviation from the optical limit indicates either the importance of the triplet part of the J = 1 components or the need for additional configuration singlet levels in the model to describe these mixed levels. If only pure spin-exchange excitation of these levels occurs, and only the $1\,^3P > $ LS-component in Eqn. 2 is excited,

$$\text{Exchange Lim r'} = \alpha^2/\beta^2. \tag{4}$$

EXPERIMENT, RESULTS, OBSERVATIONS AND CONCLUSIONS

Experiment

The apparatus used to determine r and r' is comprised of a high resolution electron energy loss spectrometer with double hemispherical energy selectors in both the gun and the analyzer sections and has been detailed in Guo *et al* [3]. An important feature of this spectrometer is the absence of "wings" in the instrumental profile, often seen in spectrometers with single hemispherical analyzers. This characteristic enabled us to resolve the weak metastable energy loss features from the stronger allowed transitions in Ne. The spectrometer operated at an energy resolution of 25 - 40 meV (FWHM) with an electron current ranging from 3 - 20 nA. The electron energy loss features were fitted using a well-tested multi-Gaussian linear least-squares program [1] to unfold the contributions of overlapping features. The resulting line intensities were used to evaluate the values of r and r' and their error constituted the statistical error on these intensities plus an additional error of <2% (in quadrature) of the determination of our analyzer transmission across the spectrum.

Results and Observations

A sample of the data for $E_0 = 30$eV is shown in the figures below. From Fig. 1, we observe r to be close to the statistical weight ratio value of 5. This indicates that excitation of the metastable levels of Ne is LS-coupling dominated within an experimental error of approximately 3% when all data sets at all θ in this figure are averaged. Interestingly, the Multi-Configuration Ground-State Relativistic Distorted-Wave calculation indicates a large departure from r=5 for $\theta<40°$. However, for a more rigid test of theory, one needs more precise values of r over the full angular range.

In Fig. 2, it is clearly observed that *only* the Distorted Wave Approximation [8] agrees with the experimental data in the small θ limit (dipole-limit; this observation is made possible because of the small experimental r' errors). However, this is only because mixing coefficients of α = 0.964, β = 0.266 were used, which are obtained experimentally from the optical measurements of Tsurubuchi *et al* [9,1]. The remaining models use theoretical target wave functions obtained by minimizing differences in the theoretical energies of the target levels with those obtained experimentally [10]. The use,

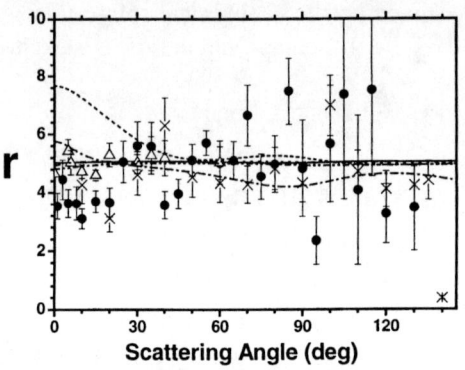

FIGURE 1. Values of r at E_0 = 30eV. Legend: *Experiments:* • Present Results; Δ Khakoo *et al.* [1]; × Register et al. [4] ; *Theoretical Models:* —— 15-state, Semi-relativistic R-Matrix method using "Super-Structure" target wave functions [5]; — — 5-configuration, Unitarized First-Order Many-Body Theory [6]; - · · - Relativistic Distorted-Wave Approximation with single-configuration ground-state wave functions [7]; ---- Relativistic Distorted-Wave Approximation with multi-configuration ground-state wave functions [7]; . . . Distorted-Wave Approximation with Hartree-Fock wavefunctions[8]; - · - Distorted-Wave Approximation with 15-state, Super-Structure wavefunctions [8].

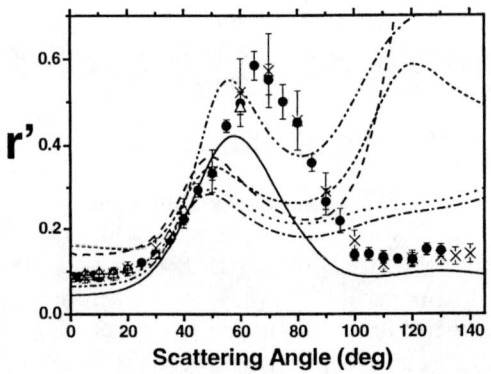

FIGURE 2. Ratio r' taken at E_0=30eV. Legend is the same as for Fig. 1. See text for discussion.

164

in the Distorted Wave Approximation [8], of either Hartree-Fock or "Super-Structure" target wave functions gives essentially identical results, indicating that it is the intermediate-coupling expansion of these wave functions that is most important. Also, we observe that the R-Matrix model gives the best qualitative agreement with the present measurements. This theory differs from the others in that it is a close-coupling theory whereas the others are perturbative-type theories. Coupling between different open channels is very important especially for the tightly-bound $2p^5$ core of the excited Ne states and this effect is probably dealt with most effectively using close-coupling models. We also note the increase in r' at $\theta=70^o$ which probably indicates increased spin-exchange processes according to Eqn. 4. The perturbative models become more accurate with increasing E_0 and we find the distorted-wave models become reliable by $E_0=100eV$.

Conclusions

The present investigations show that DCS ratios for the rare gases provide significant insights into the importance of accurate target wave functions as well as other model-dependent effects. It would be interesting to compare with other versions of close-coupling calculations to see the sensitivity of the results for a non-perturbative calculation. Additionally, we intend to carry out electron-metastable atom coincidence measurements to obtain more precise values of r so that we can investigate the level of relativistic effects in the scattering dynamics.

ACKNOWLEDGMENTS

Funded by the National Science Foundation. Private Communications with several theoretical groups (references 5–8) are gratefully acknowledged. Collaborations with Prof. Albert Crowe (Newcastle U., UK., under a NATO Grant) and Prof. Timothy J. Gay (U. Nebraska. Lincoln) are also gratefully acknowledged.

REFERENCES

1. Khakoo, M. A., Beckmann, C. E., Trajmar S., and Csanak G., J. Phys B: At. Mol. Opt. Phys. 27 3159-3161 (1994).
2. Cowan, R.D. *Theory of Atomic Spectra,* Berkeley, CA: University of California Press (1981).
3. Guo X., Khakoo, M.A., Mathews, D.J., Mikaelian, G., Crowe, A., Kanik, I., Trajmar, S., Zeman, V., Bartschat, K. and Fontes, C.J., J. Phys. B. At. Mol. Opt. Phys. 33, 1895 (2000).
4. Register, D. F., Trajmar, S., Steffensen, G., Cartwright, D. C., Phys. Rev. A29, 1793-1805 (1984).
5. Bartschat, K., Private Communication (2000).
6. Fontes, C. J., Private Communication (2000).
7. Stauffer, A. D. and Srivastava, R., Private Communication (2001).
8. Madison, D.H., Private Communication (2001).
9. Tsurubuchi, S, Watanabe, K. and Arikawa, T., J. Phys. Soc. Japan 59, 497 (1990).
10. Moore, C. E., *Atomic Energy Levels,* NBS Circular No. 467, Washington DC.: US Gov't Printing Office (1952).

Excitation of the 2p$_9$ State of Rare Gases by Electron Impact

Seiji Tsurubuchi and Hirotsugu Kobayashi

*Dept. of Applied physics, Faculty of Technology, Tokyo University of Agriculture and Technology
Koganei, Tokyo 184-8588, Japan*

Abstract. Emission cross sections for the np$\rightarrow$ns transition (Ne: n=3, Ar: n=4, Kr: n=5) have been systematically studied. The 2p$_9$ $\rightarrow$ 1s$_5$ transition in Paschen's notation is of particular interest because the 2p$_9$ state is a pure triplet state and well represented by ^{3}D$_3$ in the LS-scheme. Theoretical excitation cross sections of the 2p$_9$ state show almost ~E^{-3} dependence on electron energy E, whereas, experimentally obtained results descend more slowly. A time resolved line-intensity measurement was carried out in order to clarify the discrepancy and to have more insight into the excitation mechanism of the 2p$_9$ state. It was found that a considerable part of the population of the 2p$_9$ state is due to the cascade transitions from the higher lying Rydberg states and that the descent of the level excitation cross section of the 2p$_9$ state at high energies is slower than expected by the DWA theories.

INTRODUCTION

Emission cross sections of rare gases are of importance in many related subjects such as discharge phenomena, laser physics, astrophysics and so on. The 2p$_9$$\rightarrow$1s$_5$ transition in Paschen's notation is of particular interest because the 2p$_9$ state is a pure triplet state and well represented by the electronic term ^{3}D$_3$ in the LS-scheme, i.e. the spin exchange step is necessary to excite the 2p$_9$ level from the ground state.

For Ne, emission cross sections for the 2p$_9$ $\rightarrow$ 1s$_5$ transition was measured by Feltsan and Zapesochnyi [1], Sharpton et al [2], Gay et al [3], Chilton et al [4] and by Tsurubuchi et al [5]. To the authors' knowledge, however, almost no theoretical report on the total excitation cross section is available for the 2p$_9$ state. Puech and Mizzi [6] gave a semi-empirical expression for the related collision process.

In the case of Ar, the situation is improved, especially in the field of theoretical works, which includes the non-relativistic distorted wave (DWA) calculation of Bubelev and Grum-Grzhimailo [7], the semi-relativistic distorted-wave (SRDW) calculation of Madison et al [8] and the relativistic distorted-wave (RDWA) calculation of Kaur et al [9]. Experimental works on the emission cross section for the 2p$_9$ $\rightarrow$ 1s$_5$ transition are given by Feltsan and Zapesochny [10], Ballou et al [11], Tsurubuchi et al [12], Gay et al [3], and by Chilton et al [13]. Chutjian and Cartwright gave the differential and integrated excitation cross sections for the 2p$_9$ state by an energy loss measurement [14].

CP604, *Correlations, Polarization, and Ionization in Atomic Systems*
edited by D. H. Madison and M. Schulz
© 2002 American Institute of Physics 0-7354-0048-2/02/$19.00

For Kr, experimental cross sections were given by Gay et al [3], Feltsan [15], Bogdanova and Yurgenson [16], Chilton et al [17] and Trajmar et al [18] as well as the relativistic distorted-wave (RDWA) calculation of Kaur et al [9]. We carried out a time resolved measurement of line intensity to have more insight into the excitation mechanism of the $2p_9$ state.

EXPERIMENTAL

A schematic illustration of the experimental setup is given in Figure 1. A collision chamber is made of stainless steel with diameter of 30 cm and is pumped by a turbo-molecular pump below $\sim 10^{-9}$ Torr before operation. Target gas was introduced through a needle valve. An electron beam was modulated by a fast rectangular pulse generator with 100 kHz repetitions. The fall off time of the electron beam was less than 7 nanoseconds.

The decaying light emission was wavelength selected by a monochromator and monitored with a digital oscilloscope and a multi-channel analyzer. An ordinal photon counting method was used for detection of emitted signals.

If we have ratios between population of the level excitation and that of the cascading, we have the level-excitation cross section of $2p_9$ state by use of equation (1). In general, the level excitation cross section Q_j of level j is related to the apparent cross section Q_j' of level j by

$$Q_j = Q_j' \{ f_j + \sum_k f_k (\tau_j / \tau_k) \}, \qquad (1)$$

where f_j is the fractional population of the state j immediately following excitation of

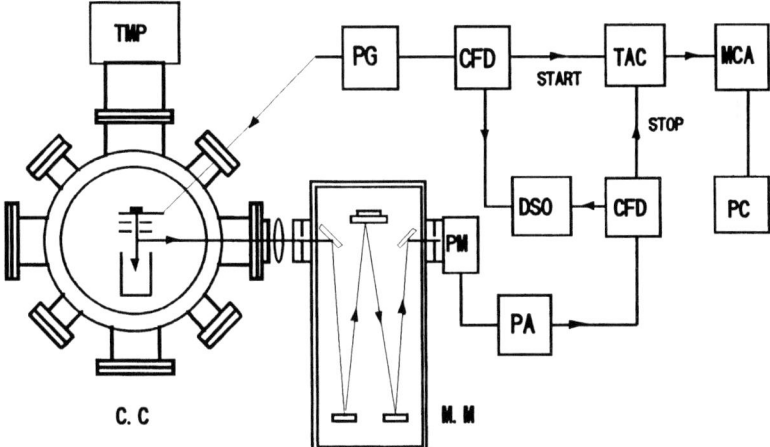

FIGURE 1. A schematic illustration of the experimental setup. CC: collision chamber, PG:pulse generator, TMP:turabo-molecular pump, MM: visible monochromator, PM: photomultiplier, PA:fast preamplifier, CFD:constant fraction discriminator, TAC: time to amplitude convertor, MCA: multichannel analyzer, DSO: digital storage oscilloscope, PC: computer

167

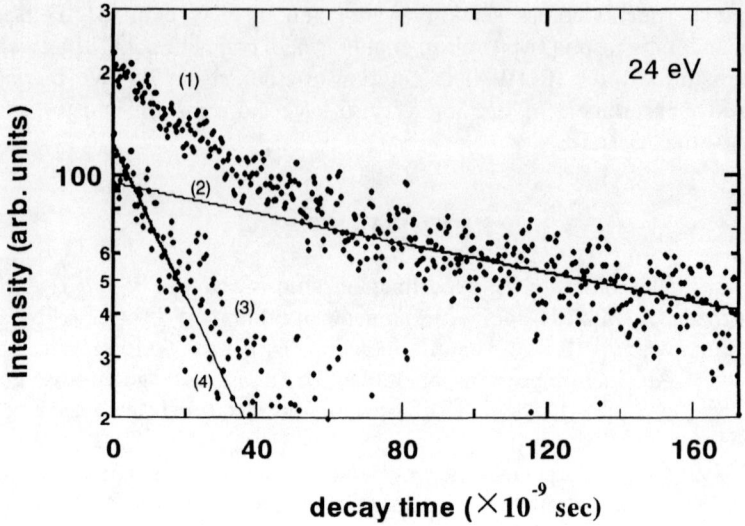

FIGURE 2. A decay curve of the $2p_9 \rightarrow 1s_5$ transition (640.2 nm line) of Ne. The exponential curve (2) is the least squares fit to the tail of plot (1). The plot (3) is the result obtained by subtracting (2) from (1). A solid line (4) was drawn by the method of the least squares fit to the plot (3).

the atom and τ_j is the lifetime of level j [19,20]. A typical decay curve is given in Figure 2. In fact, as seen in the figure, the effective 2-level analysis is possible, i.e. all cascade transitions are apparently compressed into one decaying curve with a composite decay time τ_k.

RESULTS AND DISCUSSION

Figures 3 to 5 illustrate emission cross sections for the $2p_9 \rightarrow 1s_5$ transition and level excitation cross sections of the $2p_9$ state of Ne, Ar, and Kr together with some cascade cross sections to the $2p_9$ state. It is interesting to notice that the energy dependence of the cascade cross sections are all very similar to those of the $2p_9 \rightarrow 1s_5$ emission cross sections. We thought initially that the slow descent of the $2p_9 \rightarrow 1s_5$ emission cross sections at high energies might be due to the cascade population transfer to the $2p_9$ state from the higher lying Rydberg states. In the present work, the related cascade emission lines measured are the 576.4 nm ($4d[7/2]_4 \rightarrow 3p[5/2]_3$) and 503.7 nm ($5d[7/2]_4 \rightarrow 3p[5/2]_3$) lines of Ne, the 549.5 nm line ($6d[7/2]_4 \rightarrow 4p[5/2]_3$) and 522.1 nm line ($7d[7/2]_4 \rightarrow 4p[5/2]_3$) of Ar, and the 645.6 nm line ($6d[7/2]_4 \rightarrow 5p[5/2]_3$) of Kr. However, what we need is the knowledge of the direct level excitation cross section of the $2p_9$ state which is obtained by correction for these cascade effects.

It is worthwhile to point out here again that the time-resolved measurements give us information of how much degree of cascade effect is included in the $2p_9 \rightarrow 1s_5$ emission

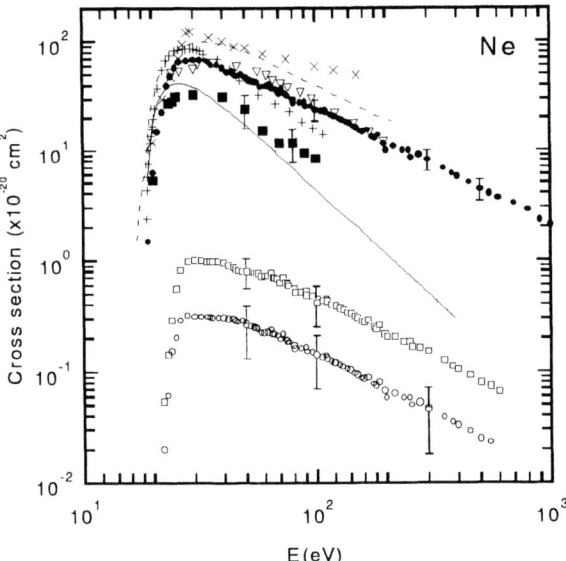

FIGURE 3: Ne, 640.2 nm line: present result (●), Chilton et al (---), Feltsan and Zapesochny (×), Gay et al (+); cascade cross section: present result for 576.4 nm line (□), 503.7 nm line (○); Level excitation cross section: present result (■),Chilton et al (▽), Puech and Mizzi (solid line).

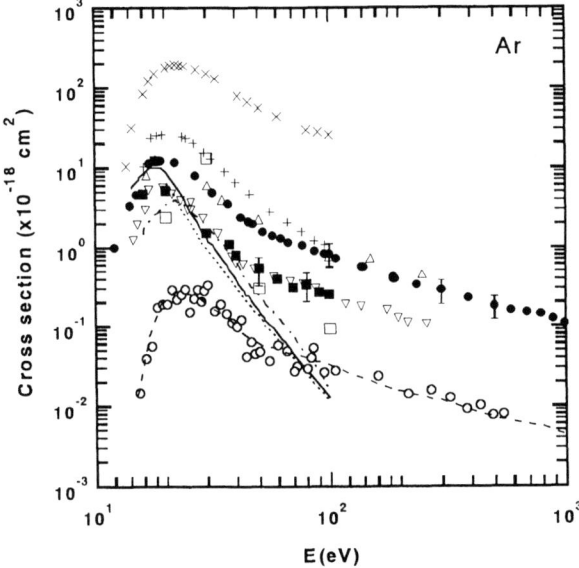

FIGURE 4: Ar, 811.5 nm line: present result (●), Feltsan and Zapesochny (×), Gay et al (+), Chilton et al (△); Level excitation cross section: present result (■), Chutjian and Cartwright (□), Chilton et al (▽), SRDW: Madison et al (solid line), RDWA: Kaur et al (dotted line), DWA: Buvelev and Grum-Grzhimailo (﹍); Cascade cross section: present result, 549.5 nm line (○).

169

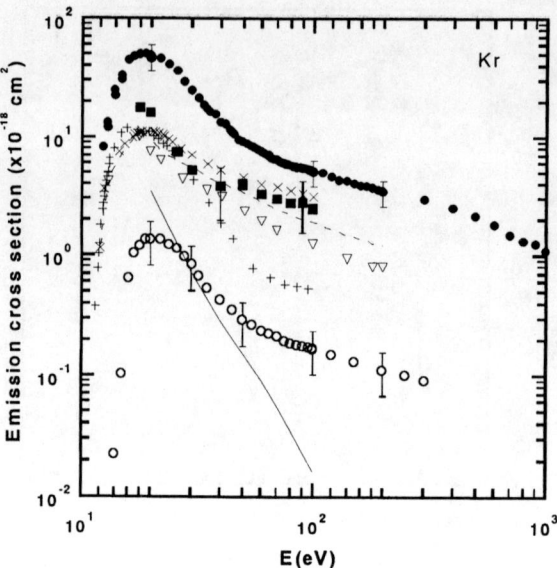

FIGURE 5 Emission cross sections for the Kr 811.3 nm line: present result (●), Feltsan (×), Gay et al (+), Chilton et al (------). Level excitation cross section of the $2p_9$ state: present result (■), Chilton et al (▽), RDW calculation: Kaur et al (solid line); Cascade cross section to the $2p_9$ state: 645.8 nm line (○).

cross section and that we have eventually the level excitation cross section for each target gas.

In the case of Ne, no integrated cross section is reported but it may be useful to notice here that a DWA calculation of Baerveldt et al for $2p^5 3p$ J=3 ($2p_9$) shows a higher DCS for the forward-scattering angles [21]. The high energy behavior of the level excitation cross section seems to be well represented by a semi-empirical result given by Puech and Mizzi [6].

The $2p_9$ state of Ar can be reached in second order through the $(3p^5 np)$ J=1 states with two direct transitions and the discrepancy may be attributed to a second order effect as pointed out by Madison et al [8]. It is worthwhile to notice that the DCS of the $2p_9$ state of Ar shows a steep rise at 100 eV in the range of small scattering angles [14], while the theoretical results show rapid decrease after making a hump [8,9].

In the case of Kr, it was experimentally found that the fraction of the cascade transition around the maximum is bigger than those at high energies in the $2p_9 \rightarrow 1s_5$ emission cross section.

CONCLUSIONS

We performed a time-resolved intensity measurement of the $2p_9 \rightarrow 1s_5$ line of rare gases and confirmed that the descent of the level excitation cross section of $2p_9$ state at high energies is slower than expected by the DWA calculations. A further consideration from experimental point of view should be necessary.

ACKNOWLEDGMENTS

The assistance of Mr. Masamitsu Hyodoh in taking some of the data is gratefully acknowledged. The present work has been partly supported by Grant-in-aid for Scientific Research (B) from the Japan Society for the Promotion of Science.

REFERENCES

1. P. V. Feltsan, I. P. Zapesochnyi and M. M. Povch, *Ukr. Fiz. Zh.* **11**, 1222-1230 (1966)
2. F. A. Sharpton, R. M. St. John, C. C. Lin and F. E. Fajen, *Phys. Rev.* A **2**, 1305-1323 (1970)
3. T. J. Gay, J. E. Furst, K. W. Trantham and W. M. K. P. Wijayaratna, *Phys. Rev.* A **53**, 1623-1629 (1996)
4. J. E. Chilton, M. D. Stewart, Jr. and C. C. Lin, *Phys. Rev.* A **61**, 052708 (2000)
5. S. Tsurubuchi, K. Arakawa, S. Kinokuni and K. Motohashi, *J.Phys. B: At. Mol. Opt. Phys.* **33**, 3713-3723 (2000)
6. V. Puech and S. Mizzi, *J. Phys. D: Appl. Phys.* **24**, 1974-1985 (1991).
7. V. E. Bubelev and A. N. Grum-Grzhimailo, *J. Phys. B: At. Mol. Opt. Phys.* **24**, 2183-2199 (1991)
8. D. H. Madison, C. M. Maloney, and J. B. Wang, *J. Phys. B: At. Mol. Opt. Phys.* **31**, 873-893 (1998).
9. S. Kaur, R. Srivastava, R. P. McEachran and A. D. Stauffer, *J. Phys. B: At. Mol. Opt. Phys.* **31**, 4833-4852 (1998).
10. P. V. Feltsan and I. P. Zapesochny, *Ukr. Fiz. Zh.* **12**, 633-639 (1967).
11. J. K. Ballou C. C. Lin and F. E. Fajen, *Phys. Rev.* A **8**, 1797-1807 (1973).
12. S. Tsurubuchi, T. Miyazaki and K. Motohashi, *J. Phys. B: At. Mol. Opt. Phys.* **29**, 1785-1801 (1996).
13. J. E. Chilton, J. B. Boffard, R. S. Schappe, and C.C. Lin, *Phys. Rev.* A **57**, 267-277 (1998).
14. A. Chutjian and D. C. Cartwright, *Phys. Rev.* A **28**, 2178-2193 (1981).
15. P. V. Felttsan, *Ukr. Fiz. Zh.* **12**, 1425-1429 (1967).
16. I. P. Bogdanova and S. V. Yurgenson, *Opt. Spectrosc.* **62**, 424-425 (1987).
17. J. E. Chilton, M. D. Stewart, Jr. and C. C. Lin, *Phys. Rev.* A **62**, 032714 (2000).
18. S Trajmar, S. K. Srivastava, H. Tanaka, H. Nishimura, and D.C. Cartwright, *Phys. Rev.* A **23**, 2167-2177 (1981).
19. R. B. Key and C. G. Simpson, *J. Phys. B: At. Mol. Opt. Phys.* **21**, 625-637 (1988).
20. R. J. Anderson, R.H.Hughes, J.H. Tung, and S.T. Chen, *Phys. Rev.* A **8**, 810-815 (1973).
21. S. Kaur, R. Srivastava, R. P. McEachranm and A.D. Stauffer, *J. Phys. B: At. Mol. Opt. Phys.* **31**, 157-174 (1998).

Alignment of Photoions for the Autoionization Analysis of Doubly Excited Rare Gas Valence Shells

Bernd Zimmermann[*], Sven Kammer[*], Sascha Mickat[*],
Karl-Heinz Schartner[*], Holger Liebel[§], Arno Ehresmann[§],
Hans Schmoranzer[§]

[*]I.Physikalisches Institut, Justus-Liebig-Universität Giessen, 35392 Giessen, Germany
[§]Fachbereich Physik, Universität Kaiserslautern, 67663 Kaiserslautern

Abstract. Rare gas valence shell satellites excited by linearly polarized synchrotron radiation are produced exclusively through electron correlations. In the near threshold range they show a strong enhancement through the autoionization of doubly excited Rydberg states. Studying the fluorescence decay of the excited photoions we derived the alignment parameter A_{20}, which depends on the dynamics of the decay process. Here we report on the population of the Ne^+ $2p^4$ ^2P_J and ^4P_J finstructure states. While the ^2P_J states show interference between direct population and autoionization, the ^4P_J states to the contrary, are populated only through autoionization. This is a clear indication for singlet/triplet mixing of the more closely investigated Ne $2p^4$ (^3P) $3p$ $^2P_{3/2,1/2}$-ns,nd-Rydberg-states, since the population of the ^4P_J-states is forbidden in LS-coupling. Corresponding autoionization processes were studied for the heavier rare gases in order to reveal systematic features.

INTRODUCTION

Photoionization of rare gases was studied using the method of photon-induced fluorescence spectroscopy (PIFS) [1]. In the threshold range doubly excited states enhance the satellite production. High resolution experiments at synchrotron radiation facilities of the third generation are needed to resolve the different Rydberg series and their finestructure splittings.

The photoexcited ions are produced exclusively through electron correlations. Their theoretical treatment by e.g. many body methods is an interesting problem which has to be solved for the understanding of complex systems [2]. We intend to supply sufficiently detailed data for a sensitive test of such methods.

The alignment of the photoion contains information about the partial wave composition of the emitted electron, which either can be a photoelectron or an electron from the autoionization process [3].

CP604, *Correlations, Polarization, and Ionization in Atomic Systems*
edited by D. H. Madison and M. Schulz
© 2002 American Institute of Physics 0-7354-0048-2/02/$19.00

EXPERIMENT

The experiments were carried out at the U125/1-PGM beamline at BESSY II. Linearly polarized photons from the undulator are focussed into our differentially pumped gas target cell containing rare gases at a pressure of 30 µbar.

Two 1m-normal-incidence monochromators are oriented with their optical axis under 90° with respect to the photon beam direction (Fig. 1). One of them is equipped with a 600 lines/mm grating and a position sensitive detector for the visible range. In addition a Wollaston prism separates the fluorescence intensities with E-vector parallel and perpendicular to the E-vector of the exciting photons. From the 2D-position-sensitive spectrum cross sections σ as well as the asymmetry parameter β_{fl} can be extracted (Fig. 2). The second spectrometer uses a 1200 lines/mm grating and an open microchannel-plate as position sensitive detector for the VUV range.

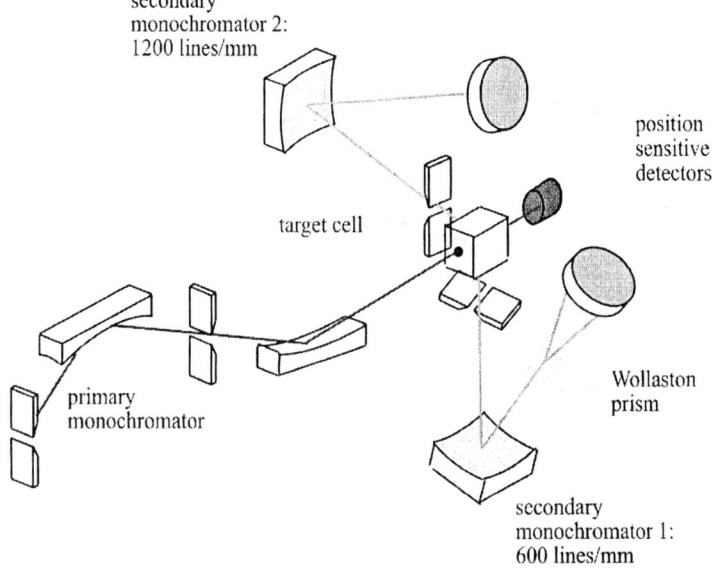

secondary monochromator 2: 1200 lines/mm

position sensitive detectors

target cell

Wollaston prism

primary monochromator

secondary monochromator 1: 600 lines/mm

Figure 1. Two-monochromator-setup for photon-induced fluorescence spectroscopy at the U125/1-PGM beamline at BESSY II.

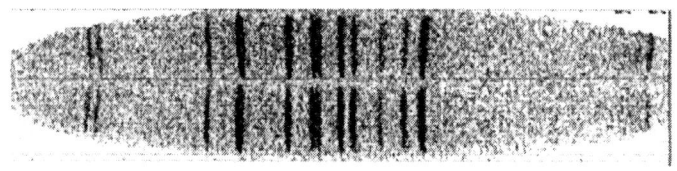

Figure 2. 2D-position-sensitive detection of Ne$^+$ satellite fluorescence intensities with E-vector parallel (upper part) and perpendicular (lower part) to the E-vector of the exciting photons separated by a Wollaston prism.

RESULTS

The autoionization of series of doubly excited Rydberg states produces a rich resonance structure of photon-induced rare gas satellites. Measuring the polarization fraction Π we can derive the anisotropy parameter β_{fl} for the transition $J_i - J_f$ that is connected to the alignment parameter A_{20} (Eq. 1) of the photoion with total angular momentum J_i. α is the kinematic structure parameter.

$$A_{20}(J_i) = \beta(J_i - J_f)/\alpha \qquad (1)$$

The alignment parameter delivers a simple access to the partial wave analysis of the in this case unobserved photo-electron, since the partial wave intensities add incoherently. Three partial waves are allowed for the total angular momentum of the emitted electron $j_{el} = 1 + s$, with $j_{el} = J_i-1, J_i, J_i+1$ [3].

Here we report measurements for Ne$^+$ $2s^2$ $2p^4$ (^{3}P) $3p$ (^{2}P$_J$)- and (^{4}P$_J$)-satellites. The (^{2}P$_J$)-satellites show interference between direct ionization and autoionization (Fig. 3), while the (^{4}P$_J$)-satellites are populated only through the autoionization of doubly excited Rydberg states (Fig. 4).

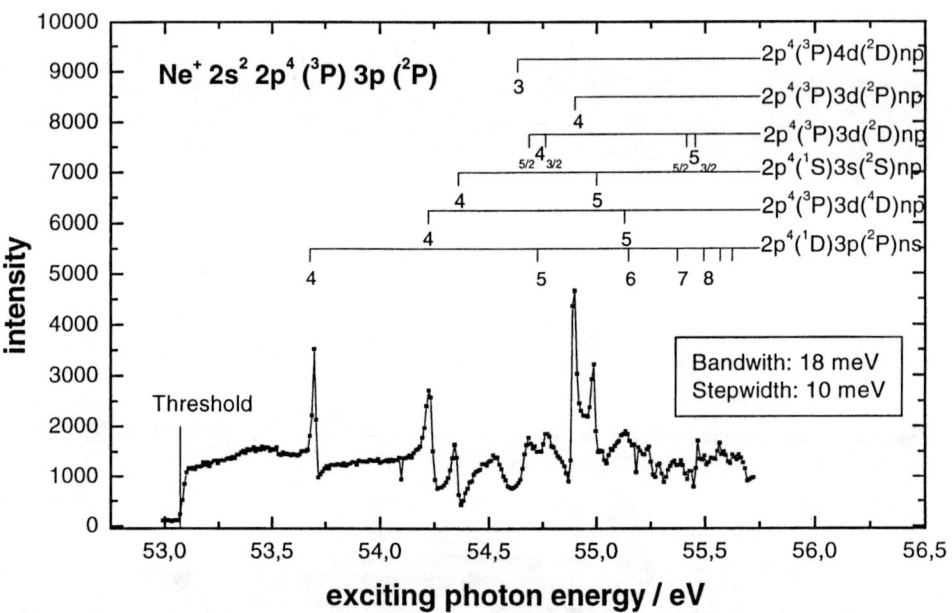

Figure 3. Fluorescence intensity of the Ne$^+$ $3p$ (^{2}P)-satellite up from the threshold at 53.07 eV in dependence of the exciting photon energy.

For the Ne$^+$ 3p ($^4P_{3/2}$)-satellite we observe a l-dependent alignment transfer from the autoionization of the doubly excited Rydberg states, which can be assigned as the Ne $2s^2$ $2p^4$ (^3P) 3p (^2P) ns- and nd-series (Fig. 4) [4]. β_{fl} for both the transitions $J_i = 3/2$ to $J_f = 1/2$ and the $J_i = 3/2$ to $J_f = 3/2$ is shown. A_{20} (for $J_i = 3/2$) values derived from both transitions agree with each other as they should. For the ns resonances the emission of a $s_{1/2}$ electron partial wave dominates. For the nd series the measured β_{fl} parameter leads to the conclusion that either a $d_{5/2}$ electron partial wave or the combination of a $s_{1/2}$ and $d_{3/2}$ electron partial wave contributes (Fig. 5). For this case no reliable prediction can be made, unless the additional information of the partial wave contribution is supplied, either through a measurement of the orientation parameter O_{10} of the photoions or through calculations.

For the J = 5/2 finestructure component of the ^4P-satellite, A_{20} shows no dependence on the Rydberg electron on the doubly excited state. Here the $s_{1/2}$ partial wave cannot exist and we find a preference of the $d_{5/2}$ electron partial wave.

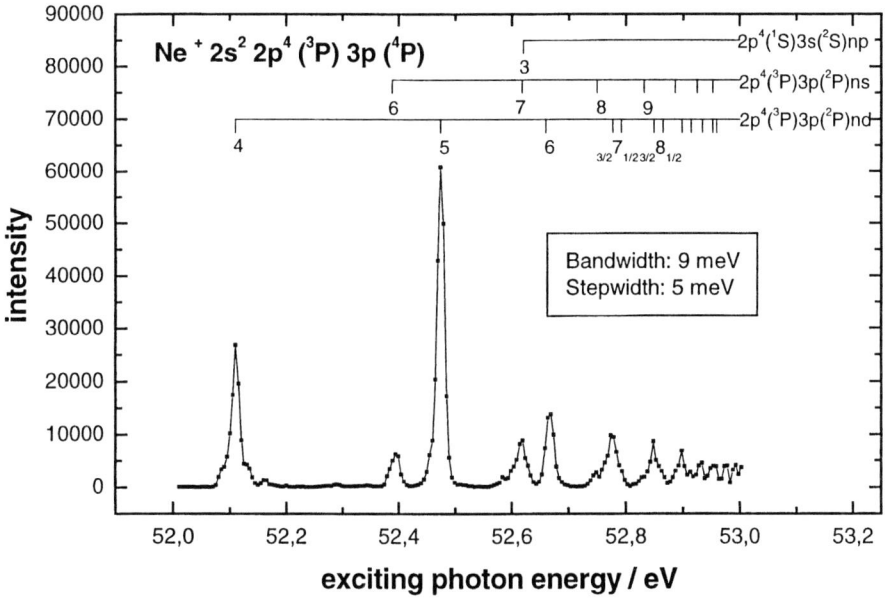

Figure 4. Fluorescence intensity of the Ne$^+$ 3p (^4P)-satellite up from the threshold at 52.08 eV in dependence of the exciting photon energy.

Our observations for satellites of corresponding 4p- or 5p-configurations of Ar and Kr, respectively, with $J_i = 5/2$ follow the same trend [1, 5]. Within the angular momentum transfer description the preference of the $d_{5/2}$ partial wave characterizes a parity unfavoured behaviour, i.e. the electron cloud of the photoion is elongated perpendicular to the E-vector of the exciting photons. For satellites with $J_i = 3/2$ we generally notice a more complex situation like the one shown in Fig. 5 or presented in [5] and [6].

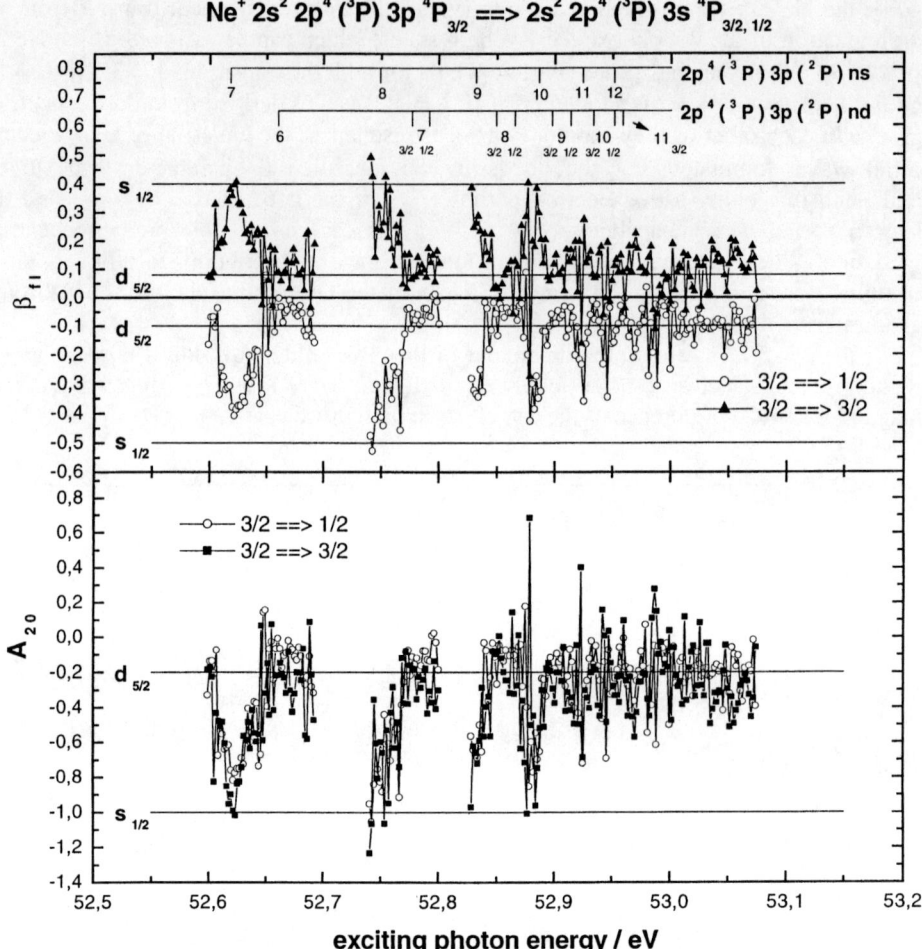

Figure 5. Resulting β_{fl} parameter and connected A_{20} parameter for the Ne$^+$ 3p ($^4P_{3/2}$) to 3s ($^4P_{3/2,1/2}$)-transitions in dependence of the exciting photon energy.

REFERENCES

[1]   Zimmermann, B., Wilhelmi, O., Schartner, K.-H., Vollweiler, F., Liebel, H., Ehresmann, A., Lauer, S., Schmoranzer, H., Lagutin, B.M., Petrov, I.D. and Sukhorukov, V.L., *J. Phys. B: At. Mol. Opt.* **33** 2467 (2000).

[2]   Lagutin, B.M., Petrov, I.D., Demekhin, Ph.V., Sukhorukov, V.L., Vollweiler, F., Liebel, H., Ehresmann, A., Lauer, S., Schmoranzer, H., Wilhelmi, O., Zimmermann, B. and Schartner, K.-H., *J. Phys. B: At. Mol. Opt. Phys.* **33** 1337 (2000)

[3]   Greene, C.H. and Zare, R.N., *Ann. Rev. Phys. Chem.* **33** 119 (1982)

[4]   Schuz, K., Dohmke, M., Püttner, R., Gutierrez, A., Kaindl, G., Miecznik, G. and Greene, C.H., *Phys. Rev. A* **54** 3095 (1996).

[5]   McLaughlin, K.W., Yenen, O. and Jaecks, D.H., *Phys. Rev. Let.* **81** 289 (1998)

[6]   Mentzel, G., Schartner, K.-H., Wilhelmi, O., Magel, B., Staude, U., Lauer, S., Liebel, H., Schmoranzer, H., Sukhorukov, V.L. and Lagutin, M., *J. Phys. B: At. Mol. Opt. Phys.* **31** 227 (1998)

Chiral Electron Pairs in Photodouble Ionization of Helium

Kouichi SOEJIMA

Graduate School of Science and Technology, Niigata University
Igarashi-ninocho 8050, Niigata 950-2181, JAPAN

Abstract. Triple differential cross sections (TDCSs) of the two ejected chiral electrons from helium were measured for several kinematic conditions using circularly polarized synchrotron radiation at the Photon Factory. The selected kinematic conditions were E=9 eV (R=2,8), E=40 eV (R=7, 19) and E=50 eV (R=24), where E is the excess energy and R is the ratio of the kinetic energy of the ejected electrons. The changes in the TDCSs with "E" and "R" are presented. The gerade (a_g) and ungerade (a_u) dipole amplitudes were determined using four parameters parametrization. The "E" and "R" dependences of the four parameters of a_g and a_u are also presented.

INTRODUCTION

The chirality in the photodouble ionization (PDI) has attracted both experimental and theoretical attention in recent years. It has become easy to observe the circular dichroism on the angular distribution (CDAD) in the PDI of free atoms because intense circularly polarized light has become available at synchrotron radiation facilities. The CDAD can be observed even for the helium target, which have spherical symmetry, when the chiral electron pairs are detected in coincidence. This chirality is caused by a handedness, which makes distinguishable a right-handed coordinate frame from a left-handed one [1]. Due to this chirality, the experiments using circularly polarized light become a new manifestation for studying electron correlation in PDI. It is noted that the experiments using linearly polarized light is also necessary for complete experimental study of determination of dynamical factor in PDI. The dynamical factor must be same for both types of experiments. Therefore, the validity of the value of the dynamical factor can be checked comparing the results for both types of experiments.

In PDI of helium, the first prediction of a strong circular dichroism (CD) has been done by Berakdar and Klar [2]. The first experimental evidence of CD was reported by Viefhaus *et al.* [3]. They showed a good example of the CD for various energy sharing between the two photoelectrons at three relative emission angles. The first CDAD in PDI of helium was reported by Mergel *et al.* in 1998 [4]. They measured the absolute five fold differential cross sections (FDCS) using COLTRIMS technique. In our group, the CDAD was measured in excess energy of 9 eV [5]. The experimental results were good agreement with convergent close-coupling (CCC)

CP604, *Correlations, Polarization, and Ionization in Atomic Systems*
edited by D. H. Madison and M. Schulz
© 2002 American Institute of Physics 0-7354-0048-2/02/$19.00

calculation [6]. Very recently, Achler *et al.* was reported the absolute FDCS in PDI of helium in the same kinetic condition as Mergel *et al.* [7]. Their careful experiments achieved excellent agreement between their results and CCC calculations. In this report, our recent experimental results of TDCS in PDI of helium in the excess energy of 40 eV and 50 eV are presented with our previous data in that of 9 eV.

EXPERIMENTAL SETUP

The experiments were performed at the helical undulator beamline BL-28A equipped with the constant deviation monochromator of the Photon Factory storage ring in Tsukuba [8]. The fundamental of helical undulator radiation provides soft X-ray with high flux and a high degree of circular polarization. Using multilayer elipsometry, the Stokes parameters S1 and S3, which describe the degree of linear and circular polarization, were measured at the photon energy of 97 eV [9]. The results were; for the right-handed mode, S1=+0.2, S2=0 and S3=+0.95 in the frame tilted by +135°±2° relative to the plane of the storage ring, and for left-handed mode, S1=+0.2, S2=0 and S3=-0.95 in the frame tilted by +44°±2°. The polarization parameters were not measured in the course of our experiments because the degree of circular polarization at the first harmonic is independent on the undulator gap.

The experimental apparatus has been previously described in detail [10]. It consists of two parallel plate energy analyzers placed in a plane perpendicular, i.e. polar at angle θ=90°, to the photon beam z-axis. One of the analyzer was placed with a fixed position, i.e. azimuthal angle ϕ=-90°, and the other was rotatable around the z-axis from ϕ=-10° to 180°. The azimuthal angle ϕ is measured counterclockwise from the storage ring plane as seen be as observer facing the photon beam. The acceptance angle and energy resolution were estimated to be $\Delta\phi$=±5.3° and ΔE=0.7 eV, respectively. The standard electronic circuit of detection system for coincidence measurement has been used.

RESULTS AND DISCUSSION

For PDI, complete information is provided by the TDCS, which can be obtained from coincidence measurement for two emitted photoelectrons. It is well known that the shape of the TDCS is dramatically changing with both geometrical, i.e. direction of e1, light polarization, and dynamical, i.e. excess energy, energy sharing, experimental conditions. In this report, the dynamical condition dependence of the TDCSs are shown. Our kinematic conditions were E=9 eV (R=2,8), 40 eV (R=7,19) and 50eV (R=24), where E is the excess energy and R is the ratio of kinetic energy of the emitted photoelectrons, i.e. R=E2/E1. The experimental results of most asymmetric energy sharing condition, E=50eV, R=24, are shown in Fig. 1. The "E" and "R" dependences of the TDCSs are presented in Fig. 2 for left circularly polarized light. It is enough to show the results only for left circularly polarized light since the TDCS for right and left circularly polarized lights have mirror symmetry relative to the fixed analyzer direction, indicated by arrows, due to the parity conservation in the perpendicular plane geometry.

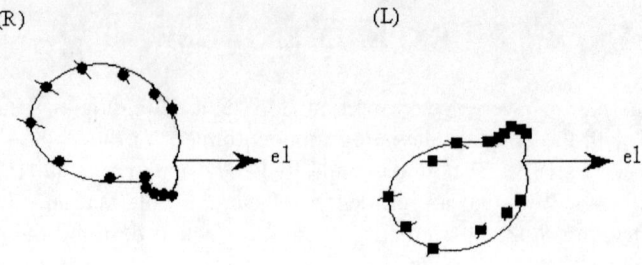

FIGURE 1. The experimental TDCS in PDI of helium for right circularly polarized light (R) and left circularly polarized light (L) in the kinetic condition of E=40 eV, R=24.

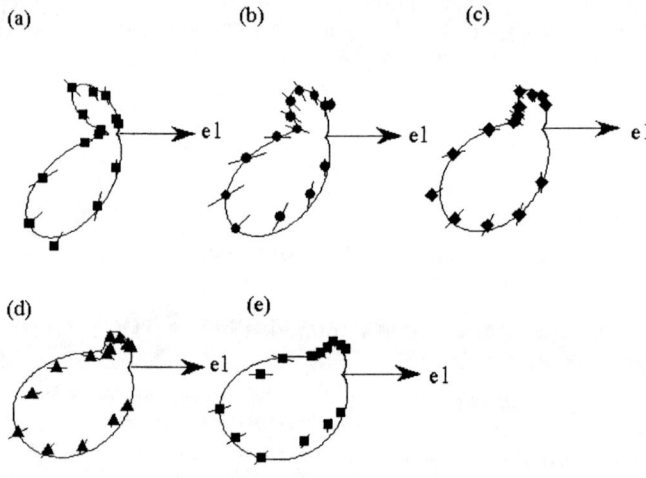

FIGURE 2. The TDCS for left circularly polarized light of E=9 eV, R=2 (a), R=8 (b), E=40 eV, R=7 (c), R=19 (d) and E=50 eV, R=24 (e). The full curves are fitted results of 4P parametrization.

Parametrization Procedure

Expansion of TDCS in terms of the angular functions is useful for a qualitative understanding of the dynamics in PDI. This approach was pioneered by Huetz and his co-workers [11]. The aim of this expansion is to separate the angular dependence of the cross section from that of the energy dependence. The TDCS can be expressed as

$$TDCS \propto |\varepsilon ü \ddot{E} \mu|^2 \qquad (1)$$

where ε denotes the electric field vector and μ represents the dipole transition moment. In Huetz formalism, μ is given by

$$\mu = M_g (\mathbf{k1} + \mathbf{k2}) + M_u (\mathbf{k1} - \mathbf{k2}) \qquad (2)$$

where M_g and M_u are the dipole amplitudes which are respectively symmetric and anti-symmetric with respect to interchanging of the two photoelectron kinetic energies, $\mathbf{k1}$ and $\mathbf{k2}$ denotes the momenta of the two photoelectrons. In our experimental geometry, equation (1) leads to following equations showing the respective contributions to TDCS of polarization insensitive (PI), linear dichroism (LD), and circular dichroism (CD) :

$$PI = TDCS_x + TDCS_y = TDCS_R + TDCS_L = (1 + \cos\phi)|M_g|^2 + (1 - \cos\phi)|M_u|^2 \quad (3)$$

$$LD = TDCS_x - TDCS_y$$
$$= \sin\phi \{(1 - \cos\phi)|M_u|^2 - (1 + \cos\phi)|M_g|^2 - 2\cos\phi \, \mathrm{Re}(M_g M_u^*)\} \qquad (4)$$

$$CD = TDCS_R - TDCS_L = \sin\phi \{2 \, \mathrm{Im}(M_g^* M_u)\} \qquad (5)$$

where $TDCS_x$ and $TDCS_y$ refer to the linear polarization along X- and Y-axis, and $TDCS_R$ and $TDCS_L$ refer to the right and left circular polarization, and ϕ is the mutual angle between two photoelectrons. It is well understood that the linearly polarized light probes $\mathrm{Re}(M_g M_u^*)$ whilst circularly polarized light probes $\mathrm{Im}(M_g M_u^*)$ from equations (4) and (5). Therefore, it is also well understood that the experiments using circularly polarized light have advantage for determination of the sign of phase difference between M_g and M_u. It is noted that the optical definition was adopted in this report, i.e. right circularly light corresponding to negative helicity. The measured TDCS for partial light polarization is given by equations (3), (4) and (5) as

$$TDCS = PI + S_1 LD + S_3 CD . \qquad (6)$$

The $|M_g|^2$ and $|M_u|^2$ are parametrized using three parameters of Γ_g, Γ_u, η as

$$|M_g|^2 = \exp(-4\ln 2(\phi - 180°)^2 / \Gamma_g^2) \tag{7}$$

$$|M_u|^2 = \eta \exp(-4\ln 2(\phi - 180°)^2 / \Gamma_u^2) \tag{8}$$

and one more parameter (Δ) is introduced for the parametrization of TDCS, which is the phase difference between M_g and M_u. Thus, the four parameters (4P) parametrization is adopted for analyzing our experimental data. From least-squares fit to the numerous experimental data, 4P was determined for respective kinematic conditions of E=9 eV (R=2,8), E=40 eV (R=7,19) and E=50 eV (R=24). The "R" and "E" dependences of 4P are shown in Fig. 3. In my analysis, Γ_g, Γ_u, and Δ are almost independent of "E" and "R". In contrast, η is strongly depend on "R" which means that $|M_u|^2$ increases rapidly for increasing "R". Then, it can be concluded that the increasing of CD in PDI is related with the increasing $|M_u|^2$ in our kinetic conditions.

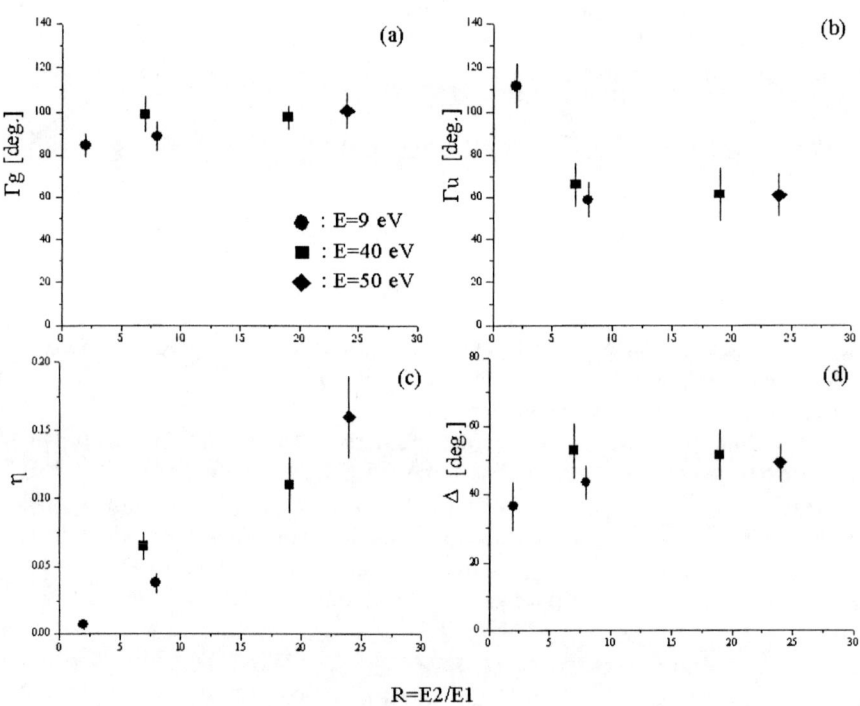

R=E2/E1

FIGURE 3. The "E" and "R" dependences of 4P of (a) Γ_g, (b) Γ_u, (c) η and (d) Δ.

ACKNOWLEDGMENTS

The author is grateful to Dr. A. Yagishita, Dr. P. Selles, and Dr. A. Huetz for fruitful discussions for PDI of helium. I also would like to acknowledge Dr. T. Koide and Dr. Y. Miyauchi for their excellent support to use helical undulator beamline BL-28A at the PF. This study was supported by a Matsuo Foundation. This work has been performed under approval of PF Advisory Committee (Proposal No.95G413, 97G352 and 99G148).

REFERENCES

1. J. S. Briggs and V. Schmidt, J. Phys. B: At. Mol. Opt. Phys. **33**, R1 (2000).
2. J. Berakdar and H. Klar, Phys. Rev. Lett. **69**, 1175 (1992).
3. J. Viefhaus et al., Phys. Rev. Lett. **77**, 3975 (1996).
4. V. Mergel et al., Phys. Rev. Lett. **80**, 5301 (1998).
5. K. Soejima et al., Phys. Rev. Lett. **83**, 1546 (1999).
6. A. Kheifets et al., J. Phys. B: At. Mol. Opt. Phys. **32**, L501 (1999).
7. M. Achler et al., J. Phys. B: At. Mol. Opt. Phys. **34**, 965 (2001).
8. Y. Kagoshima et al., Rev. Sci. Instrum. **63**, 1289 (1992).
9. H. Kimura et al., Rev. Sci. Instrum. **66**, 1920 (1995).
10. K. Soejima et al., J. Korean Phys. Soc. **32**, 368 (1998).
11. A. Huetz et al., J. Phys. B: At. Mol. Opt. Phys. **24**, 1917 (1991).

Balmer-α Polarization Studies

J.W.McConkey

Department of Physics, University of Windsor, Ontario N9B 3P4, Canada

Abstract. The polarization of Balmer-α radiation excited in collisions of electrons with atomic hydrogen is presented for an electron energy range from threshold to 1000 eV. Measurements are in good agreement with calculations carried out using either convergent-close-coupling (CCC) or R-Matrix with pseudo-states (RMPS) approaches. Cascade is demonstrated to have a significant effect. Balmer-α excitation function data are discussed and H(n=3l) data are presented. A previous measurement of the polarization of Balmer-α following dissociative excitation of H_2 by electrons is confirmed and extended.

INTRODUCTION

Atomic hydrogen still provides a fruitful testing ground for modern theories of collisional excitation, particularly when the finer details of the excitation process are being probed. Much progress has been made in recent years [1-3] particularly with regard to excitation of H(2p) but the need to obtain accurate sub-level cross section and other data for the n=3 and higher levels remains. The recent measurements were made possible by new developments in atomic source technology which resulted in atomic source densities many orders of magnitudes greater than were available some 30 years ago when many of the pioneering experiments on atomic H were being performed.

Balmer line measurements go back to the pioneering work of Ornstein and Lindeman [4]. The only reports of cross section data were by Kleinpoppen and Kraiss [10], who performed relative Balmer-α measurements and by Walker and St John [5], who obtained absolute cross section data for the first five lines of the Balmer series from 20 to 440 eV. They found reasonable agreement between their data and the First Born Approximation data [6,7]. Mahan et al [8] used a modulated excitation technique, which took advantage of the different lifetimes of the components of Balmer-α, to determine the excitation cross sections of the individual 3l sublevels up to 500 eV electron energy. They normalised their total n=3 cross section to the Born Approximation results at 500 eV.

The first measurements of the electron impact-induced H-α line polarizations were performed by Kleinpoppen et al [9,10]. Since that time the only other measurements have been the recent work of Werner and Schartner [11] who presented data for the polarization of the Balmer-α line for both electron and proton impact. In the electron impact case data were presented for electron energies from 80 eV to 3 keV. In the limited energy range where a comparison can be made with the earlier data, the Werner and Schartner results

CP604, *Correlations, Polarization, and Ionization in Atomic Systems*
edited by D. H. Madison and M. Schulz
© 2002 American Institute of Physics 0-7354-0048-2/02/$19.00

are significantly lower. Although there are many reports of calculated cross sections for the n=3 states, very few authors give magnetic sublevel or polarization data. Also, although there has been a large body of theoretical input relative to the H(n=2) states motivated by developments in electron-photon coincidence experiments, (see for example Andersen et al, [12], Andersen and Bartschat, [1]), the amount of work aimed specifically at H(n=3) is much more limited (see Bray and Stelbovics, [13]).

The present work is an attempt to redress this situation both theoretically and experimentally. Thus accurate polarization and other data, both experimental and theoretical, are presented for H (n=3) excitation and subsequent decay. In addition the earlier polarization data [14] for Balmer-α emission in electron – molecular hydrogen collisions have been extended to 300 eV incident electron energy.

EXPERIMENTAL DETAILS

A crossed beam apparatus is used in which an electron beam and the target H beam are orthogonal. Radiation, emerging from the interaction region in a direction perpendicular to both of these beams, passes through a narrow band filter (Melles-Griot, 1nm FWHM) centered on Balmer-α and through a high grade, transmission polarizing element before being focussed onto a cooled photomultiplier with a GaAs photocathode. Alternatively, radiation from the interaction region can be monitored in a direction opposite to this after it passes through a half-meter Seya-Namioka vacuum ultraviolet (VUV) monochromator equipped with a channel electron multiplier detector. Full details of this set-up have been given elsewhere, [15-18].

The polarization analyzer is used to select the radiation polarized parallel (I_{11}) or perpendicular ($I_\perp$) to the e-beam direction. The polarization is then defined in the usual way by:

$$P = [I_{11} - I_\perp] / [I_{11} + I_\perp] \tag{1}$$

The polarization efficiency of the analyzer was measured to be very close to unity and so is not included in Equation 1. The H-atom source is a special microwave discharge developed at Belfast [19]. It incorporates a 2.45 GHz cavity with Lisatano slotted line radiators immersed in a static magnetic field. The discharge was initiated in a 2.5cm diameter quartz tube within the source. The dissociation fraction in the target beam was measured by monitoring the intensity of the VUV H_2 bands near Lyman-α. An accurate knowledge of the molecular content in the interaction region is necessary in order to remove the contribution due to dissociative excitation of Balmer-α from molecular parents. Operating pressures in the source ranged from 30 to 120 mTorr. At higher pressures, depolarizing effects were observed to occur.

One effect which is likely to have a significant effect on both the measured apparent excitation cross section and the measured polarization is cascade into the n=3 sublevels from higher excited states. Starting with Equation 1 and splitting the measured intensities into directly excited and cascade components, I_D and I_C, it is easy to show [18] that the measured polarization, P_m, is related to P_D, the polarization which would be measured if only direct excitation was occurring, by

$$P_m \sim P_D/[1 + \{I_C/I_D\}\{(3 - P_D)/(3 - P_C)\}\,] \qquad (2)$$

If we assume an unpolarized cascade contribution of 10% and a value of P_D of 0.25 (typical at low energy, see later), we obtain $P_m = 0.91\ P_D$. Thus cascade can play a significant role. We have calculated the cascade contribution as a function of energy using our theoretical estimates of the cross-sections of the upper states involved together with the branching ratios for the de-exciting transitions. I_C/I_D is found to have a value of around 0.12 and stays reasonably constant as the electron energy is varied. We note that these estimates are significantly higher than ones calculated previously, [8], using the Born approximation.

RESULTS AND DISCUSSION

Polarization data, obtained as described in the previous sections, are displayed in Fig 1 along with other experimental data where available and also the present and previous theoretical results. The present theoretical results have been obtained using the convergent-close-coupling (CCC) and R-Matrix with pseudo-states (RMPS) methods. The CCC calculations include 30 S-, P-, D-, and F- states, of which 15 states are discrete states. The RMPS calculations have been described by Bartschat and Bray [20].

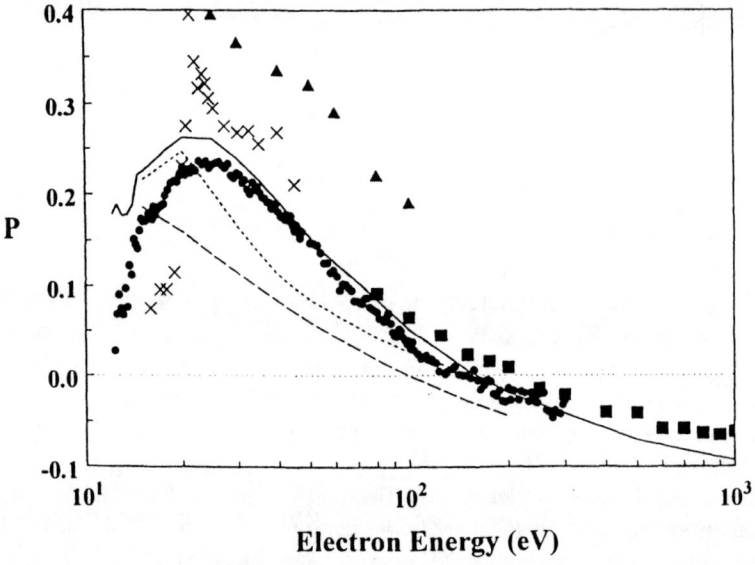

Figure 1. Balmer-α polarization as a function of electron impact energy. Filled triangles – Ref [9] Data; Filled squares –Ref [11] Data; Dots– present experimental data; X -Ref [10] Data; Solid line – Present CCC data, cascade included; long dashes – Born data, [24]; short dashes – DWPO data [24].

186

In addition we used the CCC and RMPS codes to perform standard 15-state (discrete states only) close coupling calculations (the results are labeled 15CC below) to assess the importance of channel coupling to the target continuum. Calculation of the radiation polarization from the ensemble of the n=3 levels follows from the general principles outlined, for example, by Blum [21] and will be presented in detail elsewhere.

A number of points should be noted from Fig 1. First of all there is good agreement between the present measurements and theory particularly above 40 eV. Secondly, we note that the structure observed in the early work [10] has not been reproduced. Thirdly, in the region above 80 eV, our experimental data and the CCC results are seen to lie somewhat lower than the recent data of Werner and Schartner[11]. Our data cross the zero polarization axis at 150 eV whereas their results reach this value over 50 eV higher in energy. The reason for this discrepancy is not clear. Also shown on Figure 1 are the earlier theoretical results of Syms et al [24] using Born and Distorted Wave Polarized-Orbital (DWPO) approximations. As can be seen, the Born approximation underestimates the polarization over the whole energy range. The DWPO data do show a peak at approximately the right energy and come into agreement with the present data at energies above 100 eV. There is excellent agreement between the RMPS and CCC results but the 15-state calculation is found to be inadequate. This is interesting because the 15-state calculation gives quite accurate results for the overall n=3 cross sections at higher energies. However a significant redistribution of flux occurs among the magnetic sub-states leading to the disagreement in the case of the polarization.

With the convergence of the CCC and RMPS predictions and the good agreement between experiment and theory already demonstrated, we believe that our data for n=3 excitation are reliable to high accuracy. When we compare our data with the earlier data of Walker and St John [5], we find good agreement regarding the *shape* of the Balmer-α excitation function over the whole energy range though it appears that their data were consistently too high by a factor of approximately 1.3.

In Fig 2 we present data for the excitation of the individual n=3 sublevels and compare these with the earlier data of Mahan et al [8] and with the Born approximation [6]. Ref [8] had emphasized the unusual situation where the low energy results for 3d excitation had exceeded the Born results. It is clear from Fig 2 that this anomaly has now been cleared up. Although there is considerable scatter in the results of [8], the overall level of agreement with the present data is impressive and is a testimony to the exceptional experimental efforts of Mahan et al.

Dissociative Excitation of Balmer-α

As a byproduct of the present experiment we obtained the polarization of the Balmer- α radiation excited in electron–H_2 collisions. These data are in good agreement with the earlier data of Karolis and Harting [14] and extend these data to higher energies (300eV). An interesting comparison may also be made with earlier work [22,23] where the polarization of Lyman-α following electron impact

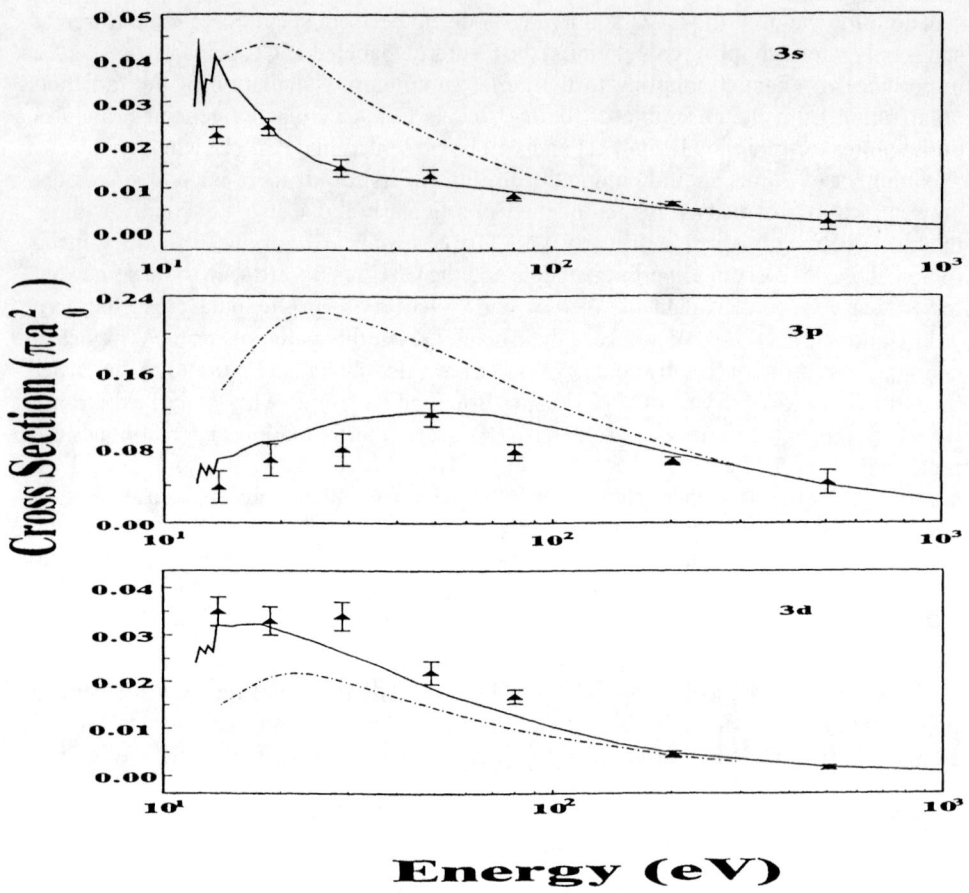

Figure 2. Cross sections for direct excitation of the H(n=3) sublevels. Solid line – Present CCC data; Dash-dot line – Born data [6]; Triangles – Mahan et al [8].

dissociative excitation of H_2 was measured. Their results are essentially identical with the present data, thus suggesting that very similar dissociation mechanisms are responsible in the two situations.

CONCLUSIONS

Data have been obtained for the polarization and excitation functions for Balmer-α excited in e-H collisions over an energy range from threshold to 1000 eV. Agreement between experiment and theory is good. Individual n=3 sublevel cross section data are also presented. Given the other successes of the CCC and RMPS theories with this target, it is reasonable to claim that they will predict the excitation of higher n states with comparable high accuracy. Although there is this level of agreement between experiment and theory as far as polarization and excitation function data are concerned, we should point out that serious disagreement still exists when coincidence measurements of the circular polarization of radiation from the 3D state are considered (see Bray and Stelbovics, [13], for details). Previous data for the polarization of Balmer-α line radiation following dissociative excitation of H_2 have been confirmed and extended.

ACKNOWLEDGEMENTS

This work has been done in close collaboration with colleagues in Windsor (W Kedzierski and A Abdellatif), at Drake University (K Bartschat), and at Flinders University (I Bray and D Fursa). Thanks are due to NSERC (Canada) for financial assistance and to the staff of the mechanical and electronic workshops at Windsor for expert technical help.

REFERENCES

1. Andersen N and Bartschat K, *"Polarization, Alignment and Orientation in Atomic Collisions"*, Springer (New York), (2000).
2. James, G. K., Slevin, J. A., Dziczek, D., McConkey J W and Bray I, Phys Rev A, **57,** 1787, (1998).
3. James, G. K., Slevin, J. A., Shemansky, D.E., McConkey, J. W., Bray, I., Dziczek, D., Kanik, I. and Ajello, J. M., Phys Rev A, **55,** 1069, (1997).
4. Ornstein L S and Lindeman H, Z Physik, **80,** 525, (1933).
5. Walker J D and St John R M, J Chem Phys, **61,** 2394, (1974).
6. Vainshtein L A, Opt Spectrosc, **18,** 538,(1965).
7. Green L C, Rush P P and C D Chandler, Astrophys J Suppl, **3,** 37, (1957).
8. Mahan A H, Gallagher A and Smith S J, Phys Rev A, **13,** 156, (1976).
9. Kleinpoppen H, Krueger H and Ulmer R, Phys Letts, **2,** 78, (1962).
10. Kleinpoppen H and Kraiss E, Phys Rev Lett, **20,** 361, (1968).
11. Werner A and Schartner K-H, J Phys B, **29,** 125, (1996).
12. Andersen N, Gallagher J W and Hertel I V, Phys Reps **165,** 1, (1988).
13. Bray I and Stelbovics A T, J Phys B, **30,** L493, (1997).
14. Karolis C and Harting E, J Phys B, **11,** 357, (1978).
15. Woolsey J M, Forand J L and McConkey J W, J Phys B, **19,** L 493, (1986).
16. Wang S and McConkey J W, J Phys B, **25,** 5461, (1992).
17. Forand J L, Woolsey J M and McConkey J W, Can J Phys, **66,** 349, (1988).
18. Kedzierski W, Abdellatif A, McConkey J W, Bartschat K, Fursa D V and Bray I, J Phys B, **34,** In Press, (2001).
19. McCullough R W et al , Meas Sci and Tech, **4,** 79, (1993).
20. Bartschat K and Bray I, J Phys B, **29,** L577, (1996).
21. Blum K, *"Density Matrix Theory and Applications"*, Plenum, (New York), (1981)
22. Malcolm I C, Dassen H W and McConkey J W, J Phys B, **12,** 1003, (1979).
23. Ott W R, Kauppila W E and Fite W L, Phys Rev A, **1,** 1089, (1970).
24. Syms R F, McDowell M R C, Morgan L and Myerscough VP, J Phys B, **8,** 2817, (1976).

Elastic Scattering Of Spin Polarised Electrons From Krypton

M. R. Went[*], R. P. McEachran[†], Birgit Lohmann[*] and
W. R. MacGillivray[*]

[*]Laser Atomic Physics Laboratory, Griffith University, Brisbane, Queensland, AUSTRALIA 4111
[†] Research School of Physical Sciences and Engineering, Australian National University, Canberra,
AUSTRALIA 0200

Abstract. Calculated values of the Sherman function for krypton at 60 and 65eV incident
electron energy show large predicted values and a rapid dependence on energy for the sign of the
asymmetry. These unexpectedly large asymmetries have prompted an experimental investigation
of this target. In this paper the current status of these measurements and further theoretical and
experimental results for the Sherman function for elastic scattering of spin polarised electrons
from krypton will be discussed.

INTRODUCTION

The scattering of spin-polarised electrons from atoms yields an immediate means of
probing spin-dependent interactions. A sensitive spin-dependent observable is the left-
right asymmetry in the scattering of the spin up/down electrons by the target. An
asymmetry in the scattering of polarised electrons elastically scattered from noble
gases can only arise from the spin-orbit interaction of the projectile electron in the
field of the atom. Electrons, with a degree of polarisation P perpendicular to the
scattering plane, impinging on a noble gas target are scattered. Measurements of the
difference in the number of electrons scattered at equal angles to the left and right of
$0°$ yields the asymmetry A such that

$$A = \frac{N_L - N_R}{N_L + N_R} = S_A P \ , \tag{1}$$

where S_A is called the asymmetry function and for elastic scattering is equal to the
Sherman function, S [1]. Equivalently if the electron polarisation can be switched
without affecting the beam or introducing experimental asymmetries, N_L and N_R can
be replaced with scattering intensities on one side but for up $N_\uparrow$ and down $N_\downarrow$ electron
polarisations respectively.

In this paper we present new calculations of the Sherman function for elastic
scattering from krypton at 60 and 65eV incident electron energy which predict large
values of the Sherman function, and a rapid dependence on energy of the sign of the
asymmetry. These unexpectedly large asymmetries at 60 and 65eV have prompted our
investigation of this process.

CP604, Correlations, Polarization, and Ionization in Atomic Systems
edited by D. H. Madison and M. Schulz

THEORY

The details of the derivation of the Dirac equations for the case of elastic scattering of electrons from closed-shell atomic systems has been given elsewhere [2]. Consequently, we will present only a brief outline of our procedure. In the relativistic Dirac treatment of electron scattering, the large and small radial components, $f_\kappa(r)$ and $g_\kappa(r)$, of the scattered electrons satisfy the coupled first-order differential equations

$$\frac{d}{dr} f_\kappa(r) + \frac{\kappa}{r} f_\kappa(r) - \alpha \left[\frac{2}{\alpha^2} - V(r) + \varepsilon \right] g_\kappa(r) - \alpha W_Q(\kappa; r) = 0 \tag{2a}$$

and

$$\frac{d}{dr} g_\kappa(r) - \frac{\kappa}{r} g_\kappa(r) + \alpha [-V(r) + \varepsilon] f_\kappa(r) + \alpha W_P(\kappa; r) = 0 \tag{2b}$$

subject to the boundary conditions

$$f_\kappa(0) = g_\kappa(0) = 0, \tag{3}$$

$$f_\kappa(r)_{r \to \infty} \to C_\kappa(k) \sin\left(kr - \frac{l\pi}{2} + \delta_\kappa(k) \right), \tag{4a}$$

and

$$g_\kappa(r)_{r \to \infty} \to C_\kappa(k) \frac{ck}{E + c^2} \cos\left(kr - \frac{l\pi}{2} + \delta_\kappa(k) \right), \tag{4b}$$

where $\delta_\kappa(k)$ is the phase shift and $\alpha = 1/c$ is the fine-structure constant. In equations (2a,b), the potential $V(r)$ is given by the sum of the usual static potential and an adiabatic dipole polarisation potential while the terms $W_P(\kappa; r)$ and $W_Q(\kappa; r)$ represent the large and small components of the non-local exchange interaction between the incident electron and the atom. For the definitions of the remaining parameters in the above equations the reader is referred to Ref [2].

Once the phase shifts have been determined the direct and spin-flip scattering amplitudes, $f(\theta)$ and $g(\theta)$, can be determined from the expressions

$$f(\theta) = \frac{1}{k} \sum_{l=0}^{\infty} \left[(l+1) T_l^+(k) + l T_l^-(k) \right] P_l(\cos\theta) \tag{5a}$$

and

$$g(\theta) = \frac{1}{k} \sum_{l=0}^{\infty} \left[-T_l^+(k) + T_l^-(k) \right] P_l'(\cos\theta) \tag{5b}$$

with

$$T_l^\pm(k) = \frac{1}{2i} \left[\exp\left(2i\delta_l^\pm(k) \right) - 1 \right]. \tag{6}$$

Here $\delta^+_l(k)$ is the so-called spin-up phase shift corresponding to $\kappa = -(l+1)$ while $\delta^-_l(k)$ is the spin-down phase shift corresponding to $\kappa = l$. Once the scattering amplitudes have been determined, the Sherman function $S(\theta)$ can be determined according to

$$S(\theta) = \frac{2\operatorname{Im}\left\{ f^*(\theta) g(\theta) \right\}}{\left| f(\theta) \right|^2 + \left| g(\theta) \right|^2}. \tag{7}$$

APPARATUS

An overview of the experimental apparatus can be seen in figure 1. Light from a diode laser of wavelength 780nm is passed through a linear polariser, liquid crystal variable retarder combination to produce circularly polarised light which produces longitudinally polarized photoelectrons from a GaAs crystal. The electrons pass through a 90° deflector and exit with transverse polarisation. The electron beam is then focused onto the gas target by an electrostatic lens system. The gas enters the system through a single capillary. Electrons which are scattered from the gas are collected by a hemispherical electron energy analyser. The analyser is able to be rotated through an angle of 30°-130° in the scattering plane. The electron polarisation may be monitored by the use of a Mott detector operating at 100keV.

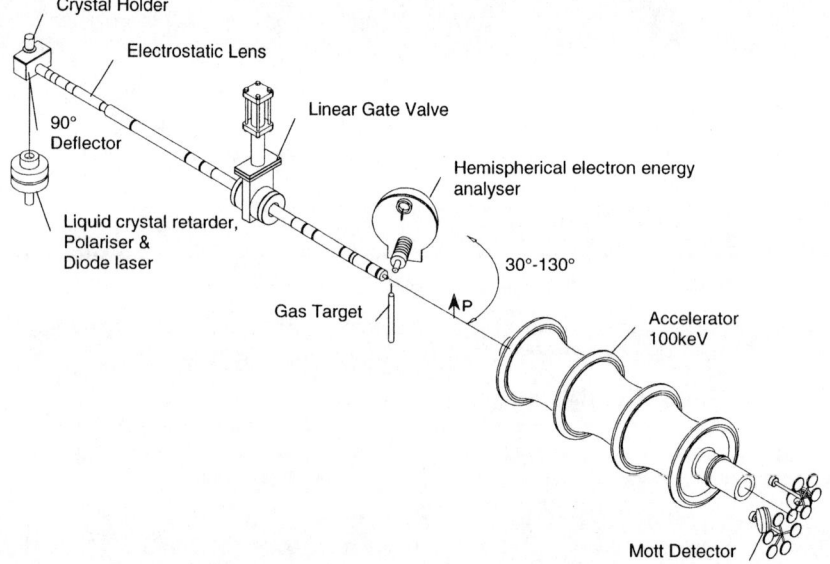

Figure 1. Schematic diagram of the experimental set-up.

192

The standard GaAs polarised electron source is used in this experiment. Polarised electrons are produced by photoemission from a negative electron affinity (NEA) surface. The NEA surface is produced by the application of caesium and oxygen to the GaAs surface which has first been heat cleaned at 660°C. The emission current for the duration of this experiment was 6-8μA with an associated 'half life' of approximately three weeks. Due to losses in the electrostatic lens system this current was reduced to 100-150nA in the interaction region.

A single hemispherical electron energy analyser detects electrons in the scattering plane. The inner and outer hemispheres of the analyser are 65mm and 125mm in diameter respectively. The entrance and exit apertures of the analyser are 1mm in diameter. The energy resolution of the analyser was determined to be 0.55eV FWHM. A set of electrostatic aperture lenses image the interaction region at the entrance to the hemispherical energy analyser; the geometrical acceptance angle of the analyser is ±2°. Electrons of appropriate energy to pass through the electron energy analyser are detected by a channel electron multiplier. Pulses from this device are passed through a simple pick-off circuit into a discriminator via a preamplifier and are counted by standard counters. The angular positioning of the analyser is achieved by a stepper motor connected to a personal computer. The analyser angular calibration was performed by comparison of measured argon elastic cross sections with previously published results [3].

RESULTS

Figure 2 shows the angular dependence of the Sherman function for elastic scattering from krypton in the energy range 50-100eV. The solid line in each case is the present calculation. As problems with the Mott detector prevented a reliable calibration measurement of the polarisation of the incident beam, the experimental results were normalised to the theory at 60eV, 70° scattering angle, yielding a spin polarisation of 20%. This same inferred polarisation was then used in the determination of the Sherman function at the other energies. At 50eV other experimental results for the Sherman function are available [4,5] for comparison (Figure 2 (a)). The latter values are obtained from polarisation measurements of the scattered electrons produced by an initially unpolarised electron beam. There is generally good agreement with Refs. [4] and [5], except at 40° where Ref. [4] appears to overestimates the Sherman function. There is also good agreement with theory across most of the angular range, except in the region 130°-150°, where theory predicts a large, varying asymmetry. At 100eV (Figure 2(d)) our experimental results are compared with results from Ref. [4], and with theory. Again there is good agreement with previous experimental data and with the present theory. At 60 and 65eV, there are no other experimental results for comparison. Figures 2(b), (c) show our data and the theoretical calculation. There is satisfactory agreement between theory and experiment at smaller angles, but the very large predicted values of the Sherman function near 110° (with a sign change in going from 60 to 65eV) are not observed in the experimental results. Given the predicted rapid variation as a function of energy, we also measured the asymmetry 2eV above and below 60eV, but found no evidence for the large values predicted by the theory. As these large asymmetry values coincide with the minimum in the differential cross section, determining S in this region becomes difficult experimentally and theoretically. Further measurements in this angular and energy range are planned.

ACKNOWLEDGMENTS

This research was supported by the Australian Research Council. MW gratefully acknowledges the support of an Australian Postgraduate Award.

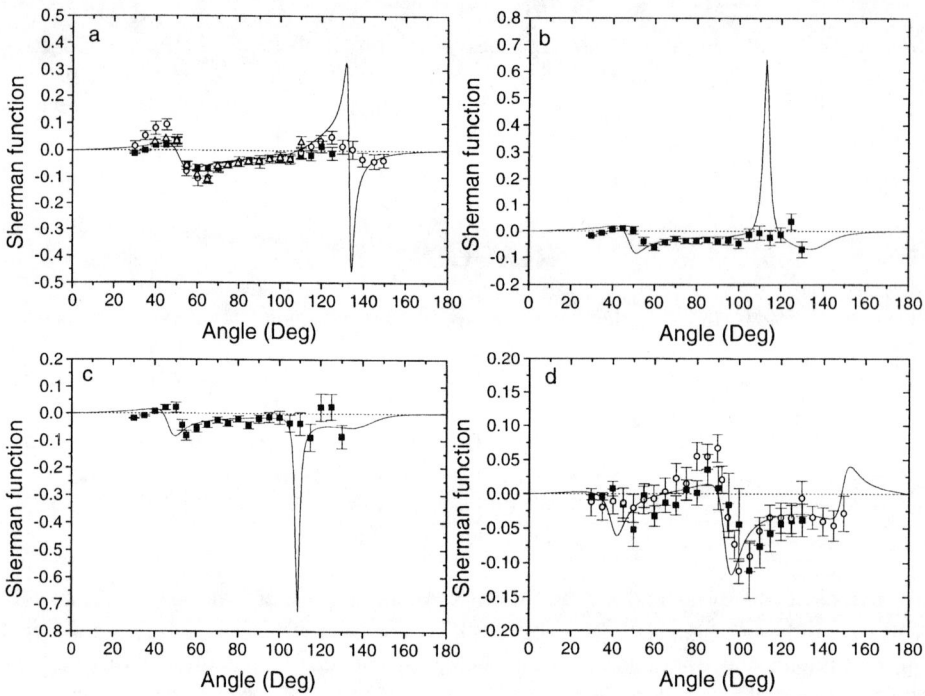

FIGURE 2. Sherman function for elastic scattering from krypton at energies of (a) 50eV, (b) 60eV, (c) 65eV and (d) 100eV. ■ Our experimental results. Electron polarisation measurements from O Ref[4], Δ Ref [5]. — Present calculations.

REFERENCES

1. Kessler, J., *Polarized Electrons*, Springer Verlag, Berlin, 1976, pp.38-48.
2. McEachran, R. P., and Stauffer, A. D., *J. Phys. B: At Mol Phys.* **19**, 3523-3538 (1986).
3. Panajotović, R., Filipović, D., Mariković, B., Pejčev, V., and Kurepa, M., *J. Phys. B: At. Mol. Phys.* **30**, 5877-5894 (1997).
4. Schackert, K., *Z. Phys.* **213**, 316-322 (1968).
5. Beerlage, M. J. M., Quing, Z., and Van der Wiel, M. J., *J. Phys. B: At. Mol. Phys.* **14**, 4627-4635 (1981).

Super Elastic Electron Scattering from Alkali Atoms

V. Karaganov, P. J. O. Teubner and M.J. Brunger

School of Chemistry, Physics and Earth Sciences,
The Flinders University of South Australia,
G.P.O. Box 2100, Adelaide, S.A. 5001, Australia.

Abstract. The results from the super elastic electron scattering experiments on lithium and potassium that have been carried out at Flinders University over the past decade are reviewed and compared with the predictions of the CCC theory of Bray. We find that there is excellent agreement between theory and experiment over a wide range of energies in lithium. The agreement in the case of potassium is very good. We present the first results for orientation and alignment parameters for electron scattering from cesium. Our preliminary results at 10 eV show that relativistic effects are very small.

As the title suggests, this paper was meant to be a review of the status of super elastic electron scattering from the alkalis. After we received the invitation to speak, we have made very significant progress in our experiments on super elastic scattering from cesium so we have decided to devote most of this talk to these very exciting experiments. It is appropriate that we present these results at the eleventh international symposium on polarization and correlation in atomic collisions because the impetus for our present study of cesium was first suggested by Dr Bartschat [1] at the ninth international symposium in Frascati four years ago.

In a conventional super elastic experiment a well collimated beam of ground state atoms lies in the scattering plane and the atoms are oriented by optical pumping to the first excited state. The optical pumping is achieved with laser radiation which is introduced normal to the scattering plane. The laser light has a well defined polarization. The interaction region is defined as the overlap between the laser beam and the atomic beam. A well collimated beam of electrons is directed at the interaction region. The energy of the electrons that are scattered from this region through an angle θ in the scattering plane is analysed by an electron spectrometer. Electrons that have gained energy equal to the energy difference between the ground and the first excited state are detected and counted as a function of the laser polarization.

The three components of the equivalent Stokes vector can be deduced from these measurements. Specifically

CP604, *Correlations, Polarization, and Ionization in Atomic Systems*
edited by D. H. Madison and M. Schulz
© 2002 American Institute of Physics 0-7354-0048-2/02/$19.00

$$P_1 = \frac{I_0 - I_{90}}{I_0 + I_{90}}$$

$$P_2 = \frac{I_{45} - I_{135}}{I_{45} + I_{135}}$$

$$P_3 = \frac{I_{RHC} - I_{LHC}}{I_{RHC} + I_{LHC}}$$

(1)

where I_ϕ represents the super elastic count rate when the laser light is polarised at an angle ϕ with respect to the scattered electron direction. The terms I_{RHC} and I_{LHC} denote the count rates for right and left handed circular polarization.

The influence of hyperfine structure on the optical pumping process can be allowed for through the parameters K and K' [2,3]. The reduced parameters $\overline{P}_i$ are defined by

$$\overline{P}_1 = \frac{1}{K} P_1$$

$$\overline{P}_2 = \frac{1}{K} P_2$$

$$\overline{P}_3 = \frac{1}{K'} P_3$$

(2)

These components $\overline{P}_i$ are related to the magnitudes and relative phase of the scattering amplitudes for the time reversed excitation process of the ^2P state from the ground state [4]. Consequently they provide an extremely sensitive test of the scattering theory.

The alkali atoms have provided a very fertile ground for the use of the super elastic scattering technique because in each case excitation of the first excited state by CW laser radiation is readily achievable. The extension of the study of super elastic scattering from alkali targets other than sodium in our laboratory has coincided with the development by Bray of the convergent close coupling (CCC) theory to describe electron scattering from the alkalis [5]. In the case of electron scattering from lithium, Karaganov and Teubner [6] have shown that the CCC provides an extremely accurate description of the scattering process for equivalent energies from 5 eV to 20 eV and over a very wide range of scattering angles from 0° to about 140°. The exemplary agreement between experiment and theory in this case has led to improvements in experimental technique and to the discovery of quite subtle yet significant sources of systematic error in the experiments which has further enhanced the status of the CCC theory. The accord that has been thereby established has prompted us to suggest that the electron lithium problem has been solved [7].

It was puzzling therefore that when the CCC theory was applied to electron scattering from potassium it seemed unable to predict the differential cross section for the excitation of the 4 ^2P state. There has been considerable interest in the theory of electron scattering from potassium and these theories were reviewed by Stockman *et al.* [8]. Although the theories differed at middle angles, none could explain the difference of about a factor of 10 in this angular range with the experimental results [9,10]. Our measurement of the orientation and alignment parameters and of the differential cross section at 54.4 eV [8] demonstrated that the problem lay with the previous experiments and not with the theories. This work also demonstrated that the frozen core approximation used by Bray to describe the target atom was appropriate at energies above the threshold for the ionization of electrons from the core. However we were able to observe significant departures from the CCC and the measured $\overline{P_3}$ at 80 eV and middle angles [11]. We attribute these small aberrations to problems with the frozen core approximation. In summary we find that the CCC provides an excellent description of the electron impact excitation of the 4 ^2P state in potassium over the energy range from 4 eV to 80 eV and over a wide range of scattering angles from 0° to 140°. In particular we found no evidence of the influence of relativistic effects.

Relativistic effects both in the target description and in the scattering process were predicted by Bartschat [1] in the electron-cesium problem. It is clear that the next step in the theory is to develop a relativistic theory that is applicable over a wide range of energies. Zeman *et al.* [12] have developed a relativistic distorted wave calculation for the cesium problem. Bartschat [13, 14] is developing a theory that is a 40 state Breit-Pauli R matrix calculation, where relativistic effects were accounted for through the one-electron terms of the Breit-Pauli Hamiltonian. A core potential was used to describe the interaction between the projectile, the outer electron and the 54 core electrons. Bartschat and Bray [15] have applied CCC to electron scattering from cesium using a non relativistic model.

The possibility of relativistic effects influencing the scattering process in electron scattering from cesium complicates the technique used in the superelastic scattering experiments. Hall *et al.* [16] have analysed both the optical pumping process and super elastic scattering for a rubidium target in which they proposed relativistic effects would be apparent. Their analysis introduces the spin flip cross section ρ_{00} and an additional optical pumping parameter K'' which can be deduced from the polarization of fluorescence radiation after excitation by circularly polarized light.

In addition to the pseudo Stokes parameters P_i given by equation (1), the quantity

$$r(\theta) = \frac{I_0(\theta)}{I_{90}(\theta)} \tag{3}$$

is introduced, where I_0, I_{90} are the super elastically scattered signals when the laser polarisation is parallel and perpendicular to the scattering plane. In this case the laser radiation is introduced in the scattering plane whereas in equation (1) the laser radiation is introduced normal to the plane.

The equivalent relationships to equation (2) now become more complex; the four equations are given by Hall *et al.* [16].

The apparatus used in the cesium experiments has been described previously [6, 8, 11]. The cesium beam was produced in a resistively heated oven to a temperature of about 140°C which resulted in a beam density of about 10^9 cm^{-3} in the interaction region. It has been established that the influence of radiation trapping on our results can be ignored at this beam density. The cesium beam was collimated to a Doppler width of about 15 MHz and the atoms were optically pumped by an external cavity diode laser (TuiOptics DL100) tuned to a wavelength of 852.346 nm and using peak power of 50 mW. The laser was tuned to the $6\,{}^2S_{1/2}\ F = 4$ to $6\,{}^2P_{3/2}\ F = 5$ resonance line. The laser radiation was polarised using a quarter wave plate followed by a linear polarizer both under computer control. The relevant HFS splitting in the ${}^2P_{3/2}$ state is 213 MHz so the influence of hyperfine structure trapping on this experimental technique is very small. A standard saturated absorption technique was used for long term stabilisation of the laser system.

The electron beam was derived from our standard gun and produced a 500 nA beam at 8.5 eV with an energy spread of 0.3 eV. The electron beam energy was calibrated against the *b* feature in krypton. The electron spectrometer that analysed the energy of the scattered electrons was a retarding field analyzer with a resolution < 0.3 eV. The scattering angle was varied by rotating the gun about the axis perpendicular to the scattering plane.

The four simultaneous equations (Hall *et al.* [16]) can be solved for $\overline{P}_i$ and ρ_{00} provided optical pumping parameters, P_i and r are known. Ideally each of the parameters P_i and r should be measured during the same run; that is for the same scattering angles. This complicates the experimental technique considerably but these conditions can be satisfied by the application of suitable optical components. In the present experiments the parameter r was determined from a separate set of experiments to the parameters $\overline{P}_i$. We deduce from these results that ρ_{00} is small and indeed currently we are unable to distinguish it from zero. Future experiments will reduce the influence of systematic errors so that we can better test theoretical predictions for this parameter. With respect to the present study we assume $\rho_{00} = 0$. This assumption is justified by noting that it is predicted to be < 10% by the R matrix theory and therefore will contribute only minimally to the values $\overline{P}_i$. Under such conditions equations [1, 2] are used to interpret the results of our superelastic experiment on cesium.

The results of the present study are shown in Figure 1 where we compare the measured values of the reduced Stokes parameters at an equivalent energy of 10 eV with the predictions of the RDW of Zeman *et al.* [12], the Breit-Pauli R matrix calculation of Bartschat and Fang [13, 14] and the CCC theory of Bartschat and Bray [15]. Clearly the CCC accurately predicts the values of $\overline{P}_3$ over the whole angular range. The R matrix calculation appears to break down at middle angles but we note that it is superior to CCC in the prediction of $\overline{P}_1$. Both theories provide impressive predictions for the values of $\overline{P}_2$. That the RDW calculation is inferior to the CCC and R matrix theories is perhaps not surprising at this incident energy because distorted wave theories in general are better at higher energies.

The fact that the non-relativistic CCC theory does so well when it is compared with the experiments perhaps reflects the fact that the relativistic effects are small at this energy. Based on the results shown in Figure 1 there would seem to be very little to be gained in making the CCC relativistic.

ACKNOWLEDGMENTS

The research reported in this paper was supported by the Australian Research Council in the form of several Large Grants. We thank Dr. A. Sidorov for pointing out the merits of TuiOptics laser system, Dr. K. Bartschat for several essential discussions on the theory of electron cesium collisions and Drs. I. Bray and R. McEachran for supplying tabulated results of their theories.

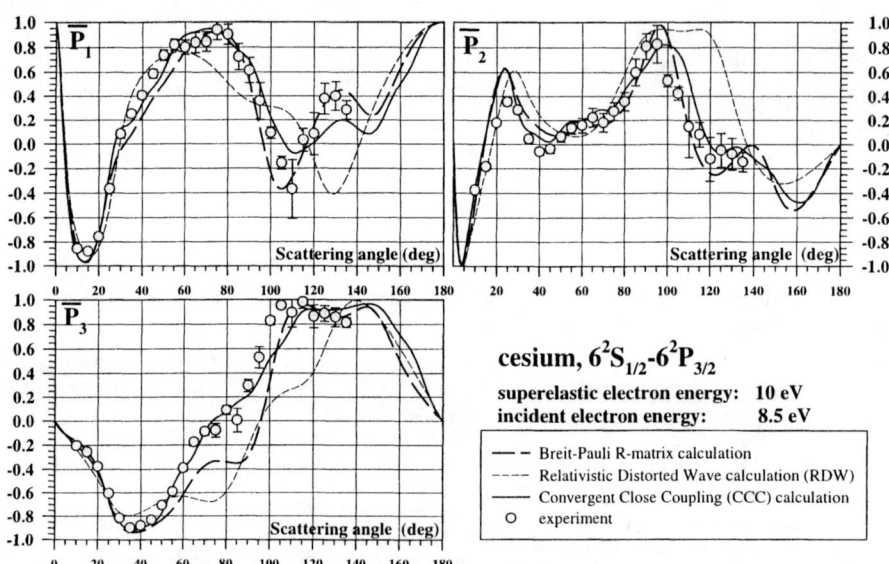

FIGURE 1. Reduced Stokes parameters for electron scattering from cesium. The experimental results are compared with the three theories that are explained in the text.

REFERENCES

1. Bartschat. K., *private communication*, (1997).
2. Farrell, P.M., MacGillivray, W.R. and Standage, M.C., *Phys. Rev. A* **44**, 1828 (1991).
3. Meng, X-K., MacGillivray, W.R. and Standage, M.C., *Phys. Rev. A* **45**, 1767 (1992).
4. Andersen, N., Gallagher, J.W. and Hertel, I.V., *Phys. Rep.* **165**, 1 (1988).
5. Bray, I., *Phys. Rev. A* **49**, 1066 (1994).
6. Karaganov, V. and Teubner, P.J.O., in *Photonic, Electronic and Atomic Collisions*, edited by Aumayr, F. and Winter, H-P., World Scientific, Singapore, 1998, p291.
7. Teubner, P.J.O., in *The Physics of Electronic and Atomic Collisions*, edited by Itikawa, Y., Okuno, K., Tanaka, H., Yagashita, A. and Matsuzawa, M., A.I.P. Conference Proceedings 500, Melville, New York, 2000, p256.
8. Stockman, K.A., Karaganov, V., Bray, I. and Teubner, P.J.O., *J. Phys. B: At. Mol. Opt. Phys.* **32**, 3003 (1999).
9. Buckman, S.J., Noble, C.J. and Teubner, P.J.O., *J. Phys. B: At. Mol. Phys.* **12**, 3077 (1979).
10. Vuskovic, L. and Srivastava, S.K., *J. Phys. B: At. Mol. Phys.* **13**, 4849 (1980).
11. Stockman, K.A., Karaganov, V., Bray, I. and Teubner, P.J.O., *J. Phys. B: At. Mol. Opt. Phys.* **34**, 1105 (2001).
12. Zeman, V., McEachran, R.P. and Stauffer, A.D., *Z. Phys. D* **30**, 145 (1994).
13. Bartschat, K. and Fang Y., *Phys. Rev. A* **62**, (2000).
14. Bartschat. K., *private communication*, (2001).
15. Bray, I. and Bartschat, K., *Phys. Rev. A* **54**, 1723 (1996).
16. Hall, B.V., Sang, R.T., Shurgalin, M., Farrell, P.M., MacGillivray, W.R. and Standage, M.C., *Can. J. Phys.* **74**, 977 (1999).

Simultaneous Ionization–Excitation:
A Challenge for Theory and Experiment

Klaus Bartschat and Yanghua Fang

Department of Physics and Astronomy, Drake University
Des Moines, Iowa 50311, USA

Abstract. As a highly correlated process, simultaneous electron-impact ionization–excitation presents a major challenge to experimentalists and theorists alike. With the rapid advance of computational power, it has recently become possible to account for both first-order and second-order effects in the interaction of a "fast" projectile with the target, combined with a convergent close-coupling-type description of the initial bound state and the interaction between a "slow" ejected electron and the residual ion. Examples of our recent work on this problem are presented, with particular emphasis on the role of auto-ionizing resonances in ionization–excitation of helium, as well as preliminary results for the more complex calcium target.

INTRODUCTION

Electron-impact ionization of atoms has been of great interest for many years. In most cases, experimental efforts concentrated on the measurement of ionization probabilities, with the most detailed information being obtained from triple-differential cross-section (TDCS) studies, for which the kinematics of the incident, scattered, and ejected electrons are fully defined. Figure 1 shows the generic scheme of such an experiment [1]. Note that the residual ion may end up in an excited state, which subsequently can decay by photon emission.

Even if the projectile and target spins are irrelevant, the full information about an ionization–excitation process such as

$$e_i(\boldsymbol{k}_0) + \mathrm{He}(1s^2)^1\mathrm{S} \rightarrow \mathrm{He}^+(2p)^2\mathrm{P} + e_e(\boldsymbol{k}_1) + e_s(\boldsymbol{k}_2)$$
$$\downarrow$$
$$\mathrm{He}^+(1s)^2\mathrm{S} + h\nu \ (30.4\,\mathrm{nm}) \tag{1}$$

can only be obtained in a triple-coincidence experiment between the two outgoing electrons and the photon which, in turn, also needs to be analyzed with respect to its polarization. In (1), $\boldsymbol{k}_0$ is the momentum of the incident electron while $\boldsymbol{k}_1$ and $\boldsymbol{k}_2$ denote the momenta of the scattered and ejected electrons, respectively.

CP604, *Correlations, Polarization, and Ionization in Atomic Systems*
edited by D. H. Madison and M. Schulz

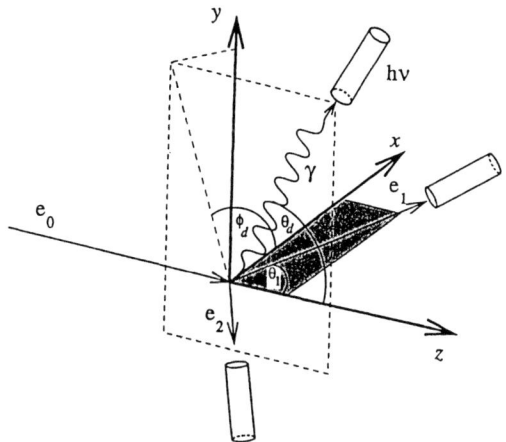

FIGURE 1. Scheme of an $(e, 2e\gamma)$ process for ionization–excitation.

Although the two outgoing electrons are, in principle, indistinguishable, the faster (slower) one of these is usually referred to as the scattered (ejected) electron.

Although pioneering triple-coincidence setups between one electron and two photons [2] or all three outgoing electrons in double-ionization studies [3] have been developed, the two-electron–one-photon coincidence experiment described above has not been performed to date. Instead, the investigations have been simplified in the following ways:

- Only the emitted photon is observed to perform an (optical) measurement of the angle-integrated ionization–excitation cross section [4,5] and, if the polarization or the angular distribution of the light is determined, to obtain additional information about the magnetic sublevel cross sections for the residual ions [6,7]. Such experiments yield no information about the outgoing electrons and, therefore, the phases between the scattering amplitudes.
- Only one electron is observed to obtain double-differential cross sections (DDCS) [8] or, in the decay of autoionizing states [9], information about the magnetic sublevels of such a state.
- One electron and the photon are observed in coincidence. This problem exhibits exactly the same symmetry properties as the electron–photon coincidence setup for excitation [1,10–12], although the number of independent parameters may be larger than in the excitation-only case, owing to the incoherence associated with the undetected second electron.
- Both electrons are observed in coincidence to obtain triple-differential cross sections (TDCS) [13,14]. Such studies yield the most (and in some cases the complete) information about these electrons, but they do not reveal any details regarding the magnetic sublevels of the residual ion.

SUMMARY OF THE PWB2–RMPS METHOD

Parallel to the difficulties involved in the experimental investigation of simultaneous ionization–excitation, these processes also represent major challenges in the theoretical description. However, with the rapid advance of computational power, it has recently become possible [15–19] to include both first-order and second-order effects in the interaction of a "fast" projectile with the target, combined with a convergent close-coupling-type description of the initial bound state and the interaction between a "slow" ejected electron and the residual ion.

The principal ideas of our method were outlined in detail by Fang and Bartschat [17], and the similarities and the differences between our method and that of Marchalant *et al.* [15,16] were discussed in [18]. Briefly, we describe a fast ionizing projectile with incident momentum k_0 by a plane wave and obtain the ionization amplitude for two outgoing electrons with momenta k_1 and k_2, respectively, as

$$f(k_2, k_1, k_0) = f^{B1}(k_2, k_1, k_0) + f^{B2}(k_2, k_1, k_0),$$ (2)

where f^{B1} and f^{B2} are the first-order and second-order contributions to the scattering amplitude. We use the same close-coupling expansion to represent the initial target state $|\psi_0\rangle$ and the continuum state $|\Psi_f^-(k_2)\rangle$ describing the "slow" ejected electron in the field of the residual ion, and hence we expect similar convergence properties for both terms. This is advisable, since the principal effect of the second-order term in the calculation of observable quantities comes from interference with the first-order amplitude in the bilinear product of scattering amplitudes.

There are several advantages associated with the present method. Most importantly, we extended the general (non-relativistic) R-matrix code of Bartschat [20] to include the second-order term. Hence, the program can handle more complex targets than helium, and the use of an R-matrix expansion allows for very efficient calculations in cases where the ejected-electron–residual-ion interaction is dominated by resonances. Furthermore, using an explicit numerical integration scheme to calculate the second-order contributions allows us to investigate the angular range in the intermediate state where these contributions come from, and the integration scheme can be systematically refined.

RESONANCE EFFECTS IN
IONIZATION–EXCITATION OF HELIUM

In our calculations for ionization–excitation of $He(1s^2)$ resulting in $He^+(2s,2p)$, we used a 23-state model including ten bound states and 13 continuum pseudo-states, with the latter chosen to represent the effect of coupling to the (double) ionization continuum. In order to allow for a proper treatment of autoionizing resonances with configuration $(4\ell n\ell')$ [19], it was necessary to choose all ten bound states of He^+ to be physical states, in contrast to previous work [17,18] where the

$n=4$ states were chosen as pseudo-states to simulate the coupling to the remaining members of the bound spectrum. We found that changing the character of the $n = 4$ states from pseudo to physical only had a minor effect on the results outside of the resonance region explicitly associated with these states.

Although the $(2\ell 2\ell')$ resonances below the threshold for ionization with excitation to the 2s state and the 2p state of He^+ were studied experimentally about a decade ago by Lower and Weigold [21] and by McDonald and Crowe [22], until recently [19] the possible effect of higher-lying resonances had been completely ignored in all published comparisons of theoretical predictions and experimental data for the simultaneous ionization–excitation process. An analysis of the He^+ spectrum, on the other hand, reveals that such resonances might have a major effect for three of the experimentally chosen kinematical situations in the work of Dupré et al. [13], Avaldi et al. [14], and Rouvellou et al. [23], namely those with nominal slow-electron energies of 10 eV and 9.25 eV, respectively. In the ionization–excitation channel to the $n = 2$ final states of He^+, these energies lie between the $n = 3$ and $n = 4$ thresholds, i.e., one can expect a wealth of autoionizing resonances, particularly of the form $(4\ell n\ell')$, to partly decay into the $n=2$ final ionic states.

Figure 2 shows an example of the ejected-electron energy and angle dependence of the triple-differential cross section for simultaneous ionization–excitation of He $(1s^2)^1$S to the $n= 2$ (2s+2p) states of He^+, as predicted by the PWB2-RMPS model mentioned above. For an incident projectile energy of 365.8 eV and an observation angle of $\theta_1 = 4.5°$ for the fast outgoing electron of energy $E_1 \approx 291.2$ eV [23], the results are plotted in the vicinity of the nominal slow-electron energy of 9.25 eV. Clearly, the potential effect of resonances in an experiment with finite energy resolution should not be ignored.

To illustrate this point further, the left part of Fig. 3 presents the ejected-electron energy dependence of the triple-differential cross section in the above kinematical situation, this time for a fixed slow-electron detection angle near the first maximum. In addition to the PWB2-RMPS results summed to convergence over the partial-wave angular momenta of the ejected electron, we show the contributions from the partial waves with total (residual ion plus ejected electron) angular momentum $L=1$ and $L=2$, respectively, i.e., the dominant contributions for relatively low incident projectile energies, and also the first-order PWB1-RMPS results. In the right part of Fig. 3, we then show PWB2-RMPS predictions for slightly different ejected-electron energies in the vicinity of the nominal energy of 9.25 eV, and the result obtained after convoluting with a Gaussian energy profile of 400 meV (FWHM). [Note that Rouvellou et al. [23] quote a "coincidence resolution of 750 meV." The earlier Dupré et al. [13] experiment was performed with a resolution of 5 eV, indicating that their results at nominal slow-electron energies of 5 eV and 10 eV could be strongly affected by resonances, as well as by an overall energy dependence in the background.] Already at 400 meV, however, we clearly see that the theoretical energy chosen for comparison can have significant effects on both the shape and the magnitude of the predictions, and the angular dependence also depends on the convolution procedure. Since the experimental data for this case are not absolute,

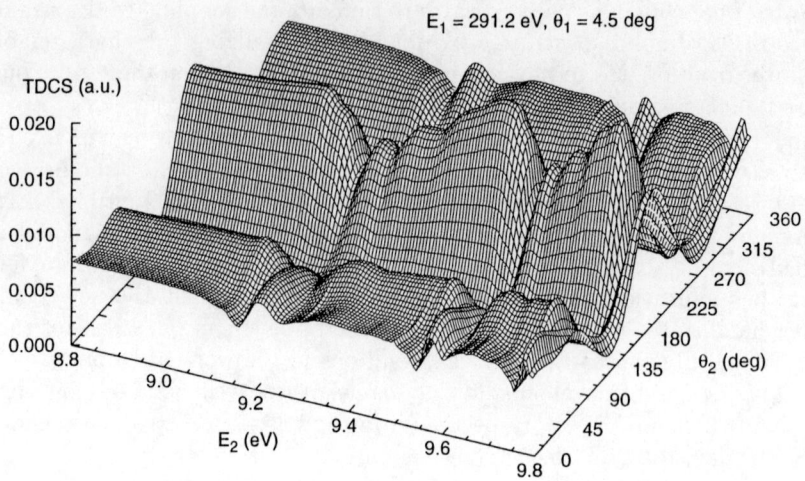

FIGURE 2. Ejected-electron energy and angle dependence of the triple-differential cross section, in atomic units $a_0^2/(2 \cdot \text{Ryd} \cdot \text{sr}^2)$, for simultaneous ionization–excitation of He $(1s^2)^1S$ to the $n = 2$ $(2s+2p)$ states of He$^+$. For an incident projectile energy of 365.8 eV and an observation angle of $\theta_1 = 4.5°$ for the fast outgoing electron of energy $E_1 \approx 291.2$ eV, PWB2-RMPS results are plotted as a function of the detection angle θ_2 for slow-electron energies between 8.8 eV and 9.8 eV.

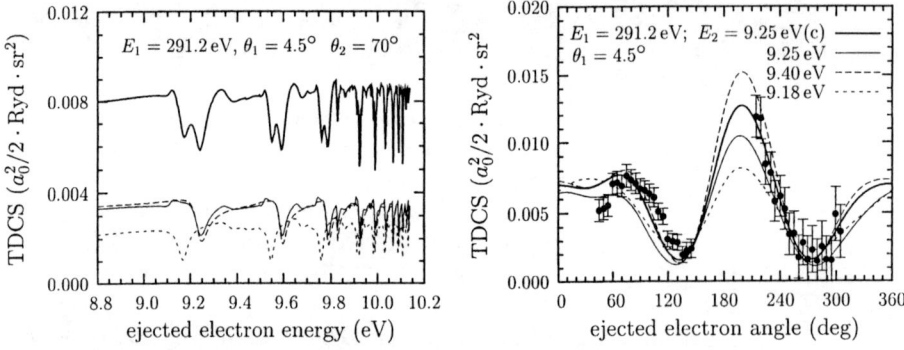

FIGURE 3. Ejected-electron energy and angle dependence of the triple-differential cross section for simultaneous ionization–excitation of He $(1s^2)^1S$ to the $n = 2$ $(2s+2p)$ states of He$^+$. For an incident projectile energy of 365.8 eV and an observation angle of $\theta_1 = 4.5°$ for the fast outgoing electron of energy $E_1 \approx 291.2$ eV, the PWB2-RMPS results are plotted as a function of the ejected-electron energy for a fixed detection angle $\theta_2 = 70°$ (left) and as a function of θ_2 for different slow-electron energies in the vicinity of 9.25 eV (right). In addition to the PWB2-RMPS results (thick solid line), the left figure also shows the partial contributions from $L = 1$ (long-dashed) and $L = 2$ (short-dashed), as well as the first-order PWB1-RMPS results (thin line). The curve marked (c) was obtained after convoluting the 9.25 eV results with a 400 meV (FWHM) Gaussian energy distribution. The experimental data are from Rouvellou *et al.* [23].

206

they were normalized to provide a good visual fit to the theoretical predictions after convolution. Nevertheless, changes within a factor of two are readily conceivable if predictions for slightly different slow-electron energies were compared to experiment.

IONIZATION–EXCITATION OF CALCIUM

From an experimental point of view, a very attractive target for studies of ionization–excitation is calcium [24]). The reason is two-fold: 1) for comparable kinematical situations, one can expect larger cross sections in Ca than in He due to the large $(4p^2)$ correlation contribution in the $Ca(4s^2)$ ground state; 2) the photon emitted in the $4p \to 4s$ transition of Ca^+ has a wavelength of approximately 395 nm, which is significantly easier to handle experimentally than the VUV radiation originating from the $2p \to 1s$ transition of He^+.

We therefore applied our method to explore both channel-coupling and second-order effects in ionization–excitation of calcium. For an incident electron energy of 400 eV and a detection angle of 6° for the fast electron, Figure 4 shows the TDCS for leaving the Ca^+ ion in either the (4s) or the (4p) state. For a fixed energy of ≈ 386 eV of the scattered electron, this corresponds to ejected-electron energies of 8 eV and 5.1 eV, respectively. Comparing results from 11-state (7 physical, 4 pseudo), 7-state, and 1-state models of e–Ca^+ scattering shows that the relatively large cross section for leaving the ion in its ground state is strongly affected by channel coupling (the RM1 results differ substantially from RM7 and RM11), while the effect of including the second-order projectile–target interaction is small. On the other hand, the smaller ionization–excitation cross section strongly depends on the order of the interaction included. Fortunately, the similarity of the 11-state and the 7-state predictions indicates reasonable convergence with the number of states included in the close-coupling part of the model.

For the same kinematical situation described above, Figure 5 exhibits the predicted charge-cloud parameters $L_\perp$ and γ for ionization–excitation to the $Ca^+(4p)$ state. We note a very strong dependence of $L_\perp$ on the order of the projectile–target interaction, particularly for ejected-electron angles in the range up to $0° - 120°$, where the size of the cross section is substantial. This suggests an attractive case for an experimental study.

SUMMARY

We have presented results from detailed second-order calculations of electron-impact ionization–excitation of helium and calcium. The helium results indicate the need to account for possible resonance effects when comparing experimental and theoretical data. Finally, we hope that the calcium predictions will stimulate new experimental work in this area, with the ultimate goal of a complete $(e, 2e\gamma)$ ionization experiment.

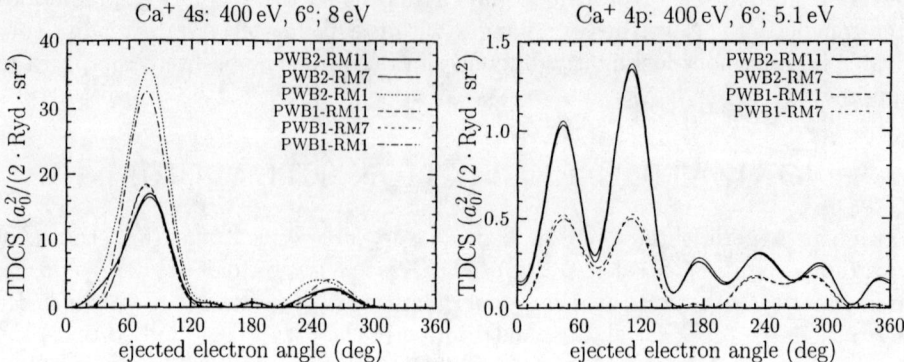

FIGURE 4. TDCS for ionization and ionization–excitation of Ca $(4s^2)^1$S to Ca$^+(4s)$ and Ca$^+(4p)$. For an incident projectile energy of 400 eV and an observation angle of $\theta_1 = 6°$ for the fast outgoing electron, the results are plotted as a function of the detection angle θ_2 for the slow electron with energy 8 eV (4s) and 5.1 eV (4p). The theoretical models are indicated in the legend, with PWB1 and PWB2 referring to first-order and second-order treatments of the fast electron, and the number after RM to the states included in the close-coupling expansion.

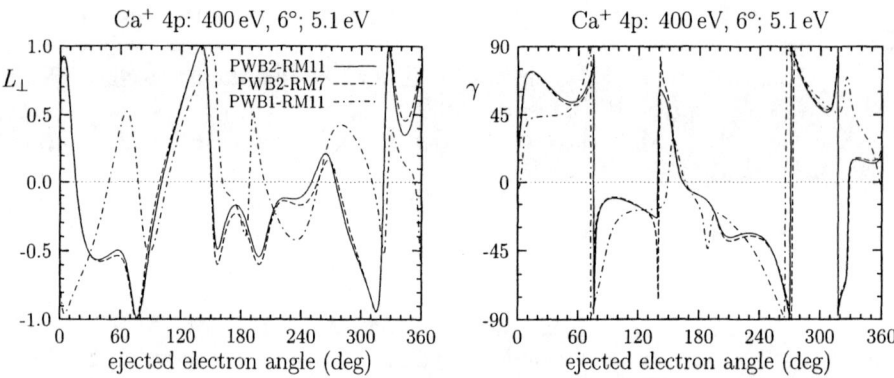

FIGURE 5. Ejected-electron energy and angle dependence of the charge-cloud parameters $L_\perp$ and γ for simultaneous ionization–excitation of Ca $(4s^2)^1$S to Ca$^+(4p)$. For an incident projectile energy of 400 eV and an observation angle of $\theta_1 = 6°$ for the fast outgoing electron, the results are plotted as a function of the detection angle θ_2 for the slow electron with energy 5.1 eV.

ACKNOWLEDGMENTS

The authors would like to thank Albert Crowe for stimulating discussions. This work was supported by the United States National Science Foundation under grant PHY-0088917.

REFERENCES

1. N. Andersen and K. Bartschat, *Polarization, Alignment, and Orientation in Atomic Collisions*, Springer, New York, 2000.
2. A.G. Mikosza, J.F. Williams, and J.B. Wang, Phys. Rev. Lett. **79**, 3375 (1997).
3. I. Taouil, A. Lahmam-Bennani, A. Duguet, and L. Avaldi, Phys. Rev. Lett. **81**, 4600 (1998).
4. E.W.O. Bloemen, H. Winter, T.D. Märk, D. Dijkkamp, D. Barends, and F.J. de Heer, J. Phys. B **14**, 717 (1981).
5. J.J. Forand, K. Becker, and J.W. McConkey, J. Phys. B **18**, 1409 (1985).
6. A. Götz, W. Mehlhorn, A. Raeker, and K. Bartschat, J. Phys. B **29**, 4699 (1996).
7. H. Merabet, M. Bailey, R. Bruch, I. Bray, D.V. Fursa, J.W. McConkey, and P. Hammond, Phys. Rev. A **60**, 1187 (1999).
8. R. Müller-Fiedler, K. Jung, and H. Ehrhardt, J. Phys. B **19**, 1211 (1986).
9. B. Matterstock, R. Huster, B. Paripas, A.N. Grum-Grzhimailo, and W. Mehlhorn, J. Phys. B **28**, 4301 (1995).
10. R. Schwienhorst, A. Raeker, K. Bartschat, and K. Blum, J. Phys. B **29**, 2305 (1996).
11. P.A. Hayes and J.F. Williams, Phys. Rev. Lett. **77**, 3098 (1996).
12. M. Dogan, A. Crowe, K. Bartschat, and P.J. Marchalant, J. Phys. B **31**, 1611 (1998).
13. C. Dupré, A. Lahmam-Bennani, A. Duguet, F. Mota-Furtado, P.F. O'Mahony, and C. Dal Cappello, J. Phys. B **25**, 259 (1992).
14. L. Avaldi, R. Camilloni, R. Multari, G. Stefani, J. Langlois, O. Robaux, R.J. Tweed, and G.N. Vien, J. Phys. B **31**, 2981 (1998).
15. P.J. Marchalant, C.T. Whelan, and H.R.J. Walters, J. Phys. B **31**, 1141 (1998).
16. P.J. Marchalant, J. Rasch, C.T. Whelan, D.H. Madison, and H.R.J. Walters, J. Phys. B **32**, L705 (1999).
17. Y. Fang and K. Bartschat, J. Phys. B **34**, L19 (2001).
18. Y. Fang and K. Bartschat, J. Phys. B **34**, 2747 (2001).
19. Y. Fang and K. Bartschat, Phys. Rev. A **64**, 020701(R) (2001).
20. K. Bartschat, Comp. Phys. Commun. **25**, 219 (2000).
21. J. Lower and E. Weigold, J. Phys. B **23**, 2819 (1990).
22. D.G. McDonald and A. Crowe, Z. Phys. D **23**, 371 (1992).
23. B. Rouvellou, S. Rioual, A. Pochat, R.J. Tweed, J. Langlois, G.N. Vien G N, and O. Robaux, J. Phys. B **33**, L599 (2000).
24. A. Crowe, private communication (2001).

Double photoemission studies
at metal surfaces

N. Fominykh, J. Henk, J. Berakdar, P. Bruno

Max-Planck-Institut für Mikrostrukturphysik, Weinberg 2, D-06120 Halle (Saale), Germany

Abstract   In the present work, we take a closer look at the theoretical model of double-electron photoemission (DPE) we have introduced in [1,2]. A local wave-vector-dependent approximation of the Coulomb interaction between the photoelectrons gives a connection of DPE to the one-step model of single photoemission (SPE). Calculations of DPE photocurrent for Cu(001) and Cu(111) surfaces are provided to illustrate the interplay between scattering and correlation, as well as the manifestation of the single-particle properties in angular and energy sharing distributions.

1.   INTRODUCTION

In DPE the single photoabsorption event leads to the simultaneous emission of two photoelectrons. These are detected with energy and momentum resolution, the simultaneity of their creation is controlled by the time-of-flight technique [3]. To our knowledge, a theoretical investigation of DPE from surfaces has been undertaken in two different frameworks. In the first one [4], the DPE matrix element was evaluated for the jellium-like semi-infinite solid and screened Coulomb interaction between outgoing electrons. It was shown that there exists a certain condition for the diffraction of the correlated electron pair and a certain DPE selection rule. The second approach [1,2] was oriented towards utilization of the up-to-date level of single photoemission (SPE) calculations [5] in order to be able to use realistic surface band structures and densities of states. Accurate treatment of single-particle properties requires a manageable form of the interaction between the two photoelectrons. We designed a method in which the screened Coulomb potential was incorporated in a non-perturbative way into the one-particle Green's functions, provided by an *ab initio* SPE computer code [5]. The screening length λ is a parameter derived from a Thomas-Fermi approach. In this way the two-electron photocurrent can be approximately expressed through the one-particle matrix elements, formally resembling the ones encountered in one-step model of SPE (see e.g. [6]). Our first numerical results [2] revealed the fingerprints of the electronic correlation, in particular the appearance of Coulomb correlation hole, and confirmed the expected qualitative trends. In the present work we continue these investigations and calculate the DPE photocurrent from Cu(001) and Cu(111) surfaces. In Sec.2 we discuss DPE angular distributions (ADs), Sec.3 deals with energy sharing distributions (ESDs), Sec.4 summarises the results.

CP604, *Correlations, Polarization, and Ionization in Atomic Systems*
edited by D. H. Madison and M. Schulz
© 2002 American Institute of Physics 0-7354-0048-2/02/$19.00

2.   ANGULAR CORRELATION PLOTS

The experimentally defined two-electron final state provides six independent variables: four angles of emission $(\theta_1, \phi_1), (\theta_2, \phi_2)$ and two kinetic energies E_1, E_2. Two particular situations will be considered here. At a given energy and (linear) polarization of the photon the DPE probability will be given: (i) as a function of the energy difference $(E_1 - E_2)$ at fixed total energy $E_{tot} = E_1 + E_2$ and fixed angles $(\theta_1, \phi_1), (\theta_2, \phi_2)$, or (ii) as a function of (θ_2, ϕ_2) for fixed (θ_1, ϕ_1) and E_1, E_2. The first mode is referred to as energy sharing distribution (ESD), the second one is the DPE angular distribution (AD).

The scattering of the electrons from the lattice and their mutual repulsion most easily can be observed in the DPE ADs. In general, the DPE intensity depends on the difference in emission angles according to two competing tendencies. This angle should be small enough to provide sufficient interaction, and it should be large enough to avoid strong Coulomb repulsion. Therefore the region of high DPE intensity looks roughly like a 'ring' around the fixed-electron direction (see Fig.1 and it's caption). This general pattern is mixed with the diffraction effects due to scattering of the electrons from the lattice. The mean angle with respect to the first electron, at which the second electron is preferentially emitted, depends on the value of the screening length λ and, hence, on the material. For the lack of space we are not visualizing this trend here: The larger is

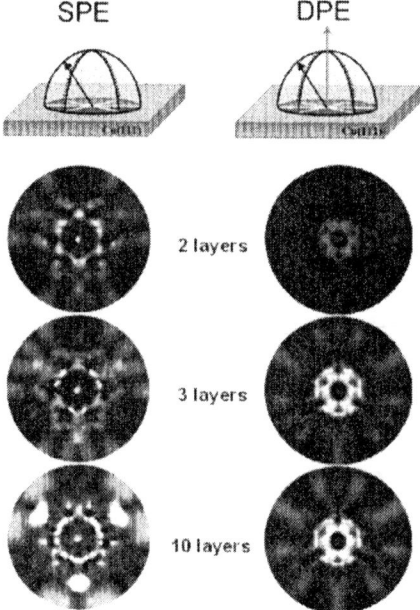

Figure 1 DPE ADs from Cu(111) surface (right column) in the form of stereographic projection. The direction of emission of the 'fixed' electron is perpendicular to the surface, as shown by a grey arrow on the sketch above. $\hbar\omega = 45$ eV, $E_1 = E_2 = 15$ eV, the light is polarized normally to the surface. For comparison, the left column shows the SPE ADs. The normalized intensity from zero to one is shown by black-and-white scale (the normalization is the same within each column and differs by five orders between the columns). The angle of emission, which is varied, is shown by a black arrow in both sketches.

211

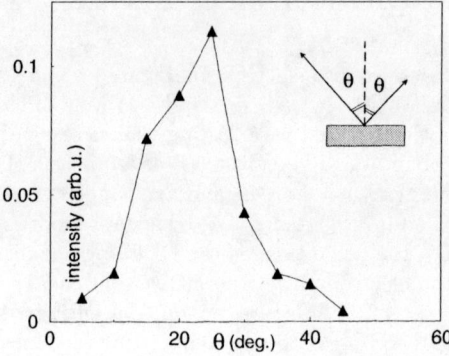

Figure 2 Cu(001) DPE intensity, averaged over all possible energy sharings between photoelectrons for given ω and E_{tot}. $E_{tot} = 33\,eV$, $\omega = 45\,eV$, angle θ changes simultaneously for both electrons and serves as an abscissa of the plot. Here and below angles ϕ_1, ϕ_2 are set to zero.

the region where the electrons can interact and decline their trajectories, the wider will be the 'ring' of high intensity. The depth of generation of correlated pairs, as compared to the depth of generation of photoelectrons in SPE process, is illustrated by showing the photocurrents from 2, 3 and 10 monolayers of Cu(111). One can see that most of the DPE signal is formed at a depth up to the third monolayer. SPE and DPE intensities differ by five orders of magnitude, that is in agreement with experimental observation. In Fig.2 we are interested in finding the 'most efficient' final state geometry, i.e. the angle between two electrons, which gives the highest DPE rate at a given photon energy and E_{tot}. The following set up is chosen (Fig.2, inset): electrons are directed symmetrically, each making an angle θ with the normal. All combinations of E_1 and E_2 are taken into account, which satisfy $E_{tot} = 33\,eV$. One can indeed observe the single peak of intensity at $\theta = 25^0$ as a result of the interplay between the above trends.

3.   DPE ENERGY SHARING DISTRIBUTIONS

An ESD shows, how favourable for photoemission is one or another *sharing* of the total energy between two electrons, the latter being dependent on the momentum transfer due to the Coulomb interaction. Contrary to the SPE process, in DPE the individual surface-parallel momenta of the photoelectrons $\mathbf{k}_{\|1,2}$ are, in general, not conserved. It is the sum of the parallel components $(\mathbf{k}_{\|1} + \mathbf{k}_{\|2})$ that is conserved (*modulo* a surface reciprocal lattice vector) upon photoemission. However, our approximate expression of the two-electron photocurrent through the single-particle matrix elements [1,2] implies the conservation of the individual $\mathbf{k}_{\|1}$ and $\mathbf{k}_{\|2}$. Thus, only those initial states are participating in the optical transition, that are characterized by the same irreducible representation of the translation group as the asymptotic final states. This means that the $\mathbf{k}_{\|}$-resolved density of states is a meaningful quantity for the rough analysis of the DPE ESDs. Finally, being the optically excited flux, ESD reflects the features of the dipole transition and contains a kinematical factor $\sqrt{E_1 \cdot E_2}$ (in the non-relativistic case). The latter explains the zero value of the ESD at the edges, where either E_1 or E_2 is equal to zero. So, each point of the ESD is related to certain E_1, E_2 and corresponding $\mathbf{k}_{\|}$-resolved densities of initial states. This relation is not straightforward, since one has to integrate the DPE current over a certain energy range of the initial states [2], as defined

212

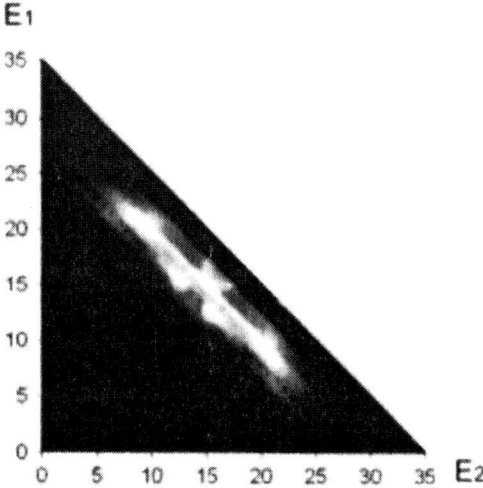

Figure 3 DPE intensity (black-and-white contrast) as a function of E_1, E_2. Electrons are emitted at 30^0 symmetrically with respect to the surface normal. The photon energy $\omega = 45\,eV$ and the Fermi energy $E_F \sim 5\,eV$ define the maximal total energy $E_{tot} \sim 35\,eV$.

by the energy conservation. In Fig.3 we present the DPE current from Cu(001) as a function of E_1, E_2 for fixed photon energy (p-polarized light) and fixed angles of emission. The difference of this plot and the experimental ones (see example in [3]) is due to the disregard of the secondary processes. The appreciable DPE signal (white region) is obtained in the region $E_{tot} = 25...35\,eV$. The d-band of copper ($\sim$3 eV band below the Fermi level) supplies the initial state electrons; pairs with $E_{tot} < 25\,eV$ and in the narrow black band just below the $E_{tot} = 35\,eV$ boundary are not accessible due to the low density of the initial states. Fine details on the picture arise from the Coulomb interaction and multiple scattering at the crystal potential.

The next aspect, that we would like to clarify, is the role of the screening length λ. It is the only parameter in the model, which describes how effectively the range of interaction between two outgoing electrons is *cut-off* by the ground state electrons of the surface. The screening length is the measure of the interaction, for if it is strongly screened, the electrons may be treated as non-interacting, in which case the DPE probability becomes zero [4]. The question arises as to whether the screening length changes only the magnitude of the ESDs distributions, or it's role is dynamical, i.e. influences their shapes too. In Fig.4 we plot two series of ESDs, where both trends are shown. Fig.4(a) corresponds to E_{tot}=33 eV, $\omega = 45$ eV, $\theta_1 = -\theta_2 = 20^0$, $\lambda = 1.0, 2.66, 4.0\,a_0$ for the curves 1 (long-dashed), 2 (solid) and 3 (dashed), respectively. Curve 2 corresponds to the Thomas-Fermi value of λ for Cu (used throughout if not stated otherwise). One can see that the value of the screening length alters not only the magnitude, but also the shape of ESD. In Fig.4(b) the values of λ and the geometry are the same except for the angles of emission: $\theta_1 = -\theta_2 = 40^0$. It may be inferred, that for the larger angle between electrons the ESD shapes are less sensitive to screening and the role of the screening length is only scaling.

213

It was shown in [4], that there exists a propensity rule, stating that DPE current is zero if the scalar product $(\mathbf{k}_1 + \mathbf{k}_2) \cdot \hat{\epsilon} = 0$, where $\hat{\epsilon}$ is the polarization vector of light. It was deduced as a consequence of the dipole approximation and the approximation for the DPE final state, used in that model. In the simple case, whatever the initial two-particle state Φ_2 is, if each electron in the final state is represented by a plane wave, the above scalar product factors out of the matrix element:

$$M_{\mathbf{k}_1,\mathbf{k}_2} = \int d^3\mathbf{p}\, d^3\mathbf{p}' \, \langle \mathbf{k}_1, \mathbf{k}_2 | \hat{\epsilon}(\mathbf{p}_1 + \mathbf{p}_2) | \mathbf{p}, \mathbf{p}' \rangle \, \langle \mathbf{p}, \mathbf{p}' | \Phi_2 \rangle = \hat{\epsilon}(\mathbf{k}_1 + \mathbf{k}_2) \cdot \tilde{\Phi}_2(\mathbf{k}_1, \mathbf{k}_2) \; , \; (1)$$

where $\tilde{\Phi}_2(\mathbf{k}_1, \mathbf{k}_2)$ stands for a Fourier transform of the initial state. This remains valid also when the interaction between the final state electrons is present, but disregarded is the final-channel coupling to the host system.

In the present framework the final states are represented by the so-called time-reversed LEED (Low Energy Electron Diffraction) states, whose essential feature is that they are eigenstates of the in-plane lattice translations. Irrespective of the form of the initial state, the matrix element built from the two Bloch sums in the final state reads:

$$M_{\mathbf{k}_1,\mathbf{k}_2} = \int \left\langle \sum_{\mathbf{g}} a_{\mathbf{g}} e^{-i(\mathbf{k}_1 + \mathbf{g})\mathbf{r}} \cdot \sum_{\mathbf{g}'} b_{\mathbf{g}'} e^{-i(\mathbf{k}_2 + \mathbf{g}')\mathbf{r}'} \middle| \hat{\epsilon} \cdot (\mathbf{p}_1 + \mathbf{p}_2) \middle| \mathbf{p}, \mathbf{p}' \right\rangle \times$$

$$\times \; \langle \mathbf{p}, \mathbf{p}' | \Phi_2 \rangle \, d^3\mathbf{p}\, d^3\mathbf{p}' = \tilde{\Phi}_2(\mathbf{k}_1, \mathbf{k}_2) \cdot \sum_{\mathbf{g}\mathbf{g}'} a_{\mathbf{g}} b_{\mathbf{g}'} \, \hat{\epsilon} \cdot (\mathbf{k}_1 + \mathbf{k}_2 + \mathbf{g} + \mathbf{g}'). \; (2)$$

Here $a_{\mathbf{g}}$ and $b_{\mathbf{g}'}$ are the coefficients in the Bloch expansion of the LEED states, vectors $\mathbf{g}$ and $\mathbf{g}'$ are the (surface) reciprocal lattice vectors. In general, the last sum in Eq.(2) cannot be zero for any choice of $\mathbf{k}_1$, $\mathbf{k}_2$, $\hat{\epsilon}$, and the selection rule described before does not hold. On the other hand, the plane-wave representation of the LEED state becomes adequate when going to higher kinetic energies, meaning that the selection rule becomes

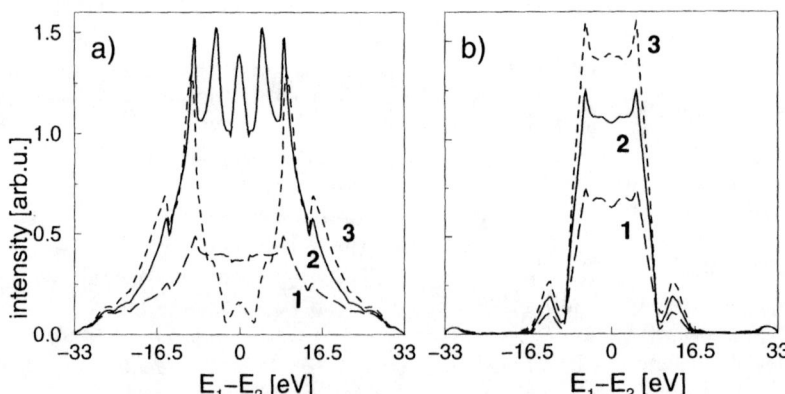

Figure 4 DPE EDS from Cu(001) for different values of the screening length (see text for details).

applicable. To estimate this, we performed a 'numerical experiment', deliberately decreasing the number of 'beams' (plane waves, participating in the expansion of the LEED states) from convergency limit (~40 beams per state) to just one beam. The latter case would simulate the situation, when the selection rule does apply. The value, which should vanish in this case, is the DPE intensity in the middle of the ESD ($E_1 = E_2$) for the symmetric geometry (Fig.5, inset). Fig.5 indeed demonstrates the downfall of the middle-point value of ESD from a finite value to zero when the number of beams is decreased from 37 to 1. Note the saturation of the curve at ~20 beams per state.

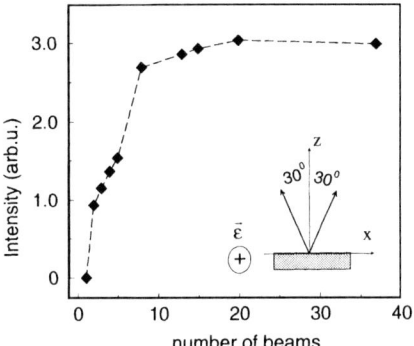

Figure 5   Cu(001) DPE current at $E_1 = E_2 = 15\,eV$, $\omega = 45\,eV$, $\theta_1 = -\theta_2 = 30^0$, $\hat{\epsilon}$ parallel to y-axis, calculated for different number of plane waves in the expansion of the LEED states.

This conclusion is supported by another setup, Fig.6, in which ω and E_{tot} are concurrently growing. The value of the ESD at $E_1 = E_2$ is maximum at E_{tot}=13 eV and decreases towards higher energies. The variety of ESD shapes in Fig.6 is guided by several factors: (i) the interaction between electrons depends on their kinetic energies; (ii) the Bloch spectral functions of the initial states are 'scanned' differently along different ESDs; (iii) the number of relevant 'beams' changes with energy.

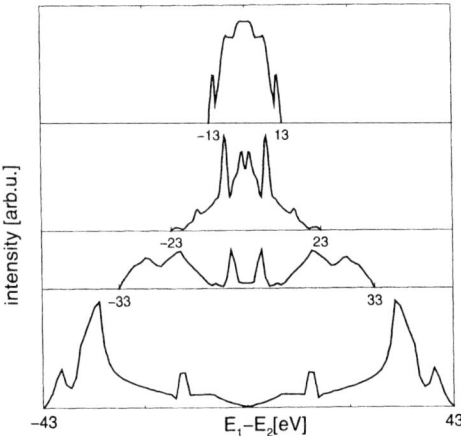

Figure 6   ESDs from Cu(001) for ω=25, 35, 45 and 55 eV and E_{tot}=13, 23, 33 and 43 eV, respectively. For all cases the ejection angles are $\theta_1 = -\theta_2 = 30^0$, $\hat{\epsilon}$ parallel to x-axis.

215

4.   SUMMARY

We have presented numerical results of DPE calculations from copper surface. Some of them, like the manifestation of Coulomb interaction in the angular distribution of the DPE cross section, can be understood from simple qualitative reasoning. Another, like the role of initial state densities, the screening and the non-plane-wave character of the electronic states are not obvious, have a strong effect on the DPE spectra and require extensive numerical investigations. The analysis of the DPE distributions is complicated by the fact that none of these features can be singled out and traced separately. Nevertheless, the present framework remains a direct and useful tool for the exploration of correlation-accented techniques.

REFERENCES

1. N. Fominykh, J. Henk, J. Berakdar, P. Bruno, H. Gollisch, R. Feder,
   Solid State Communications, 113 (12), 665-669, (2000)
2. N. Fominykh et. al., in "Many-particle spectroscopy", 461-470, Kluwer (2001).
3. R. Herrmann, S. Samarin, H. Schwabe, J. Kirschner, Phys. Rev. Lett. 81 (1998) 2148.
4. J. Berakdar, Phys. Rev. B 58 (1998) 9808.
5. J. Henk, T. Scheunemann, S. V. Halilov, E. Tamura, and R. Feder (1999)
   omni – fully relativistic electron spectroscopy calculations.
6. S. V. Kevan (ed.), Angle-resolved Photoemission, Elsevier, Amsterdam (1992).

Polarization in Collisions between Ultra-cold Sodium Atoms

H.G.M. Heideman, P. van der Straten and A. Amelink

Debye Institute, Utrecht University, PO Box 80000, 3508 TA Utrecht, The Netherlands

Abstract. We have studied polarization effects in the photoassociation of two colliding ultra-cold Na-atoms. In a first step the two colliding atoms absorb a laser photon to associate to a singly excited molecule in a bound state of the 0_g^- or 1_g potential. In a second step the excited molecule absorbs a second photon and is further excited to a bound level in a doubly excited potential (0_u^- or 1_u which are autoionizing). By measuring the produced ions with parallel and perpendicular polarizations of the two laser beams used possible polarization effects can be studied. We find strong polarization effects if a 0_g^- state is used as the intermediate state and very little or no effects at all if a 1_g state is used.

INTRODUCTION

The study of collisions between cold atoms ($T < 1$ mK, corresponding to relative velocities of the order of a few m/s) has been made possible by the introduction of laser cooling and trapping techniques. When two very slow atoms collide in the presence of a radiation field they may combine to produce a bound, electronically excited molecule. This photoassociation process may be detected as a decrease of the fluorescence of the trapped atomic sample [1]. Alternatively the excited molecule may be excited by a second photon to a higher lying state which lies in the ionization continuum or which is autoionizing via a crossing with the continuum at short internuclear range. By measuring the ions formed as a function of the frequencies of the lasers both the first and the second step of this process may be studied. This is shown schematically in the Figs. 1a and 1b, respectively. In the recent past several experiments of this type have been performed and valuable information has been obtained on the potential curves of the states involved [2, 3].

In the present experiment we have studied the above described two-step process as a function of the polarization of the two lasers used. When a sample of isotropically oriented molecules is excited by polarized light the excited molecules will in general be aligned with respect to the polarization direction of the exciting light beam [4]. This is a consequence of the fact that the absorption probability is dependent on the orientation of the molecular axis with respect to the polarization direction of the light. Therefore, if a sample of cold atoms moving with random relative velocity directions is irradiated with polarized light, atom pairs with favorable relative velocity directions will preferentially absorb a photon and combine to an excited molecule. Immediately after its formation this molecule will have its axis aligned along the relative velocity direction of the two colliding atoms before the absorption. Similarly the probability for the second photon to

CP604, Correlations, Polarization, and Ionization in Atomic Systems
edited by D. H. Madison and M. Schulz
© 2002 American Institute of Physics 0-7354-0048-2/02/$19.00

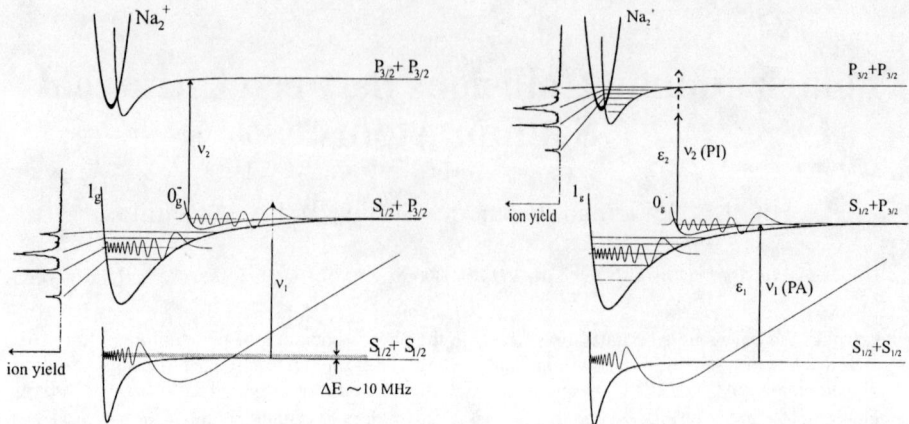

FIGURE 1. Schematic diagram showing ground, singly-excited and doubly-excited potentials of the Na_2 molecule. Two ground state atoms colliding slowly absorb a red detuned photon to produce a singly-excited, translationally cold molecule in a well-defined rovibrational level. By absorption of a second photon the singly-excited molecule may be further excited to a doubly-excited (autoionizing) state. By monitoring the ion production as a function of the frequency of the first (left panel) or second (right panel) photon, the rovibrational levels in the singly- or doubly-excited potential may be probed, respectively.

be absorbed in the second step will depend on the relative orientation of the molecular axis in the intermediate excited state with respect to the polarization direction of the second photon. As a result the production rate of the ions formed in the autoionization process is expected to depend on the relative orientation of the polarization directions of the two successive photons that have been absorbed.

In our experiment we have chosen as the intermediate state the $(v, J) = (1, 2)$ rovibrational level in the 0_g^- or 1_g potential which asymptotically connect to the $S_{1/2} + P_{3/2}$ dissociation limit of Na_2. The final state reached upon the second photon absorption is one of the bound levels in either the 0_u^- or the 1_u potential connecting to the $P_{3/2} + P_{3/2}$ dissociation limit.

EXPERIMENT

The experiments are performed in a so-called "dark-spot" magneto-optical trap (MOT) [5] which leaves almost all atoms in the lowest, $f = 1$, hyperfine state. The trap is continuously loaded from a Zeeman-tuned, laser cooled, atomic beam [6]. The trapping beams are turned on and off at a 100 kHz rate with a duty cycle of 50%. During the trap off periods two probe beams are focused in the MOT at the region with the highest density of trapped atoms. One of the lasers (the PA-laser) is detuned to the red of the atomic $S_{1/2} + P_{3/2}$ resonance while the other laser (the PI-laser) is detuned to the blue. The frequency of the PA-laser is adjusted to drive a transition from the colliding ground state atoms to a specific rovibrational level in the 0_g^- or 1_g potential. The frequency of the PI-laser is scanned to drive the transitions from the chosen inter-

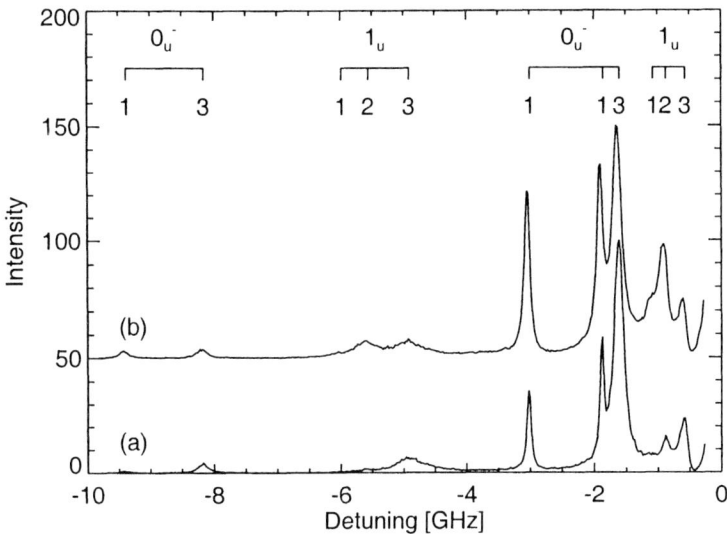

FIGURE 2. Spectra of rovibrational lines within the highest two vibrational bands in the 0_u^- and 1_u doubly-excited potentials. The rovibrational levels are excited via the two-step mechanism depicted in Fig. 1b with the PA-laser fixed at the transition to the $(v,J) = (1,2)$ rovibrational level in the 0_g^- potential. The laser light is linearly polarized, where in spectrum (a) the polarizations of the PA- and PI-laser are parallel to each other and in spectrum (b) they are perpendicular to each other. The baseline of spectrum (b) has been shifted for clarity. Both spectra have been normalized to each other by setting the intensity for the $0_u^-(J = 3)$ peak equal to 100.

mediate state to bound levels in the 0_u^- or 1_u potentials which are autoionizing at short internuclear range. By collecting the produced ions an ion spectrum is measured where the peaks correspond to the bound levels in the doubly excited potentials. The whole process is shown schematically in Fig. 1b. To facilitate the study of polarization effects the light of the PA- and PI-lasers is linearly or circularly polarized. The ion spectra are measured in two cases: (i) With the (linear or circular) polarizations oriented parallel to each other and (ii) With the polarizations oriented at right angles to each other.

RESULTS

Figure 2 shows the results of our measurements when linearly polarized light is used for the two successive excitation steps. The frequency of the first (PA) laser is fixed at the transition to the $(v, J) = (1, 2)$ rovibrational level in the 0_g^- potential, whereas the second (PI) laser is scanned through the bound levels in the 0_u^- or the 1_u doubly excited potentials. The identifications of the various lines in the spectra have been made using the dipole selection rules. In the spectra we have identified two vibrational levels of each molecular symmetry, 0_u^- and 1_u, as indicated. Within each vibrational band

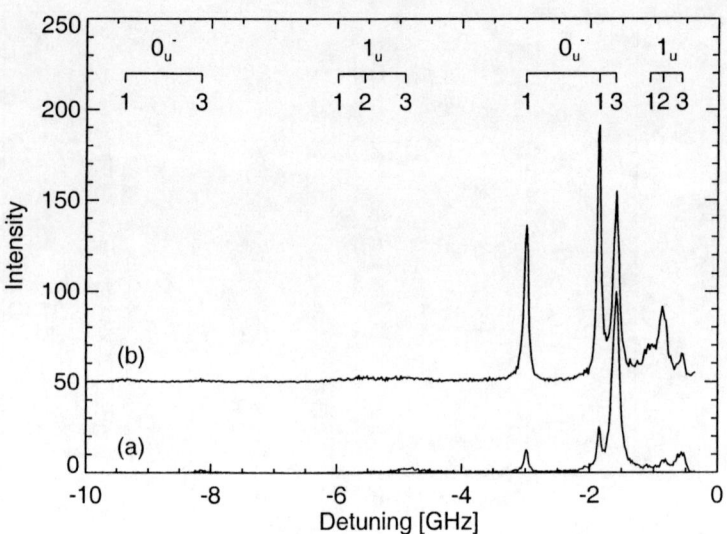

FIGURE 3. Same as Fig. 2, but now with the laser light circularly polarized.

rotational lines are observed with quantum numbers J = 1, 2 or 3. (Remember that the intermediate state is a $(v, J) = (1, 2)$ level). For the vibrational levels of the 0_u^- state the $J = 2$ rotational line is clearly missing. This is a result of the selection rule $\Delta J \neq 0$ for a $J = 0 \rightarrow J' = 0$ transition. For some hitherto unexplained reason the $J = 1$ line of the second vibrational band of the 0_u^- state seems to have split in two lines. In panel (a) the two polarizations are oriented parallel and in panel (b) they are oriented at right angles to each other. When comparing the two panels in Fig. 2 it is immediately clear that the relative intensities of the various peaks differ appreciably in the two cases, indicating that strong polarization effects are present. Similar results have been obtained for circular polarization (see Fig. 3). Also in that case strong polarization effects are observed. The results have been summarized in Table 1. All spectra have been normalized to each other by setting the intensity of the 0_u^- ($J = 3$) peak equal to 100. This strong peak appeared to vary negligible when the orientation of the two polarizations was changed from parallel to perpendicular. In the column 6 and 9 of the table the ratios of the peak intensities for parallel and perpendicular orientation of the two polarizations are given for linear and circular polarization, respectively. We see that for transitions from the intermediate to the final state with $\Delta J = +1$ (R-branch) the polarization effect is small (the parallel/perpendicular ratio ≈ 1) whereas for transitions with $\Delta J = 0$ (Q-branch) and $\Delta J = -1$ (P-branch) they are strong. The strange $J = 1$ states in the 0_u^- potential appear to deviate in this respect.

Similar experiments as described above have been performed, but with one of the rovibrational levels of the 1_g state chosen as the intermediate state. In this case very little or no polarization effects at all were observed.

TABLE 1. Intensities of the rotational levels within the highest two vibrational bands of the 0_u^- and 1_u symmetry. For each polarization combination the intensities are normalized with respect to the $\Delta J = +1$ transition of the highest vibrational state of the 0_u^- symmetry. The intensity of this peak appeared to change negligible, when the orientation of the polarizations was changed from parallel to perpendicular. The results are averages over three scans for each combination.

v	state	ΔJ	lin∥lin	lin⊥lin	lin∥lin/lin⊥lin	$\sigma-\sigma$	$\sigma^+-\sigma^-$	$\sigma-\sigma/\sigma^+-\sigma^-$
$v_{max}-1$	0_u^-	-1	1.06	4.78	0.22	0.28	2.79	0.10
		$+1$	6.60	6.45	1.02	2.51	2.21	1.13
$v_{max}-1$	1_u	-1	0.16	2.04	0.08	0.16	3.98	0.04
		0	1.89	20.53	0.09	0.05	8.95	0.006
		$+1$	23.02	24.48	0.94	9.18	10.11	0.91
v_{max}	0_u^-	-1	24.28	59.66	0.41	9.49	77.86	0.12
		-1	39.25	70.60	0.56	14.06	119.08	0.12
		$+1$	100	100	1	100	100	1
v_{max}	1_u	-1	6.93	26.16	0.26	1.12	25.93	0.04
		0	14.38	55.11	0.26	6.69	54.69	0.12
		$+1$	23.82	23.56	1.01	13.06	16.48	0.79

DISCUSSION

At first sight it is surprising that polarization or orientation effects do exist at all for the above described processes. Since the rotational spacing is of the order of 1 GHz the excited molecule in the intermediate state may make several rotations during its lifetime which is of the order of 10 ns. In this case one would expect the electronic angular momentum precessing around the molecular axis to have completely lost all memory of its initial orientation by the time of the second photon absorption. However, this is only true if the electronic angular momentum indeed precesses around the molecular axis. The 0_g^- state has zero angular momentum component along the molecular axis and hence there is no rotation around this axis. Therefore the electronic angular momentum remains fixed in space during the rotation of the molecule at right angles with the plane of rotation. The 1_g state on the other hand does have an electronic angular momentum component along the molecular axis and therefore the electronic angular momentum vector carries out precessions around the rotating axis and the memory of its initial orientation is lost by the time of the second photon absorption. This is most likely the reason why we fail to observe significant polarization effects when we choose one of the rovibrational states in the 1_g potential as the intermediate state.

ACKNOWLEDGMENTS

This work is supported by NIST, the US Office of Naval Research, Stichting FOM and the Debye Institute. P.v.d.S. and A.A. are in particular grateful to Kevin Jones and Paul Lett for their hospitality at NIST and for making their apparatus available for the present measurements.

REFERENCES

1. J. Weiner, V.S. Bagnato, S. Zilio and P.S. Julienne, Rev. Mod. Phys. **71**, 1 (1999)
2. K.M. Jones, S. Maleki, L.P. Ratliff and P.D. Lett, J. Phys. B: At. Mol. Opt. Phys. **30**, 289 (1997)
3. A. Amelink, K.M. Jones, P.D. Lett, P. van der Straten and H.G.M. Heideman, Phys. Rev. **61**, 042707 (2000)
4. R.E. Drullinger and R.N. Zare, J. Chem. Phys. **16**, 67 (1969)
5. W. Ketterle, K.B. Davies, M.A. Joffe, A. Martin and D.E. Pritchard, Phys. Rev. Lett. **70**, 2253 (1993)
6. W.D. Phillips and H.J. Metcalf, Phys. Rev. Lett. **48**, 596 (1982)

New developments in the problem of a complete experiment for photoinduced Auger processes

N.M. Kabachnik

*Institute of Nuclear Physics, Moscow State University, 119899 Moscow, Russia, and
Fakultät für Physik, Universität Bielefeld, 33615 Bielefeld, Germany*

Abstract. A theoretical analysis of the problem of a complete experiment, i.e. experimental determination of all Auger amplitudes and their relative phases for a particular transition, is given. It is shown, that recently found relations between intrinsic parameters which determine the angular distribution and spin polarization of Auger electrons make it impossible to realize a complete experiment by measuring the characteristics of Auger electrons only. Additional information about the polarization state of the ion in the final state is necessary.

CONCEPT OF A COMPLETE EXPERIMENT FOR AUGER PROCESSES

A set of measurements is referred to as realizing a ' complete' experiment if its analysis provides the parameters (usually transition matrix elements), which give a complete quantum mechanical description of the process. A concept of a complete experiment for Auger decay was formulated in [1] within the framework of the conventional two-step model of the Auger process. An Auger decay is considered as a quantum transition from a well-defined initial ionic state, characterized by its energy, angular momentum (J_i), and parity (π_i), to a certain final state of the residual ion (J_f, π_f) and the Auger electron in the continuum. The initial ionic state is prepared in the first step of the Auger process: ionization by photon or particle impact. In the following we consider photoionization but all conclusions are also valid for particle impact ionization. Since the initial and final ionic states have certain angular momenta and parities, it is convenient to expand the Auger-electron wavefunction in partial waves. Then the Auger decay may be described in terms of a limited number of complex matrix elements (Auger amplitudes) $M_{lj} \equiv \langle J_f \pi_f, lj \| O \| J_i \pi_i \rangle$ where l and j are the orbital and total angular momenta of the Auger electron and O is the transition operator.

For a particular transition $J_i \rightarrow J_f$ the number of Auger amplitudes is limited by the angular momentum and parity selection rules. In the general case the total number of amplitudes is $2J_i + 1$. For example, for the transition $N_4 O_{2,3} O_{2,3} (J_f = 2)$ there are four different electron continuum channels: $s_{1/2}$, $d_{3/2}$, $d_{5/2}$, and $g_{7/2}$ and correspondingly four complex Auger amplitudes. The moduli of the amplitudes and relative phase shifts form a set of the $4J_i + 1$ parameters to be determined experimentally for a complete characterization of the Auger decay. In practice, there are many transitions for which the number of possible decay channels is less than the maximum number. If $J_f < J_i$ then the number of amplitudes reduces to $2J_f + 1$, thus only $4J_f + 1$ parameters need to be determined experimentally. In particular, if $J_f = 0$ only one channel is possible, $j = J_i$, and the Auger decay is characterized by a single amplitude. This particular case is widely used in studies of inner-shell alignment and orientation.

Recently several attempts have been made to perform a complete experiment in photoinduced autoionization and Auger decay using various particular schemes for its realization [2]–[6]. In this paper we discuss, however, not the particular experiments but some general limitations that arise from newly found relations between parameters describing the Auger decay.

INTRINSIC PARAMETERS OF AUGER DECAY

The experimental information about the photoinduced Auger process can be gained from various measurements that involve Auger electrons. Consider first experiments in which only Auger electrons are detected. The first parameter which depends on the Auger amplitudes is, naturally, the intensity of the line in the Auger spectrum, measured in the 4π geometry or in an angular-resolved experiment at the "magic" angle: $I_0 \approx \sum_{lj} |M_{lj}|^2$. In photoionization the produced ion is aligned along either the photon beam direction (unpolarized or circularly polarized photons) or along the polarization vector of linearly polarized photons. Thus the Auger electrons are emitted nonisotropically and their angular distribution may be presented as [7, 8]

$$I(\vartheta) = \frac{I_0}{4\pi} \left(1 + \alpha_2 \mathcal{A}_{20} P_2(\cos\vartheta)\right), \tag{1}$$

where ϑ is the angle between the alignment axis (chosen as z axis) and the direction of the Auger electron emission, $P_2(x)$ is the second Legendre polynomial. The anisotropy is determined by the coefficient $\beta = \alpha_2 \mathcal{A}_{20}$ which is the product of the alignment parameter $\mathcal{A}_{20}$ and the *intrinsic* anisotropy parameter α_2. The former depends on the photoionization amplitudes only, while the latter depends on the characteristics of the Auger decay: the initial and final state quantum numbers and Auger amplitudes. Since the alignment $\mathcal{A}_{20}$ can be measured in an independent experiment (for example, by measuring the anisotropy of an Auger transition to the $J_f = 0$ final state, where α_2 is known), measurements of the angular distribution give the second experimental parameter α_2 that depends on the Auger amplitudes.

One can also measure the spin polarization of Auger electrons. For example, for photoionization with circularly polarized light one gets the following expressions for the components of the Auger-electron spin polarization [8]–[11]

$$P_{x'} = \frac{\xi_1 \mathcal{A}_{10} \sin\vartheta}{1 + \alpha_2 \mathcal{A}_{20} P_2(\cos\vartheta)}, \tag{2}$$

$$P_{y'} = \frac{\xi_2 \mathcal{A}_{20} \sin 2\vartheta}{1 + \alpha_2 \mathcal{A}_{20} P_2(\cos\vartheta)}, \tag{3}$$

$$P_{z'} = \frac{\delta_1 \mathcal{A}_{10} \cos\vartheta}{1 + \alpha_2 \mathcal{A}_{20} P_2(\cos\vartheta)}. \tag{4}$$

In these equations $(P_{x'}, P_{y'}, P_{z'})$ are the spin polarization components in the frame with z' axis oriented along the direction of Auger emission, so that $P_{z'}$ is a longitudinal component. The other two are transverse components: $P_{x'}$ is perpendicular to the emission direction in the reaction plane, while $P_{y'}$ is perpendicular to the reaction plane. Similar to the alignment, the orientation parameter of the ion $\mathcal{A}_{10}$ can be determined from independent experiments. Thus the measurement of the spin polarization provides three additional intrinsic parameters ξ_1, ξ_2 and δ_1 that depend on Auger amplitudes. Due to the dipole character of photoionization, orientation and alignment of the produced ion are the only possible state multipoles in a noncoincidence experiment (only the Auger electron is detected). Therefore, only the four intrinsic parameters introduced above (and the total intensity) may be obtained from such experiments.

In a more elaborate experiment that includes the coincident angular-resolved detection of photoelectron and Auger electron, more intrinsic parameters are accessible. In general, for a fixed direction of photoelectron emission $(\vec{n}_1)$, the angular distribution of the Auger electrons can be presented in the following form [8]

$$I(\vartheta, \varphi; \vec{n}_1) = C \left(1 + \sum_{k=2,4,\ldots} \alpha_k \sum_q \sqrt{\frac{4\pi}{2k+1}} \mathcal{A}_{kq}(J_i, \vec{n}_1) Y_{kq}(\vartheta, \varphi) \right), \qquad (5)$$

where $Y_{kq}(\vartheta, \varphi)$ are the spherical harmonics. In coincidence experiment higher-order state multipoles $(2 < k \leq 2J_i)$ are formed and therefore higher-order intrinsic parameters are accessible from the angular-correlation function. Quite analogously, one can consider, at least in principle, an experiment where the spin polarization of Auger electrons is measured in coincidence with the photoelectron. Here also higher multipoles will contribute and the corresponding intrinsic parameters of higher orders will determine the spin polarization. The longitudinal spin polarization is determined by δ_k (k odd), while the transverse spin polarization is determined by ξ_k (k even and odd). The maximum number of measurable intrinsic parameters for a given transition (plus the intensity) is $4J_i + 1$ [1], i.e. it is equal to the maximum number of unknown real parameters characterizing the set of Auger amplitudes. Thus if the intrinsic parameters were independent the measurements of angular distribution and spin polarization of Auger electrons would constitute a complete experiment. However, as it has become clear now, the intrinsic parameters are not independent.

RELATIONS BETWEEN INTRINSIC PARAMETERS

First evidence that the intrinsic parameters may be related to each other has been obtained in an attempt to extract matrix elements from the measured anisotropy and spin polarization of Auger electrons for the transition $N_4O_{2,3}O_{2,3}$ $^3P_1$ in Xe [5]. It has been shown that the four intrinsic parameters describing the transition are connected and the explicit relation has been found. Earlier, an analogous relation was found for the process of photoionization [12]. Since then several relations between intrinsic parameters for some particular cases have been derived [13, 14, 15]. Below we discuss them and their implications to the problem of the complete experiment.

225

Case 1: $J_i = 1/2$

Consider the case of initial angular momentum $J_i = 1/2$. In this case the Auger electron emission is isotropic. The $J_i = 1/2$ state cannot be aligned and the corresponding intrinsic parameter α_2 is zero [7]. By the same reason the spin component perpendicular to the reaction plane, the so-called dynamical spin polarization, is also zero ($\xi_2 = 0$) [9, 16]. On the other hand, the $J_i = 1/2$ vacancy can be oriented, for example, when it is created by absorption of circularly polarized light ($\mathcal{A}_{10} \neq 0$). In this case, due to polarization transfer, the emitted Auger electrons may be spin-polarized in the reaction plane, the polarization being determined by the two parameters ξ_1 and δ_1, see equations (2,4).

The decay of the $J_i = 1/2$ state is described by only two Auger amplitudes, corresponding to the emission of Auger electrons with the total angular momentum $j = J_f \pm 1/2$. Therefore, only three real values (two moduli of the amplitudes and the phase difference) should be determined from the experimental data for a complete characterization of the Auger decay in this case. Thus, if ξ_1 and δ_1 would be independent parameters, their measurement combined with the measurement of the total Auger yield might have realized the complete experiment. However, as we have shown recently [13], δ_1 and ξ_1 are related by a simple equation

$$\delta_1 = (-1)^{l+J_f+1} 2\xi_1 - 1.\tag{6}$$

(We define ξ_1 as in [14], i.e. it has an opposite sign with respect to the definition in [13].) This relation is very general. It is valid for any Auger transition from the initial state with $J_i = 1/2$, being independent of any dynamical model of Auger decay, particular choice of the wavefunctions, correlations, excitation mechanism etc. Its existence makes the complete experiment impossible provided only parameters of the Auger electrons are measured.

Case 2: $J_i = 1$

If an electron from an inner shell of a closed-shell atom is photoexcited to one of the unoccupied orbitals, a resonance is formed with $J_i = 1$ which predominantly decays via emission of resonant Auger electrons. The Auger decay in this case is characterized by three complex amplitudes (5 parameters), and beside the intensity, four intrinsic parameters α_2, ξ_1, ξ_2 and δ_1 exist that determine within the two-step model the angular distribution and spin polarization of resonant Auger electrons. As it was shown in [13] a similarity of this process to the nonresonant photoionization permits to transform the relation found by Cherepkov for photoionization [12] and apply it to the case of resonant Auger emission. The following relation exists between the intrinsic parameters

$$3(\xi_1)^2 + (2\xi_2)^2 = \frac{1}{\sqrt{2}}(1 - \sqrt{2}\alpha_2)\left[\sqrt{2} + \alpha_2 + \sqrt{3}\delta_1\right].\tag{7}$$

Relation (7) is valid for any resonant Auger transitions from the initial state with $J_i = 1$ independent of the decay dynamics and the parity of the final state. We see again that the number of independent observables is less than necessary for a complete experiment.

Case 3: $J_i = 3/2$ and $J_i = 5/2$

In general four (for $J_i = 3/2$) and six (for $J_i = 5/2$) Auger amplitudes determine the intrinsic parameters I_0, α_k, δ_k and ξ_k ($k \leq 2J_i$) in this case. However, only five of them $I_0, \alpha_2, \delta_1, \xi_1$ and ξ_2 are accessible in noncoincidence measurements. A general relation between the intrinsic parameters is still unknown although we believe that it exists. A particular case of transitions to the final state $J_f = 1$ was analyzed recently both experimentally and theoretically [5, 14]. In this case only three Auger amplitudes (5 parameters) describe the transition and it seemed to be possible to extract them from the experiment. However, we have proved that the intrinsic parameters are not independent in this case and obtained the following relation for $J_i = 3/2$

$$\left[\alpha_2 - \sqrt{5}\left(\delta_1 + (-1)^l \xi_1\right)\right]^2 + (2\xi_2)^2 = (1 + \alpha_2)\left[5 - \sqrt{5}\left(\delta_1 - 2(-1)^l \xi_1\right)\right]. \quad (8)$$

A similar equation has been found for $J_i = 5/2$ [14]. Due to these relations, measurements of the above four intrinsic parameters do not constitute a complete experiment. One could expect that the two higher-order parameters δ_3 and ξ_3 available in coincidence experiments would improve the situation. However, as follows from the next section they are not independent also.

Case 4: $J_i > J_f$

We have proved recently [15] that the anisotropy parameters α_k and the longitudinal spin-polarization parameters δ_k are related by the following equation valid for $J_i > J_f$

$$\sum_{k=0,even} (J_iM, J_i - M \,|\, k0)\,\alpha_k = \sum_{k=1,odd} (J_iM, J_i - M \,|\, k0)\,\delta_k. \quad (9)$$

Here $(J_iM, J_i - M \,|\, k0)$ are Clebsch-Gordan coefficients, and equation (9) is valid for each $M \geq J_f + 1/2$. We have found also a very simple physical explanation of this equation based on the conservation of the projection of the total angular momentum in Auger decay [15].

For a particular case considered above ($J_i = 3/2$, $J_f = 1$) equation (9) gives

$$\sqrt{5}\,(1 + \alpha_2) = 3\delta_1 + \delta_3. \quad (10)$$

This relation determines the value of δ_3 if δ_1 and α_2 are known. Therefore, experimental determination of δ_3 is not necessary, or better to say it will not give additional information. Strong arguments suggest that another equation exists that connects the anisotropy α_k and the transverse spin-polarization parameters ξ_k. However, it is still unknown.

CONCLUSIONS

The discovered relations between intrinsic parameters show that they are not independent and therefore, at least for the discussed particular cases the complete experiment cannot be realized by studying the properties of the Auger electrons only. One needs additional information, for example, about the polarization state of the residual ion that can be obtained by measurements of the angular distributions or/and polarization of fluorescence. In this connection more complicated experiments that involve the coincident detection of the Auger electrons and the fluorescence in combination with a polarization analysis of the latter [2, 3], look very promising. Nevertheless a thorough theoretical analysis is necessary in view of the discussed relations.

Although we know now that in many cases the intrinsic parameters describing the Auger decay are related, the general expression that would unify the found particular relations is still missing. Moreover, equations (6)-(8) have been derived purely mathematically, and it is not clear why they exist. To find the general relation and the physical reason of its existence is a challenge for theoreticians.

ACKNOWLEDGMENTS

I am greatly indebted to the Deutsche Forschungsgemeinschaft (DFG) and the University of Bielefeld for financial support. Numerous discussions with U. Heinzmann, N. Cherepkov, A. Grum-Grzhimailo, and B. Schmidtke are gratefully acknowledged.

REFERENCES

1. Kabachnik, N.M., and Sazhina, I.P., *J. Phys. B* **23**, L353 (1990).
2. West, J.B., Ross, K.J., and Beyer, H.J., *J. Phys. B* **31**, L647 (1998).
3. Ueda, K., West, J.B., Ross, K.J., Beyer, H.J., and Kabachnik, N.M., *J. Phys. B* **31**, 4801 (1998).
4. Ueda, K., Shimizu, Y., Chiba, H., Sato, Y., Kitajima, M., Tanaka, H., and Kabachnik, N.M., *Phys. Rev. Lett.* **83**, 5463 (1999).
5. Schmidtke, B., Khalil, T., Drescher, M., Müller, N., Kabachnik, N.M., and Heinzmann, U., *J. Phys. B* **33**, 5225 (2000).
6. Grum-Grzhimailo, A.N., Dorn, A., and Mehlhorn, W., "Towards a Complete Experiment for Auger Decay," in *Complete Scattering Experiments*, edited by U. Becker, and A. Crowe, Kluwer Academic/Plenum Publishers, New York, 2001.
7. Kabachnik, N.M., and Sazhina, I.P., *J. Phys. B* **17**, 1335 (1984).
8. Balashov, V.V., Grum-Grzhimailo, A.N., and Kabachnik, N.M., *Polarization and Correlation Phenomena in Atomic Collisions. A Practical Theory Course*, Kluwer Academic/Plenum Publishers, New York, 2000, ch. 3.
9. Klar, H., *J. Phys. B* **13**, 4741 (1980).
10. Kabachnik, N.M., and Lee, O.V., *J. Phys. B* **22**, 2705 (1989).
11. Lohmann, B., *Habilitation thesis*, Westfälisher Wilhelms-Universität Münster, 1998.
12. Schmidtke, B., Drescher, M., Cherepkov, N.A., and Heinzmann, U., *J. Phys. B* **33**, 2451 (2000).
13. Kabachnik, N.M., and Grum-Grzhimailo, A.N., *J. Phys. B* **34**, L63 (2001).
14. Schmidtke, B., Khalil, T., Drescher, M., Müller, N., Kabachnik, N.M., and Heinzmann, U., *J. Phys. B* **34** to be published (2001).
15. Kabachnik, N.M., and Sazhina, I.P., *J. Phys. B* submitted.
16. Kabachnik, N.M., *J. Phys. B* **14**, L337 (1981).

Recent Developments in the Theory of Electron Emission

B. Lohmann\$†, R. Srivastava‡, U. Kleiman†, and K. Blum†

\$: *Fritz-Haber-Insitut der Max-Planck Gesellschaft, Faradayweg 4-6,
D-14195 Berlin/Dahlem, Germany*

†: *Westfälische Wilhelms-Universität Münster, Institut für Theoretische Physik,
Wilhelm-Klemm-Str. 9, D-48149 Münster, Germany*

‡: *Department of Physics, University of Roorkee, Roorkee-247667, India*

Abstract. We present our results for resonant Auger emission after deep inner shell electron impact excitation. A full *ab initio* relativistic distorted wave approximation is applied where intermediate coupling between the different configuration state functions of the resonantly excited states and exchange interaction with the continuum has been taken into account. While several investigations have been published on electron impact excitation of the outer electronic shells, for a subsequent Auger emission, we present, to our knowledge for the first time, data on alignment and orientation for deep inner shell excitation. Opposite to the case of photoexcitation, the calculation of alignment and orientation is no longer purely analytic but is a complicated function of the electron impact energy. As example we consider the angular distribution and spin polarization of the resonantly excited $Ar^*(2p_{3/2}^{-1} \rightarrow 4s_{1/2})$ $L_3M_{2,3}M_{2,3}$ Auger spectrum after electron impact excitation. Within a two-step model we are able to determine all dynamic parameters for the primary excitation process, i.e. alignment and orientation, as well as angular distribution and spin polarization parameters describing the Auger decay dynamics. At present, no experimental alignment or orientation data are available. We hope that our calculation may stimulate related experiments.

INTRODUCTION

During the last decade a number of investigations have been focused on the Auger emission after photoionization and photoexcitation, respectively [1, 2, 3, 4, 5, 6, 7]. This has been mainly due to the fact of the availability of third generation synchrotron beam sources. While this type of research, on one hand, improved the understanding of Auger emission, it, on the other hand, disregarded the fact that, in the beginning of angle- and spin resolved experiments of Auger emission electron beam sources have been used. We refer here, for instance, to the first angle resolved experiments by Cleff *et al* [8, 9] and to the first angle- and spin resolved

CP604, *Correlations, Polarization, and Ionization in Atomic Systems*
edited by D. H. Madison and M. Schulz
© 2002 American Institute of Physics 0-7354-0048-2/02/$19.00

Auger emission experiments by Hahn *et al* [10] and by Merz and Semke [11].

The advantage of observing the angular distribution and spin polarization using photon beam techniques is that the number of parameters to determine is generally restricted by the dipole approximation [5]. Moreover, for the case of photoexcitation and applying the well observed two-step model [12], alignment and orientation become constant numbers. Such experiments and its predictions with respect to the spin polarization vector have been discussed extensively [6, 13].

On the other hand, the alignment and orientation parameters after electron impact excitation have been investigated in somewhat more detailed manner in the light of coincidence and non-coincidence experiments [14, 15] but in a rather limited manner in context of studying Auger emission processes (see e.g. Mehlhorn [12] and Kaur and Srivastava [16]). However, even such studies have been confined to electron excitation from outer atomic shells, only and their subsequent decay via Auger emission has been to a state with total angular momentum $J = 0$ in which case the angular distribution parameter of Auger emission becomes a constant number indenpendent of the matrix elements.

Therefore, it is our aim to consider the process of investigating the angle- and spin resolved Auger emission after electron impact excitation of a deep inner shell. In contrast to the deep inner shell photoexcitation case, the parameters of alignment and orientation are no longer independent of the excitation matrix elements. Moreover, they become a function of the electron impact energy. Even more important is the fact that, in some sense, the emission of an Auger electron after electron impact excitation can be seen as a special case of nowadays $(e, 2e)$ experiments [17, 18, 19, 20, 21]. Thus, the investigation of Auger emission experiments after electron impact excitation yields more information on $(e, 2e)$ experiments, too.

THEORY

Let us consider the following two-step process

$$e + A \longrightarrow A^* + e'$$
$$\searrow A^{+*} + e_{res} \; .$$

(1)

In the first step, after a primary electron impact excitation the exciting electron is not detected while another electron is excited from a deep inner shell into a Rydberg state. After a certain lifetime, this Rydberg state decays via resonant Auger decay and the emitted Auger electron is eventually detected. The validity of this approach has been proved in a variety of experiments (e.g. see the review by Mehlhorn [12] and refs. therein).

For the remainder of this paper we will consider the angle- and spin resolved resonant Auger emission of argon atoms after electron impact excitation. In particular, we consider the process

$$e + \text{Ar}\,(^1\text{S}_0) \longrightarrow \text{Ar}^*(2\text{p}_{3/2}^{-1}4\text{s}_{1/2})_{J=1} + e'$$

$$\searrow \text{Ar}^{+*}(3\text{p}_{1/2,3/2}^{-2}4\text{s}_{1/2}) + e_{res}\ . \tag{2}$$

Due to the application of a two-step model, the set of dynamic parameters describing the excitation/scattering and the subsequent emission process, can be factorized into parameters of orientation and alignment, and angular distribution and spin polarization parameters, respectively. The alignment and orientation parameters contain the information about the electron impact excitation while the latter yield information about the Auger decay dynamics, respectively.

In contrast to the case of photoexcitation the total set of parameters describing the excitation–emission process is not limited by the dipole approximation. Therefore, the maximum rank of irreducible measurable quantities is given by the general restriction $K \leq 2J$, only [22]. However, the general equations of angular distribution and spin polarization after electron impact excitation still remain rather complicated. Explicit expressions are given by Lohmann [23].

For the considered case of deep inner shell argon 2p → 4s excitation, we are focusing on $J = 1$ intermediate excited states. Therefore, quantities of a maximum rank of $K = 2$ can occur, only. In order to simplify the discussion we assume an excitation process with either longitudinally or unpolarized electrons for the remainder of this work, only. This simplification yields the advantage that the expressions for angular distribution and spin polarization remain the same as has been derived for the case of photoexcitation [13]. In particular, we obtain the angular distribution as

$$I(\theta) = \frac{I_0}{4\pi}\left(1 + \alpha_2\,\mathcal{A}_{20}\,P_2(\cos\theta)\right)\ , \tag{3}$$

and the cartesian components of the spin polarization vector as

$$p_x(\theta) = \frac{\xi_1\,\mathcal{O}_{10}\,\sin\theta}{1 + \alpha_2\,\mathcal{A}_{20}P_2(\cos\theta)}\ , \tag{4}$$

$$p_y(\theta) = \frac{-\frac{3}{2}\,\xi_2\,\mathcal{A}_{20}\,\sin(2\theta)}{1 + \alpha_2\,\mathcal{A}_{20}P_2(\cos\theta)}\ , \tag{5}$$

and

$$p_z(\theta) = \frac{\delta_1\,\mathcal{O}_{10}\,\cos\theta}{1 + \alpha_2\,\mathcal{A}_{20}P_2(\cos\theta)}\ . \tag{6}$$

The orientation parameters $\mathcal{O}_{10}$ can be non-zero, only if a longitudinally polarized electron beam is used for the primary excitation process whereas the alignment parameters $\mathcal{A}_{20}$ can be different from zero even for an unpolarized electron beam.

CALCULATIONAL MODEL

An important point of our emphasis in obtaining numerical data is to apply a consistent model for the calculation of all parameters. For obtaining the numerical data we employ a relativistic distorted wave born approximation (RDWA). Detailed information about the calculation of alignment and orientation parameters may be found e.g. in [14, 15, 16], and for the calculation of angular distribution and spin polarization parameters e.g. in [13, 23]. In particular, the bound state wavefunctions of the primary, the excited intermediate, and the ionized final state are constructed using the multiconfigurational Dirac-Fock (MCDF) computer code of Grant *et al* [24, 25]. Intermediate coupling has been taken into account. The mixing coefficients have been calculated applying the average level calculation mode. For the calculation of the continuum wavefunctions electron exchange with the continuum has been accounted for. With this, the electron excitation and Auger transition matrix elements are obtained for calculating the relevant alignment and orientation, and angular distribution and spin polarization parameters respectively.

DISCUSSION

Our main emphasis is on the investigation of the alignment and orientation parameters, respectively. Both are functions of the electron impact transition matrix elements and therefore become dependent of the electron impact energy. Thus, we calculated the primary excitation cross section as a function of the electron impact energy in order to identify energy ranges where a comparatively high cross section coincides with large values for the alignment and orientation parameters, respectively. Our results for the cross section of the electronically excited $(2p_{3/2}^{-1}4s_{1/2})_{J=1}$ state of argon are plotted in figure 1 as a function of the electron impact energy.

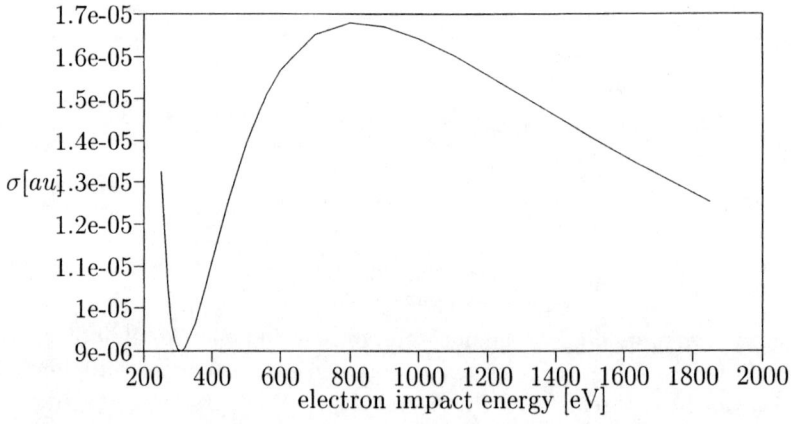

Figure 1: The cross section for the electronically excited $Ar^*(2p_{3/2}^{-1}4s_{1/2})_{J=1}$ state.

As can be seen, the cross section is comparatively large close to threshold (248.63 eV) from where it rapidly decreases to its minimum at an electron impact energy of $\sim 300eV$. Then, it continuously increases to a maximum even higher than its threshold value at an energy of $\sim 800eV$. For larger energies we find the cross section slowly decreasing.

Our results for the orientation and alignment parameters $\mathcal{O}_{10}$ and $\mathcal{A}_{20}$ are shown in figures 2 and 3.

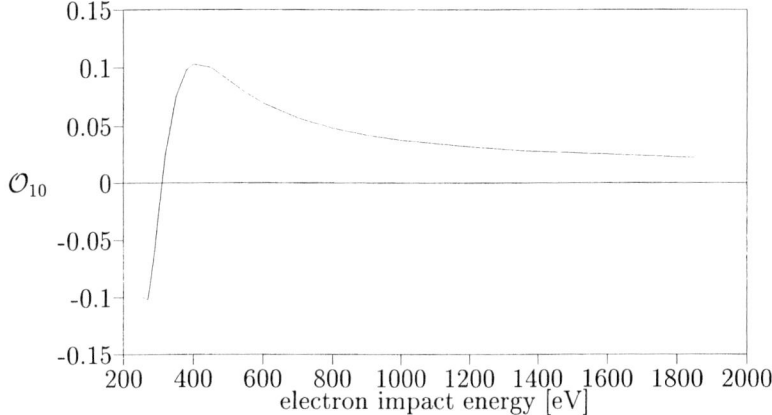

Figure 2: The orientation parameter $\mathcal{O}_{10}$ for the $Ar^*(2p_{3/2}^{-1}4s_{1/2})_{J=1}$ state after electron impact excitation. A 100% longitudinally polarized electron beam has been assumed.

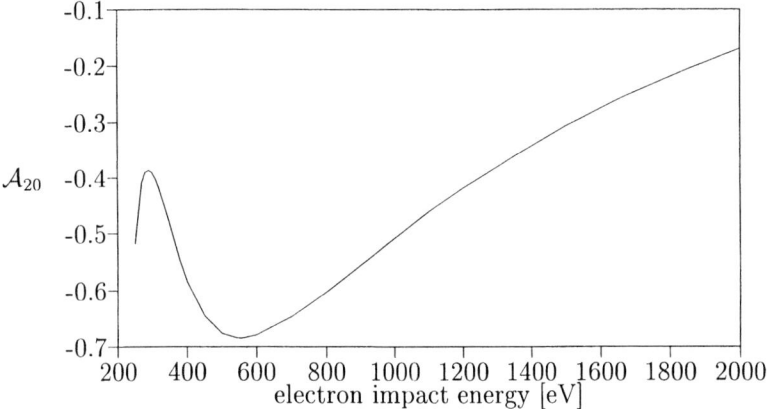

Figure 3: The alignment parameter $\mathcal{A}_{20}$ for the excited $Ar^*(2p_{3/2}^{-1}4s_{1/2})_{J=1}$ state.

Considering the alignment parameter, for the case of photoexcitation its value remains constant at $\mathcal{A}_{20} = 1/\sqrt{2}$ whereas the orientation parameter takes a value of $\mathcal{O}_{10} = \sqrt{3/2}$. As can be seen from figures 2 and 3 this is no longer the case for the case of electron impact excitation. Moreover, both parameters vary over a wide range of electron impact energy.

Though the orientation parameter varies as a function of electron impact energy and does even change sign at $\sim 300eV$, its magnitude does hardly exceed values of 10%. Close to threshold we find negative values whereas at $\sim 400eV$ the orientation takes its maximum of $\sim 11\%$ after changing sign. For larger energies the orientation smoothly decreases to small values. Note, that we assumed a 100% longitudinally polarized electron beam in the calculation.

In contrast to the photoexcitation case, the alignment has always negative magnitude but shows a dramatic behaviour. Close to threshold we find an alignment of ~ -0.5 which slightly decreases to ~ -0.4 at $300eV$ electron impact energy. Then, it increases to a maximum value of ~ -0.69 at $\sim 550eV$, from where it smoothly decreases to smaller numbers.

The numerical results for the angular distribution and spin polarization parameters for the resonant $Ar^*(4s_{1/2})L_3M_{2,3}M_{2,3}$ Auger transition are shown in table 1. We will not discuss these numbers in detail. The resonant Auger spectra have been

Table 1: The energies, relative intensities, angular distribution and spin polarization parameters for the $Ar^*(4s_{1/2})$ $L_3M_{2,3}M_{2,3}$ Auger transitions (Exchange included). (a): The leading jj coupled configuration state function has been used to identify the state, (b): States in LS coupling, †: The total intensity has been normalized to 100.

$Ar^*(4s_{1/2})$ $L_3M_{2,3}M_{2,3}$							
Final states		energy	int.	ang. & spin pol. par.			
(a)	(b)	[eV.]	I_0†	α_2	δ_1	ξ_1	ξ_2
$\lvert([3\overline{p}^2 3p^2]_2 4s^1)5/2\rangle$	$^4P_{5/2}$	213.74	5.805	-0.177	-0.257	-0.483	-0.0121
$\lvert([3\overline{p}^1 3p^3]_1 4s^1)3/2\rangle$	$^4P_{3/2}$	213.63	11.316	0.629	-1.004	-0.089	-0.0015
$\lvert([3\overline{p}^0 3p^4]_0 4s^1)1/2\rangle$	$^4P_{1/2}$	213.56	6.190	-1.149	0.153	-0.573	0.0053
$\lvert([3\overline{p}^2 3p^2]_2 4s^1)3/2\rangle$	$^2P_{3/2}$	213.21	19.359	0.699	-0.459	0.059	0.0002
$\lvert([3\overline{p}^1 3p^3]_1 4s^1)1/2\rangle$	$^2P_{1/2}$	213.07	7.945	0.617	1.172	-0.350	-0.0041
$\lvert([3\overline{p}^1 3p^3]_2 4s^1)5/2\rangle$	$^2D_{5/2}$	211.48	6.496	-0.927	-0.039	0.270	-0.2282
$\lvert([3\overline{p}^1 3p^3]_2 4s^1)3/2\rangle$	$^2D_{3/2}$	211.47	32.649	-0.222	0.646	0.842	0.0482
$\lvert([3\overline{p}^2 3p^2]_0 4s^1)1/2\rangle$	$^2S_{1/2}$	208.44	10.240	-0.711	0.406	0.815	0.00002

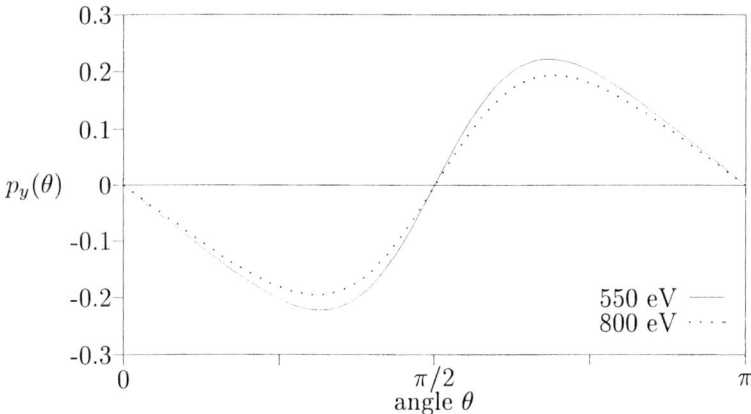

Figure 4: The p_y-component of the spin polarization vector for the resonant $\mathrm{Ar}^*(4s_{1/2})\mathrm{L}_3\mathrm{M}_{2,3}\mathrm{M}_{2,3}\ ^2\mathrm{D}_{5/2}$ final state Auger transition as a function of the Auger emission angle θ. Solid line: electron impact energy $550eV$, dotted line: electron impact energy $800eV$.

discussed e.g. by Aksela and Mursu [26]. We only mention the large angular distribution parameter α_2 for the $^4\mathrm{P}_{1/2}$ and the comparatively large dynamic spin polarization parameter ξ_2 for the $^2\mathrm{D}_{5/2}$ final states, respectively.

For our considered type of experiment the dynamic spin polarization parameter can be accessed via observation of the p_y-component of the spin polarization vector, only. Inserting α_2 and ξ_2 values for the $^2\mathrm{D}_{5/2}$ state into equation (5) we need to identify a range of electron impact energy which yields a large alignment $\mathcal{A}_{20}$ in coincidence with a comparatively large cross section. The cross section gets its maximum around $800eV$ where we still have an alignment of ~ -0.60. Though, at an energy of $500-600eV$ the cross section is still large but we are closer to the maximum of the alignment value of 0.69. Inserting both values of $\mathcal{A}_{20}$ into equation (5) we plot the p_y-component of the spin polarization vector as a function of the Auger emission angle. As can be seen from figure 4, we obtain a maximum degree of spin polarization of $\sim 19\%$ for the first, and $\sim 22\%$ for the latter alignment of the $(2p_{3/2}^{-1}4s_{1/2})_{J=1}$ excited argon state, respectively.

CONCLUSION

We investigated angle- and spin resolved Auger emission after electron impact excitation, where we focused on the alignment and orientation parameters. In contrast to the case of photoexcitation, these parameters are energy dependent for the considered case. The exact knowledge of these numbers is important for the interpretation of possible $(e, 2e)$ experiments. This is due to the fact, that the fine structure splitting of the intermediate excited state is to small to decide which electron has been scattered and which emitted, respectively. This results in the fact that experiments of this type should be carried out either as $(e, 2e)$ experiments covering their full solid angle for obtaining the full set of parameters of alignment and orientation, or at certain angles and energies with an accuracy which allows for a secure deduction of the required parameters. The investigated Auger transition to the $^2D_{5/2}$ final state shows a dynamic spin polarization of $\sim 20\%$ over a broad range of energy.

ACKNOWLEDGEMENTS

RS would like to thank the Department of Atomic Energy (DAE), Govt. of India for the financial assistance to this work. BL is grateful to the DFG and to the Auswärtiges Amt of the German Federal Republic which enabled him to attend ICPEAC and the ISPCEAC and $(e, 2e)$ satellite conferences and to present the results.

REFERENCES

1. Kabachnik, N. M., and Lee, O. V., *J. Phys. B: At. Mol. Opt. Phys.* **22**, 2705 (1989)
2. Hergenhahn, U., Kabachnik, N. M., and Lohmann, B., *J. Phys. B: At. Mol. Opt. Phys.* **24**, 4759 (1991)
3. Kuntze, R., Salzmann, M., Böwering, N., and Heinzmann, U., *Phys. Rev. Lett.* **70**, 3716 (1993)
4. Lohmann, B., Hergenhahn, U., and Kabachnik, N. M., *J. Phys. B: At. Mol. Opt. Phys.* **26**, 3327 (1993)
5. Kleiman, U., Lohmann, B., and Blum, K., *J. Phys. B: At. Mol. Opt. Phys.* **32**, 309 (1999)
6. Lohmann, B., *J. Phys. B: At. Mol. Opt. Phys.* **32**, L643 (1999)
7. Snell, G., Langer, B., Drescher, M., Müller, N., Zimmermann, B., Hergenhahn, U., Viefhaus, J., Heinzmann, U., and Becker, U., *Phys. Rev. Lett.* **82**, 2480 (1999)
8. Cleff, B., and Mehlhorn, W., *Phys. Lett.* **37A**, 3 (1971)
9. Cleff, B., and Mehlhorn, W., *J. Phys. B: At. Mol. Opt. Phys.* **7**, 593 and 605 (1974)
10. Hahn, U., Semke, J., Merz, H., and Kessler, J., *J. Phys. B: At. Mol. Opt. Phys.* **18**, L417 (1985)
11. Merz, H., and Semke, J., *X-ray and Inner Shell Processes* ed T. A. Carlson, M. O. Krause and S. T. Manson, AIP Conf. Proc. **215**, New York, 1990, pp 719
12. Mehlhorn, W., *X-ray and Inner Shell Processes*, ed T. A. Carlson, M. O. Krause and S. T. Manson, AIP Conf. Proc. **215**, New York, 1990, pp 465
13. Lohmann, B., *Aust. J. Phys.* **52**, 397 (1999)
14. Andersen, N., Bartschat, K., Broad, J. T., and Hertel, I. V., *Phys. Rep.* **278**, 107 (1997)
15. Srivastava, R., Blum, K., McEachran, R. P., and Stauffer, A. D., *J. Phys. B: At. Mol. Opt. Phys.* **29**, 3513 and 5947 (1996)

16. Kaur, S., and Srivastava, R., *J. Phys. B: At. Mol. Opt. Phys.* **32**, 2323 (1999)
17. Paripás, B., Víkor, G., and Ricz, S., *J. Phys. B: At. Mol. Opt. Phys.* **30**, 403 (1997)
18. Lohmann, B., *Aust. J. Phys.* **49**, 365 (1996) and refs. therein
19. Balashov, V. V., and Bodrenko, I. V., *J. Phys. B: At. Mol. Opt. Phys.* **32**, L687 (1999).
20. Balashov, V. V., and Bodrenko, I. V., *J. Phys. B: At. Mol. Opt. Phys.* **33**, 1473 (2000).
21. Taouil, I., *et al J. Phys. B: At. Mol. Opt. Phys.* **32**, L5 (1999)
22. Blum, K., *Density Matrix Theory and Applications*, 2nd ed, Plenum Press, New York, London, 1996
23. Lohmann, B., *Angle and Spin Resolved Auger Processes on Free Atoms and Molecules, Habilitation Thesis*, University of Münster, 1998
24. Grant, I. P., *Advan. Phys.* **19**, 747 (1970)
25. Grant, I. P., McKenzie. B., Norrington, P., Mayers, D., and Pyper, N., *Comp. Phys. Comm.* **21**, 207 (1980)
26. Aksela, H., and Mursu, J., *Phys. Rev. A* **54**, 2882 (1996)

Orientation Of Heavy Atoms Excited By Polarized-Electron Impact

C. Herting, S. Feldmann, S. Geers and G. F. Hanne

University of Münster, Physikalisches Institut, Wilhelm-Klemm-Str. 10, 48149 Münster, Germany

Abstract. The atomic orientation generated by polarized-electron impact excitation of Hg (6s6p) $^3P_1$ and Pb (6p7s) $^3P_1$ atoms has been studied experimentally using the electron-photon coincidence technique. For both atoms the orientation induced by the collision process shows a significant dependence on the spin projection of the incident electrons. The results are compared with well-established orientation propensity rules and, in the case of mercury, with a five-state Breit-Pauli R-matrix calculation. The Hg data show, in qualitative agreement with the theory, that the propensity rule is apparently violated if the electron spin is initially down. This effect is attributed to a quantum mechanical interference caused by 'intermediate coupling' within the excited state. In the Pb case the observation of the decay to $J_f = 2$ allows to extract the spin-flip probability $h^{\uparrow\downarrow}$ together with $J_\perp^{+\uparrow\downarrow}$, $P_l^{+\uparrow\downarrow}$ and $\gamma^{\uparrow\downarrow}$ when detecting photons *perpendicular* to the scattering plane.

1. INTRODUCTION

The orientation of atoms excited by electron impact, i.e. the sense of circulation of the charge cloud around the atomic core, has been of particular interest in atomic collision physics since many years. In the case of unpolarized-electron impact for example, the orientation or angular momentum transfer $J_\perp^+$ together with the alignment angle γ, the degree of linear polarization P_l^+, the height of the charge cloud h (spin-flip probability) and the differential cross section σ form a parameter set that describes the density matrix of the excited atoms. For a number of different atoms these parameters have been determined in scattered-electron polarized-photon coincidence experiments. It is a general observation for scattering systems with s → p excitation and especially for light atoms and small scattering angles that the scattered electron generates positive orientation of the atom, i.e. the orientation vector points in the direction $k_0 \times k_1$ where k_0 and k_1 represent the momenta of the incoming and scattered electrons [1]. Propensity rules have been formulated for s → p transitions along with semi-classical trajectory models to develop an intuitive insight into the excitation mechanism leading to this effect [2]. When using polarized electrons, even more information becomes accessible, since averaging over the spin quantum number of the incident electron is avoided [3]. The question arises, whether the well-established propensity rules are applicable in *polarized*-electron atomic collisions [4]. It is the purpose of this paper to provide and discuss new experimental results for *spin-resolved* angular momentum transfer $J_\perp^{+\uparrow\downarrow}$ to clarify this question.

CP604, *Correlations, Polarization, and Ionization in Atomic Systems*
edited by D. H. Madison and M. Schulz

2. THEORETICAL FRAMEWORK - BRIEF SUMMARY

The atoms are described in the 'natural coordinate system' with the incoming electron beam defining the x-axis. The z-axis is chosen to be perpendicular to the scattering plane and forms the quantization axis as defined in figure 2. The incoming electron's polarization is chosen parallel to the z-axis. Thus, the process is invariant against reflection in the scattering plane, which restricts the number of independent scattering amplitudes. In the case of a $J_0 = 0^e \rightarrow J_1 = 1^o$ excitation, this symmetry property leads to the relationship [3]

$$a(M_1, m_1, m_0) = -(-1)^{M_1 + m_0 - m_1} a(M_1, m_1, m_0) \tag{1}$$

leaving six non-vanishing complex scattering amplitudes $a(M_1, m_1, m_0)$

$$
\begin{aligned}
a(1, \tfrac{1}{2}, \tfrac{1}{2}) &= a_{+1}^{\uparrow}, & a(-1, \tfrac{1}{2}, \tfrac{1}{2}) &= a_{-1}^{\uparrow} \\
a(1, -\tfrac{1}{2}, -\tfrac{1}{2}) &= a_{+1}^{\downarrow}, & a(-1, -\tfrac{1}{2}, -\tfrac{1}{2}) &= a_{-1}^{\downarrow} \\
a(0, -\tfrac{1}{2}, \tfrac{1}{2}) &= a_0^{\uparrow}, & a(0, \tfrac{1}{2}, -\tfrac{1}{2}) &= a_0^{\downarrow}
\end{aligned}
\tag{2}
$$

The fixed quantum numbers are suppressed in the above notation. M_1 labels the excited magnetic substates and m_0 and m_1 are the spin projections of the incoming and outgoing electron. In the theoretical framework describing *polarized*-electron impact excitation the density matrix of the excited atoms can be parameterized by *two* sets of $(J_\perp^+, P_l^+, \gamma, h)$, one for spin-up ($\uparrow$) and one for spin-down ($\downarrow$) electrons, together with a factor $w_\uparrow$ weighting the relative importance of spin-up and spin-down scattering and the differential cross section σ [4]. For the spin-resolved orientation parameter $J_\perp^{+\uparrow\downarrow}$ the following relationships hold [3]:

$$
J_\perp^{+\uparrow} = \frac{\left|a_{+1}^{\uparrow}\right|^2 - \left|a_{-1}^{\uparrow}\right|^2}{\left|a_{+1}^{\uparrow}\right|^2 + \left|a_{-1}^{\uparrow}\right|^2}, \quad
J_\perp^{+\downarrow} = \frac{\left|a_{+1}^{\downarrow}\right|^2 - \left|a_{-1}^{\downarrow}\right|^2}{\left|a_{+1}^{\downarrow}\right|^2 + \left|a_{-1}^{\downarrow}\right|^2}
\tag{3}
$$

It is an important point to note that only states with $M_J = \pm 1$ contribute to the orientation of the atom, and these states are excited solely by *non-spin-flip* amplitudes.

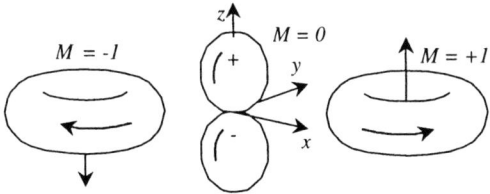

FIGURE 1. Oszillator densities corresponding to the different magnetic substates of a $J = 1^o$ state. The $M_J = \pm 1$ states have total angular momentum expectation values of $\pm \hbar$ (± 1 in atomic units), respectively.

To underline the relevance of the above mentioned expressions in a more qualitative way, figure 1 shows a sketch of the oscillator densities of the three possible magnetic substates $M_J = 0, \pm 1$ of the excited atom with $J = 1^\circ$. The $M_J = \pm 1$ states contribute $\pm \hbar$ to the net orientation. Due to the optical selection rules the corresponding decay photons observed in the z-direction have positive (σ_+) or negative (σ_-) helicity depending on whether they originated from a state with $M_J = +1$ or -1. Thus, $J_\perp^{+\uparrow\downarrow}$ is given by

$$J_\perp^{+\uparrow\downarrow} = -P_3^{\uparrow\downarrow} = \frac{I_{\sigma+}^{\uparrow\downarrow} - I_{\sigma-}^{\uparrow\downarrow}}{I_{\sigma+}^{\uparrow\downarrow} + I_{\sigma-}^{\uparrow\downarrow}} \tag{4}$$

where $I_{\sigma\pm}^{\uparrow\downarrow}$ represent the light intensities for each combination of helicity and electron spin direction. The $M_J = 0$ state does not carry orientation. It radiates like an oscillator parallel to z, thus no radiation is emitted in the z-direction. The '+' sign in the definition of $J_\perp^+$ underlines that only the part of the wave function having positive reflection symmetry contributes to this parameter. $J_\perp^+$ is equal to the total angular momentum expectation value $\langle J_z^+ \rangle$ of the part of the excited state having positive reflection symmetry.

3. EXTRACTION OF ORIENTATION - EXPERIMENTAL SETUP

In the experiment described in this section (see figure 2) the target atoms are Hg as a benchmark system in heavy atom electron collision physics, and Pb. In the case of Hg, the atoms are excited from their ground state $(6s^2)\ ^1S_0$ to $(6s6p)\ ^3P_1$. Formula (4) can be used to extract the (relative) population of the $M_J = \pm 1$ sublevels yielding directly $J_\perp^{+\uparrow\downarrow}$: With the spin of the incoming electrons switched to the desired direction, the numbers of $(6s6p)\ ^3P_1 \rightarrow (6s^2)\ ^1S_0$ decay photons having positive and negative helicity are counted by using appropriate optical filters. The photons are detected in coincidence with the scattered electron in order to restrict the observation to atoms that have scattered the electrons to a fixed direction. The same technique is applied to the Pb $(6p^2)\ ^3P_0 \rightarrow (6p7s)\ ^3P_1 \rightarrow (6p^2)\ ^3P_2$ system, and both cases represent $J = 0^e \rightarrow J = 1^o$ excitations.

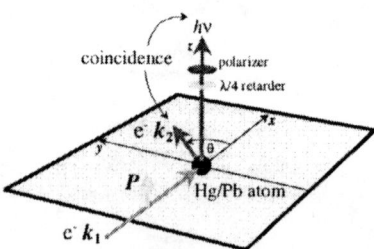

FIGURE 2. Schematic diagram of the experiment. In the 'natural frame' the incident electron beam defines the x-axis. The emitted photons are detected in the z-direction.

Figure 3 shows a sketch of the experimental setup used for the measurements. Polarized electrons emitted from a GaAs photo cathode are focused onto the scattering center and intersect with a beam of atoms emitted from the oven system. The scattered electrons are detected energy- and angle-resolved by a cylindrical-mirror analyzer (CMA). A collecting lens guides the decay photons from the scattering center through a $\lambda/4$-retarder, a linear polarization filter and an interference filter to the photomultiplier tube. The observables are the spin-resolved (e,eγ)-coincidence rates with the light analyzer set to transmit positive and negative helicity photons.

4. RESULTS AND DISCUSSION

Figure 4 (a) shows experimental results for $J_\perp^{+\uparrow\downarrow}$ of Hg $(6s^2)\,^1S_0 \rightarrow (6s6p)\,^3P_1$ at 8 eV incident energy in comparison with a five-state Breit-Pauli R-matrix calculation by Bartschat et al. [4]. In contrast to the prediction of well-established orientation propensity rules, the theoretical calculation shows a *negative* orientation for scattering of *spin-down* electrons to small angles. However, the calculated spin-averaged orientation is positive, in agreement with classical expectations for s $\rightarrow$ p transitions. Considering the complexity of the system, there is a remarkable overall qualitative agreement with the experimental data, but the predicted negative orientation for spin-down scattering is not confirmed clearly here. However, the results shown in figure 4 (b) show a significant transfer of negative orientation by spin-down electrons scattered to 10 deg at a slightly higher incident energy around 8.5 and 10.5 eV, which is an apparent violation of the propensity rule.

In the case of unpolarized electrons $J_\perp^+$ vanishes at zero scattering angle due to the axial symmetry of the process. The question arises how incident *polarized electrons* can transfer orientation to the Hg atom even at zero scattering angle.

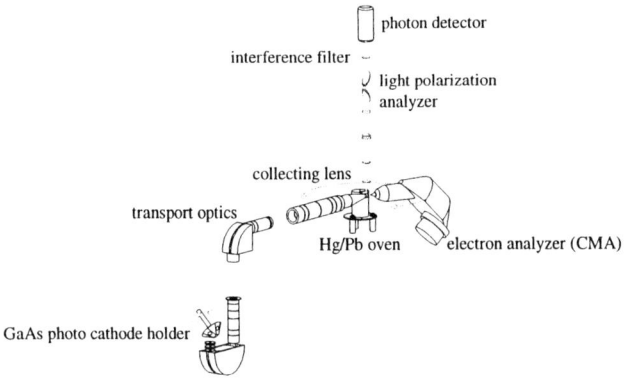

FIGURE 3. Experimental setup of the electron-photon-coincidence experiment. The laser used to extract polarized electrons from the stressed GaAs cathode is not shown.

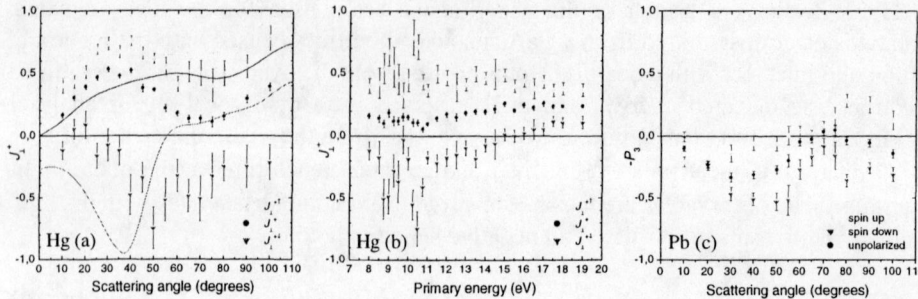

FIGURE 4. (a) Results for spin-resolved angular momentum transfer $J_\perp^+$ in Hg at 8 eV impact energy (symbols), compared with a five-state Breit-Pauli R-matrix calculation (solid lines). The top line and up triangles correspond to spin-up electrons. The bottom line together with down triangles show spin-down results. The middle line and circles give the unpolarized result. (b) $J_\perp^+$ in Hg at 10 deg scattering angle for various impact energies. The definition of the symbols is similar to (a). (c) Recent results for the spin-resolved Stokes parameter P_3 of the Pb transition at 6 eV (see text for explanation).

The excited Hg-state is best described in the 'intermediate coupling scheme' as

$$\left| {}^3P_{1M} \right\rangle_{interm.} = \alpha \left| {}^3P_{1M} \right\rangle_{LS} + \beta \left| {}^1P_{1M} \right\rangle_{LS} \tag{5}$$

where the subscript 'LS' labels pure Russel-Saunders states and α and β are real numbers. Neglecting spin-orbit terms for the continuum electron, which is a good approximation for small angle scattering, the transition matrix elements can be expressed in the LM_LSM_S-basis in terms of $F_M = f_M - g_M$ and g_M amplitudes for excitation of the LS-coupled singlet and triplet state, respectively [5]. f_M and g_M describe 'direct' and 'exchange scattering' and are independent of the initial electron spin projection. The transition matrix elements in the JM_JLS-notation of (5) can then be calculated by using standard vector coupling algebra. The differential cross sections $\sigma_M^{\uparrow\downarrow}$ for excitation with spin-up and spin-down electron are [6]

$$\sigma_{+1}^\uparrow = \left| a_{+1}^\uparrow \right|^2 = \left| \frac{\alpha}{\sqrt{2}} g_{+1} + \beta F_{+1} \right|^2, \quad \sigma_{-1}^\uparrow = \left| a_{-1}^\uparrow \right|^2 = \left| \frac{-\alpha}{\sqrt{2}} g_{-1} + \beta F_{-1} \right|^2$$

$$\sigma_{+1}^\downarrow = \left| a_{+1}^\downarrow \right|^2 = \left| \frac{-\alpha}{\sqrt{2}} g_{+1} + \beta F_{+1} \right|^2, \quad \sigma_{-1}^\downarrow = \left| a_{-1}^\downarrow \right|^2 = \left| \frac{\alpha}{\sqrt{2}} g_{-1} + \beta F_{-1} \right|^2 \tag{6}$$

For zero scattering angle, where $\frac{\alpha}{\sqrt{2}} g_{+1} = \frac{\alpha}{\sqrt{2}} g_{-1} = u$ and $\beta F_{+1} = \beta F_{-1} = v$ the orientation is given by (3)

$$J_\perp^{+\uparrow} = \frac{\sigma_{+1}^\uparrow - \sigma_{-1}^\uparrow}{\sigma_{+1}^\uparrow + \sigma_{-1}^\uparrow} = \frac{|u+v|^2 - |u-v|^2}{|u+v|^2 + |u-v|^2} = -J_\perp^{+\downarrow} \neq 0 \tag{7}$$

Thus, non-vanishing orientation at zero scattering angle is a consequence of *interference* between singlet and triplet parts of the wave function (5). Without 'intermediate coupling', i.e. in the case of $\alpha = 0$ or $\beta = 0$, $J_\perp^{+\uparrow\downarrow}$ would be zero.

In the Pb $(6p^2)\ ^3P_0 \rightarrow (6p7s)\ ^3P_1 \rightarrow (6p^2)\ ^3P_2$ process, the situation is more complicated. Since we have p → s excitation, the excited electron does not directly contribute to the orientation, but rather the remaining p-electron in combination with the angular momentum coupling mechanism can lead to this effect. Indeed, figure 4 (c) shows significant nonzero values for P_3 at 6 eV incident electron energy. The decay photons are observed in a transition $J = 1$ to $J_f = 2$. As a consequence $J_\perp^{+\uparrow\downarrow}$ is not simply given by (4). The general theory (see for example (5.2.6) of [7]) yields in this case

$$J_\perp^{+\uparrow\downarrow} = P_3^{\uparrow\downarrow}\left(\tfrac{7}{5} + \tfrac{6}{5}\tfrac{h^{\uparrow\downarrow}}{1-h^{\uparrow\downarrow}}\right) \tag{8}$$

$$\left(J_\perp^{+\uparrow\downarrow}\right)^2 = \frac{\left(P_3^{\uparrow\downarrow}\right)^2}{\left(P_3^{\uparrow\downarrow}\right)^2 + \left(5P_1^{\uparrow\downarrow}\right)^2 + \left(5P_2^{\uparrow\downarrow}\right)^2} \tag{9}$$

Interestingly, $P_3^{\uparrow\downarrow}$ now includes information about the relative spin-flip amplitude $h^{\uparrow\downarrow}$, a parameter not accessible in a transition to $J_f = 0$ when detecting the photons in z-direction [3]. (8) and (9) allow the full extraction of $J_\perp^{+\uparrow\downarrow}$ and $h^{\uparrow\downarrow}$ from the spin-resolved light polarization parameters $P_{1,2,3}^{\uparrow\downarrow}$ in addition to the parameters $P_l^{+\uparrow\downarrow}$ and $\gamma^{\uparrow\downarrow}$ accessible in a transition to $J_f = 0$. The corresponding measurements are still in progress at this time.

In conclusion, the total angular momentum of Hg atoms excited by polarized-electron impact has been measured. Incident *spin-down* electrons show a tendency to transfer *negative* orientation to the atom. Thus, orientation propensity rules are apparently violated in spin-resolved scattering from heavy atoms. This can be attributed to an interference effect caused by 'intermediate coupling' within the excited state. First experimental results for Pb atoms also show non-vanishing values $J_\perp^{+\uparrow\downarrow}$. Here, the $J = 1 \rightarrow J_f = 2$ decay photons emitted in the z-direction carry additional information about the spin-flip probability $h^{\uparrow\downarrow}$, a parameter not accessible when observing photons perpendicular to the scattering plane in a decay from $J = 1$ to $J_f = 0$.

REFERENCES

1. Andersen, N., Gallagher J. W., and Hertel, I. V., *Phys. Rep.* **165** (1988).
2. Andersen, N., and Hertel, I. V., *Comments At. Mol. Phys.* **19**, 1-34 (1986).
3. Andersen, N., Bartschat K., Broad, J. T., and Hertel, I. V., *Phys. Rep.* **279** (1997).
4. Andersen, N., Bartschat K., Broad, J. T., Hanne, G. F., and Uhrig, M., *Phys. Rev. Lett.* **76**, 208-211 (1996).
5. Hanne, G. F., *J. Phys. B: Atom. Molec. Phys.*, **9**, 805-815 (1976).
6. Hanne, G. F., *Phys. Rep.* **95** (1983).
7. Blum, K., *Density Matrix Theory and Applications*, New York: Plenum Press, 1996.

Spin Dependent Interactions in the Excitations of Atoms and Molecules

D H Yu and J F Williams

Center for Atomic, Molecular and Surface Physics,
Physics Department, University of Western Australia,
Crawley, Perth WA 6009, Australia

Abstract. Systematic studies on spin dependent interactions in the excitation, ionization of atoms and dissociative excitation of molecules have been performed with spin polarized electrons. The electron exchange, spin-orbit interaction and internal spin-orbit coupling have been characterized by the measurement of integrated Stokes parameters of the decay radiation from the excited targets for various collision processes.

INTRODUCTION

In the collisions of electrons with atoms and molecules, many fundamental forces are involved. These forces include the electron-electron and electron-nucleus Coulomb interactions, and spin-dependent interactions. The weaker spin-dependent interactions are usually masked by the stronger Coulomb interactions. The use of polarized electron can disentangle these spin-dependent interactions from the Coulomb interaction.

The spin-dependent interactions include spin-orbit interaction for the continuum electrons, electron exchange and internal spin-orbit coupling within atoms. The spin-orbit interaction for the continuum electrons is the magnetic interaction between the induced magnetic field seen by the approaching electron and the electron magnetic moment [1]. This spin-orbit interaction increases usually with increasing nuclear charge and is stronger if the electron comes closer to the nucleus. The electron exchange interaction is a direct consequence of Pauli principle which requires a completely antisymmetrized wavefunction for the description of the continuum electron and the target electrons. The exchange interaction is short range of nature and significant at low collision energies. Internal spin-orbit coupling represents the spin-orbit interaction within the atoms. It causes the atomic spin (S) and orbital angular momentum (L) to relax into fine-structure state (J).

The studies of these spin dependent interactions are motivated by both fundamental interests and practical applications. It is fundamental importance to obtain a better understanding of the roles played by different interactions, to understand the detailed collision mechanisms and dynamics in the electron-atom, electron-molecule collisions and therefore to test various theoretical models based on quantum mechanics. From the practical point of view, the spin phenomena are related to many applications. For example, the new developed Polarized Nuclear Magnetic Resonance technique

CP604, *Correlations, Polarization, and Ionization in Atomic Systems*
edited by D. H. Madison and M. Schulz

provides a better and faster imaging of lungs than the conventional magnetic resonance imaging [2]. Spin Electronics is a main subject in the coming nano-technologies. Electron spin and the related phenomena play also a crucial role in the future quantum computer.

A series of systematic studies on these spin dependent interactions has been carried out in The University of Western Australia [3,4,5,6] and the recent studies are presented here.

EXPERIMENTAL METHOD

After collision, many particles can be produced corresponding to different processes. The collision processes can be studied by detecting either scattered or emitted electrons, or the decay photon from excited targets, or both in coincidence. We detect the polarization of decay photon only but with initial spin polarized electrons which enable us to separate the effects from different spin-dependent interactions.

Spin polarized electrons are produced from photon emission of GaAs crystal. The direction of spin polarization is controlled by the helicity of the photon. The normal GaAs crystal can produce 30% electron polarization with a limit of 50%. The new strained GaAs crystal can have 75% electron polarization with a theoretical values of 100%. The electron polarization is measured by detecting the light polarization from an excited atomic state (Ne $3p^3D_3$) which is populated through pure electron exchange. In this case the circular polarization of the decay photon is proportional to the spin polarization of the incident electron [3].

The polarization of the decay photon is characterized by the well known Stokes parameters which are defined in terms of different intensities measured with different orientations of the transmission axis of the linear polarizer with respect to the electron beam direction and different intensity corresponding to different helicity of the photon [3].

The P_1 is related to the alignment of the charge cloud distribution with respect to the electron beam propagation direction. The alignment is a measure of different populations between magnetic sublevels with $M_J = \pm 1$ and $M_J = 0$, respectively. This parameter is generally not spin-dependent.

The P_2 reflects the distortion of the alignment due to spin-orbit interactions.

The P_3 measures the orientation of the charge cloud distribution and the angular momentum transfer to the atomic system through collisions. The orientation is related to the uneven populations between magnetic sublevels with $M_J = +1$ and $M_J = -1$, respectively. This parameter is mainly a measure of electron exchange effect.

Both P_2 and P_3 are directly proportional to the initial spin of the incident electron.

The detailed experimental procedure can be found elsewhere [3]. Polarized electrons are transferred through electron optics to the interaction region where it collides with the target. The decay photon from the excited state of the target is detected along the y-axis. For each electron energy and spin, different photon intensities corresponding to different settings of polarizer are recorded. These quantities are used to calculate the Stokes parameters.

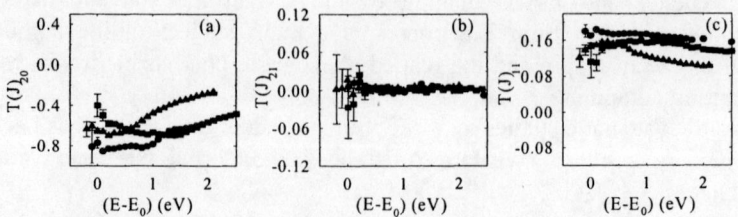

FIGURE 1. Normalized state multipoles for the J = 3 states of Ne, Kr and Xe atoms, dots: Ne; squares: Kr; triangles: Xe.

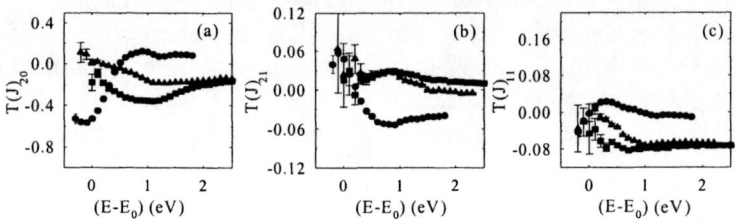

FIGURE 2. Normalized state multipoles for the J = 1 states of Ne, Kr and Xe atoms, dots: Ne; squares: Kr; triangles: Xe.

DIRECT EXCITATION OF OUTER SHELL ELECTRON OF NE, KR AND XE ATOMS

Excitation of $np^5(n+1)p$ states by polarized electrons have been studied for target atoms of neon, krypton and xenon [6]. The normalized integrated state multipoles have been determined from the measurement of the integrated Stokes parameters of the corresponding decay photons from the excited states of $np^5(n+1)p[5/2]_3$, $np^5(n+1)p'[3/2]_2$ and $np^5(n+1)p'[3/2]_1$ (n =3, 5 and 6 for Ne, Kr and Xe, respectively) within 2.5 eV of threshold. The results for the J = 3 states are shown in figure 1. The J = 3 state can be described by a pure L-S coupling state of $^3D_3$. The results show the similar energy dependence of the three state multipoles for the three atoms. The zero value of $T(J)_{21}$ are observed for the three atoms. This is consistent with the pure L-S coupling characteristics of these states. It is also an indication of unimportance of spin-orbit interaction for the continuum electrons. The similar and large $T(J)_{11}$ state multipoles for the three atoms reflect the significance of electron exchange interaction. However the situation becomes complicated for J = 1 and 2 states. As an example, the results for the J = 1 states are presented in figure 2. A few differences are noticed with respect to the results for the J = 3 states. Firstly, the $T(J)_{21}$ multipoles are non zero. Secondly, the energy dependence is different for different atoms and there is sign changing at certain energies. Thirdly, the negative values of $T(J)_{11}$ multipoles are observed. This indicates negative angular momentum transfer during the collisions. Through analysis of state multipoles with different L-S coupling states and coefficients in the description of the J = 1 state, it is found that all the different findings for the J = 1 and 2 states compared to the J = 3 states are attributed to the internal spin-orbit coupling [6]. The different behavior of the state multipole $T(J)_{20}$ contains information about the different coupling among various L-S compoents, even though it is independent on the spin of the incident electrons. The values of $T(J)_{21}$ do not necessarily correspond to the strength of internal spin-orbit interaction for these atoms. The different coupling coefficients with different signs in the description of the states can not only enhance but also cancel the total spin-orbit coupling effects. This is the reason for the lower values of $T(J)_{21}$ for Kr and Xe than Ne as shown in figure 2(b). The same mechanism is responsible for the negative angular momentum transfer as indicated in figure 2(c).

INNER SHELL IONIZATION WITH EXCITATION OF ZN ATOM

The ionization-with-excitation collision process, from the ground $3d^{10}4s^2\,^1S_0$ state to the ionic $3d^9 4s^2\,^2D_{3/2}$ state of the zinc atom has been investigated using incident polarized electrons with near threshold energies. Integrated Stokes parameters of the emitted 589.4 nm photon have been measured. The normalized state multipoles, the electric quadrupole moment of the orbital angular momentum and the magnetic dipole moment of the spin angular momentum of the excited ionic state are derived from the Stokes parameters based on L-S coupling [7]. The experimental observations show that the residual ion is aligned and oriented. The electron exchange effects in the inner-shell ionization-with-excitation process are directly demonstrated. The spin-orbit interactions are found to have no effect. The most interesting finding is that the

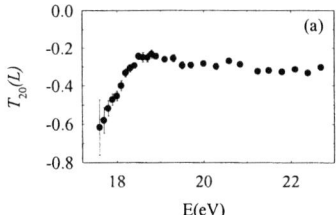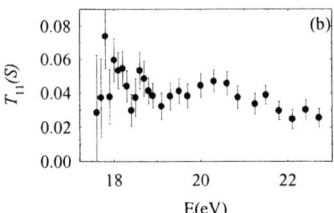

FIGURE 3. The electric quadrupole moment of the orbital angular momentum (a) and the magnetic dipole moment of the spin angular momentum (b) of the excited ionic state as a function of energies.

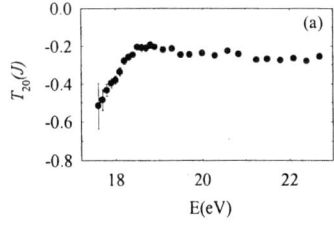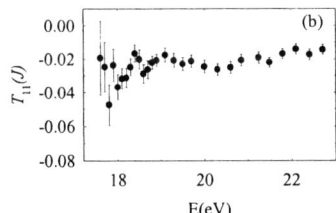

FIGURE 4. The electric quadrupole moment (a) and the magnetic dipole moment (b) of the total angular momentum of the excited ionic state as a function of energies.

combination of electron exchange and fine structure coupling enable a net negative angular momentum transfer to the atomic system. The electric quadrupole moment of the orbital angular momentum (L) and the magnetic dipole moment of the spin angular momentum (S) of the excited ionic state are indicated in figures 3 and the corresponding results for the total angular momentum (J) are shown in figure 4. Comparing the results in the figure 3(a) and 4(a), it is seen that the fine structure coupling depolarizes only the magnitude of the electric quadrupole moment (alignment) without altering the sign of the alignment. However the fine structure coupling not only depolarizes the magnitude but also changes the sign of the magnetic dipole moment (orientation) as shown in the figure 3(b) and 4(b). This results in the negative angular momentum transfer to the atomic system with spin up incident electrons.

DISSOCIATION AND EXCITATION OF H_2 MOLECULE

Dissociation and excitation of H_2 molecule has been studied by measuring polarization of the Balmer-α radiation from H(n=3) excited state after collisions with polarized electrons in the energy range from threshold to 150 eV above threshold. It was found that significant alignment (non-zero P_1) is created for H(n=3) state during the dissociation and excitation process of H_2 molecule. No explicit spin-orbit interaction has been observed because of the measured zero P_2 values. However electron exchange effects have been observed through the measured non-zero P_3 parameter for the dissociation and excitation process of H_2 molecule. The integrated Stokes parameters P_1 and P_3 are presented in figures 5(a) and 5(b), respectively, for near threshold energy range. In the figure 5(b) are shown the circular polarization of the Balmer-α radiation measured with electrons having 0%, 30% and 75% polarization.

SUMMARY AND PROSPECTS

From lighter atom (He) to heavier atom (Xe), from noble gas atoms to metal vapor (Zn) and to molecule (H_2), we have carried out systematic studies about spin-dependent interactions in various processes from direct excitation of outer shell electrons, to simultaneous ionization and excitation of either outer shell or inner shell

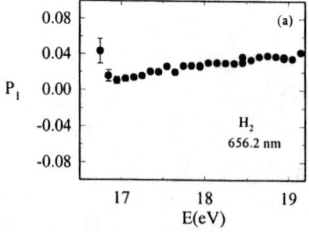 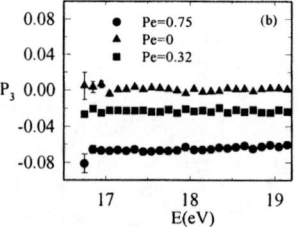

FIGURE 5. Stokes parameters P_1 and P_3 for the Balmer-α radiation as a function of energies.

electrons and to dissociation and excitation of molecules. The electron exchange has been found to be important for all the processes studied so far. The internal spin-orbit coupling within target is significant in the excitations of heavy noble gas atoms. The spin-orbit interaction for the continuum electrons has not been found to be effective in the collision processes studied here.

It is being currently planed to study resonance phenomena and molecular excitation in details by improving energy resolution of the incident polarized electrons. The electron-photon coincidence experiment is in progressing and will provide more detailed information of electron correlation. Experiment on studies of spin effects from magnetic surface has just started.

ACKNOWLEDGMENTS

This work was supported by the Australian Research Council (ARC) and the University of Western Australia. The work would not have been possible without the excellent skills in the Mechanical Workshop.

REFERENCES

1. Kessler, J, *Polarized Electrons,* 2nd ed, Springer, Berlin, 1985, pp.49-51.
2. Schwarzschild, B., *Physics Today*, **48**, no. 6, 17-18 (1995).
3. Yu, D. H., Hayes, P. A., Williams, J. F., and Furst, J. E., *J. Phys. B: At. Mol. Opt. Phys.* **30**, 1799-1812 (1997).
4. Yu, D. H., Hayes, P. A., Williams, J. F., and Furst, J. E., *Phys. Rev. Lett.* **78**, 2724-2727 (1997).
5. Hayes, P. A., Yu, D. H., and Williams, J. F., *J. Phys. B: At. Mol. Opt. Phys.* **31**, L193-200 (1998).
6. Yu, D. H., Hayes, P. A., Williams, J. F., Zeman, V., and Bartschat, K., *J. Phys. B: At. Mol. Opt. Phys.* **33**, 1881-1894 (2000)
7. Yu, D. H., Pravica, L., Williams, J. F., Warrington, N., and Hayes, P. A., *J. Phys. B: At. Mol. Opt. Phys.* **34**, (2001)

Determination of Hexadecapole Moments for the $3p^4(^1D)$ Core of Argon II Excited in Polarized e⁻-Ar Collisions

B. G. Birdsey[*], H. M. Al-khateeb[†], and T. J . Gay[*]

[*]Behlen Laboratory of Physics, University of Nebraska, Lincoln, NE 68588-0111 USA
[†]Department of Physics, Jordan University of Science and Technology, Irbid, P.O. Box 3030, 22110 Jordan

Abstract. We report on measurements of the integrated Stokes parameters of the light emitted from four well-LS coupled states of the $3p^4(^1D)4p$ manifold of ArII, following the simultaneous ionization and excitation of neutral argon by polarized electrons. As for all states, the state multipoles of J can be expanded in terms of the total orbital (L) and the total spin (S) state multipoles. By splitting each L and S state multipole into multipoles for the core and outer electron, we have experimentally obtained for the first time the normalized integrated state multipole of rank 4 (hexadecapole moment) for the $3p^4(^1D)$ core of ArII. We will comment on the Rubin-Bederson hypothesis as it pertains to this collision system [G. Csanak *et al.*, Comments At. Mol. Phys. **30**, 165 (1994)] as well as elucidate the data analysis techniques used.

INTRODUCTION

The theory of electron-atom interactions has been known in principle since the development of the Dirac equation, yet it is the tension between "what can be measured" and "what can be calculated" that has driven much of atomic collision physics to this day. Recently the convergent close coupling (CCC) technique of Bray and Stelbovics [1] has proven its efficacy in calculating electron interactions with simple atoms such as H [2], He and Na [3], and Li [4]. However the situation for the heavy noble gases is much more complicated; agreement between theory and experiment in this area ranges from adequate to dismal [5 - 8].

Our experiment is centered around the three body process of simultaneous ionization and excitation of Ar II. The specific process we examined is given by the reaction

CP604, *Correlations, Polarization, and Ionization in Atomic Systems*
edited by D. H. Madison and M. Schulz
© 2002 American Institute of Physics 0-7354-0048-2/02/$19.00

$$e^- + Ar(3p^6) \quad \rightarrow Ar^{+*}\left(3p^4\left({}^1D\right)4p\right) \quad\quad +2\ e^-$$
$$\searrow Ar^{+*}\left(3p^4 4s \text{ or } 3p^4 3d\right) + \quad \gamma \quad , \quad\quad (1)$$

where the excited states of interest in the $3p^4({}^1D)4p$ manifold are ${}^2F_{7/2}$, ${}^2F_{5/2}$, ${}^2D_{5/2}$, and ${}^2P_{3/2}$. The incident electron beam is transversely polarized and the scattered electrons are not detected. These particular residual ionic states are advantageous because they are known to be well-LS coupled [9], unlike neutral Ar. This simplifies the interpretation of our data and reduces the difficulty for theory to calculate electron-atom scattering processes [7, 8].

Our goal is to characterize the final state atomic charge cloud in terms of individual contributions from the $3p^4({}^1D)$ ionic core and the excited 4p outer electron. The shape of the charge cloud is completely described by the tensor multipoles of the excited ion's density matrix [10]. As detailed in our earlier work [11], the tensor multipoles of L can be expanded as

$$\left\langle \mathcal{J}_{KQ}^\dagger (L) \right\rangle = (2L+1) \sum_{\substack{k_c q_c \\ k_o q_o}} \sqrt{(2k_c+1)(2k_o+1)} \begin{Bmatrix} l_c & l_o & L \\ l_c & l_o & L \\ k_c & k_o & K \end{Bmatrix} \times$$
$$\left(k_c q_c, k_o q_o \middle| KQ\right)\left\langle \mathcal{J}_{k_c q_c}^\dagger (l_c) \otimes \mathcal{J}_{k_o q_o}^\dagger (l_o) \right\rangle, \quad\quad (2)$$

where the subscripts c and o refer to the multipoles associated with the core and outer electron. Notice that the multipole moments of the $\mathcal{J}_{k_c q_c}(l_c)$ and $\mathcal{J}_{k_o q_o}(l_o)$ sub-shells are generally not factorable, i.e. $\left\langle \mathcal{J}_{k_c q_c}^\dagger (l_c) \otimes \mathcal{J}_{k_o q_o}^\dagger (l_o) \right\rangle$ is not generally equal to $\left\langle \mathcal{J}_{k_c q_c}^\dagger (l_c) \right\rangle \left\langle \mathcal{J}_{k_o q_o}^\dagger (l_o) \right\rangle$. Therefore it is crucial to choose an experimental system where the multipole moments can be factored or where one can make enough observations to completely determine the coupled multipole moments. We introduce the following notation for the reduced multipole moments :

$$\ell(J,j)_{KQ,kq} = \left\langle \mathcal{J}_{KQ}^\dagger (J) \otimes \mathcal{J}_{kq}^\dagger (j) \right\rangle \Big/ \left\langle \mathcal{J}_{00}^\dagger (J) \otimes \mathcal{J}_{00}^\dagger (j) \right\rangle$$
$$\text{and } \ell(J)_{KQ} = \left\langle \mathcal{J}_{KQ}^\dagger (J) \right\rangle \Big/ \left\langle \mathcal{J}_{00}^\dagger (J) \right\rangle,$$

which will be used throughout the remainder of this work.

Because of our collision geometry, symmetry allows only a few of the multipoles of J to be non-zero [11]. Since these ionic states are well-LS coupled, L and S are good quantum numbers and the expansion of the multipoles of J into products of L multipoles and S multipoles is allowed [12]. We now invoke the Rubin-Bederson (RB) hypothesis [11], which states that if the collision time is significantly shorter than it takes for the excited system to "relax into its energy eigenstates", the collision can be considered as impulsively preparing each subsystem of the ion (the core, outer electron, and continuum electrons).

251

To determine whether the RB hypothesis should hold true it is necessary to check the appropriate time scales. The duration of the near threshold ionization/excitation process can be estimated by the time it would take an electron with the asymptotic energy of 2 eV to traverse three diameters of the residual ion, or $\sim 6 \times 10^{-16}$s. This is a factor of 3 shorter than the "Coulomb relaxation time", the time necessary to couple l_c and l_o into the total L, conservatively gauged by the largest splitting in the $3p^4(^1D)\,^2L$ manifold. The next time scale is the "fine structure relaxation time" which can be estimated from the energy splitting of the $^2F_{7/2}$ and $^2F_{5/2}$ fine-structure levels, and corresponds to 10^{-13}s. Therefore we would expect the RB hypothesis to hold in this experiment, if only marginally for decoupling L into l_c and l_o.

In our experiment, as well as the general case of atomic collisions, the fine-structure relaxation time is much longer than the collision time. According to the RB hypothesis, this implies that the subsystems of L and S can be described independently of each other, i.e. that the multipoles $t_{KQ}(L)$, $t_{KQ}(S)$ are properties of the manifold and are not properties of the individual states of the manifold. Assuming that this relationship is not then perturbed by terms that would convert angular momentum from one form to another (i.e. L to S), these multipoles of the manifold can be determined by observing a subset of the J multipoles of the manifold's individual states. The same argument allows the separation of the spin and orbital angular momentum belonging to the $3p^4(^1D)$ core and 4p outer electron. This is the cornerstone of the separation the subshell multipoles presented in this paper, where we observe the multipoles of several states in a given manifold through the integrated Stokes parameters.

In this case there are only three independent parameters to describe the orbital angular momentum distribution :

$$t_{20}(l_c),\ t_{20}(l_o),\ \text{and}\ t_{40}(l_o).$$

Expanding eq. (2) in terms of L, l_c, l_o, and these parameters we find that

$$
\begin{aligned}
t_{KQ}(L) &= F_L\left(t_{20}(l_c), t_{20}(l_o), t_{40}(l_c)\right) \\
&= \frac{c_{2L}t_{20}(l_c) + c_{3L}t_{20}(l_o) + c_{4L}t_{20}(l_c)t_{20}(l_o) + c_{5L}t_{40}(l_c)t_{20}(l_o)}{1 + c_{1L}t_{20}(l_c)t_{20}(l_o)},
\end{aligned}
$$

(3)

where the coefficients $c_{1L} \ldots c_{5L}$ depend only on L, and the function $F_L()$ is defined for use in the following section.

DATA ANALYSIS

Figure 1a presents the "raw" values of $t_{KQ}(L)$ for the states of interest. Due to the non-linear nature of eq. (3), some care must be taken in determining the values of the $t_{KQ}(L)'s$. To address this issue we inverted the equation using a terrain search

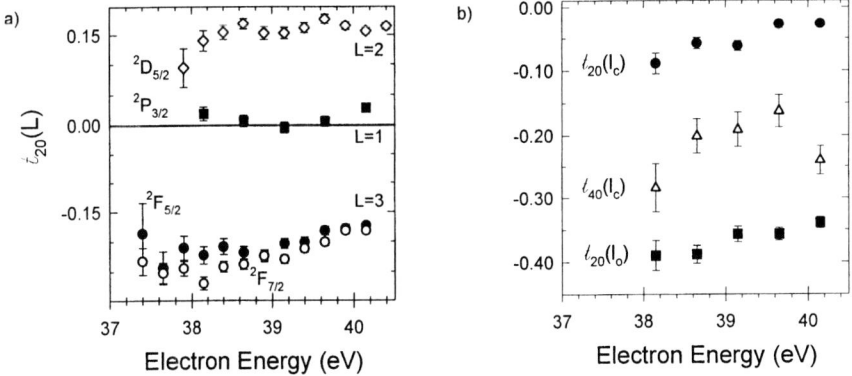

Figure 1a) The multipoles of the total angular momentum, L, generated from the relative Stokes parameters of the resonance fluorescence due to the reaction (1). **Figure 1b)** The calculated values of the multipoles of the $3p^4(^1D)$ and $4p$ subshells of the Ar^+ ion excited by electron impact.

algorithm and computed the uncertainties associated with the derived quantities (the multipoles of l_i and l_o) using a Monte-Carlo method.

The terrain search algorithm minimizes the Euclidian distance, d, between the measured vector $\{t_{20}(L=3),\ t_{20}(L=2),\ t_{20}(L=1)\}$ and the estimated vector $\{F_3\left(t_{20}(l_c),t_{20}(l_o),t_{40}(l_o)\right),\ F_2\left(t_{20}(l_c),t_{20}(l_o),t_{40}(l_o)\right),\ F_1\left(t_{20}(l_c),t_{20}(l_o),t_{40}(l_o)\right)\}$, weighted by the uncertainty in the measured values. To find the global minimum of d it is generally necessary to find the smallest local minimum in d by starting at various points in $\{t_{20}(l_c),\ t_{20}(l_o),\ t_{40}(l_o)\}$ space. Fortunately, because they are derived from angular momenta, the space of $\{t_{20}(l_c),\ t_{20}(l_o),\ t_{40}(l_o)\}$ is bounded by a finite region so it is possible to search the space exhaustively or by the more efficient method of random sampling.

Though the propagation of errors technique is the *de facto* standard for computing the uncertainty in derived quantities, given a known uncertainty in the data, this method has some significant drawbacks. It can give misleading results if the function that must be evaluated has high curvature in the region of interest. However it is possible to use Monte-Carlo methods to determine these uncertainties without any prior knowledge or prior assumptions about the function that must be determined [13]. The method relies on generating an artificial set of points that stands in as a proxy for a given datapoint and its uncertainty. Therefore the artificial set must encompass all *a priori* knowledge of data : i.e. it must be statistically indistinguishable from the parent distribution from which the data was drawn. Given the fact that we can derive values and uncertainties for the $t_{20}(L)$ from experimental data with known statistical properties, we know that the Gaussian distribution corresponds very closely to the actual parent distribution of each of the $t_{20}(L)$.

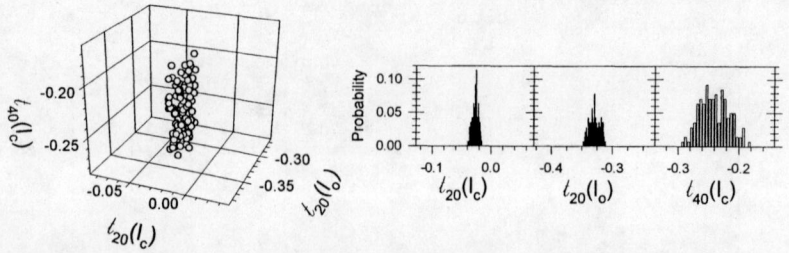

Figure 2 A representation of the distribution of the multipoles of the $3p^4(^1D)$ core and 4p outer electron generated from the set of synthetic data corresponding to the 40.2eV datapoint of Fig. 1a.

Figure 2 is an example of the typical "solution cloud" that we obtained for each energy we investigated. The top graph is a scatter plot of the local minima for the simulated set of data at an incident electron energy of 40.2 eV. The bar graphs below show the projection of this probability distribution onto the three individual axes. The mean value and width of these distributions corresponds to the solution and uncertainty in the quantities $\ell_{20}(l_c)$, $\ell_{20}(l_o)$, and $\ell_{40}(l_c)$. Each graph is unimodal, compact, and could be well described by a Gaussian function. All of these considerations point to the fact that this is a good, well defined solution.

With only moderate assurance that the RB hypothesis holds for the core and orbital angular momenta, we were somewhat concerned that the multipoles might not factor as in eq. (3), i.e. that are $\ell_{20}(l_c)$ and $\ell_{20}(l_o)$ are correlated. In this case there would be only three measurements and four independent parameters :

$$\ell\left(l_c,l_o\right)_{20,00}, \ \ell\left(l_c,l_o\right)_{00,20}, \ \ell\left(l_c,l_o\right)_{20,20}, \text{ and } \ell\left(l_c,l_o\right)_{40,20}.$$

If there was correlation between the two parameters $\ell_{20}(l_c)$ and $\ell_{20}(l_o)$, then the quantity

$$x = \frac{\ell\left(l_c,l_o\right)_{20,20} - \ell\left(l_c,l_o\right)_{20,00}\ell\left(l_c,l_o\right)_{00,20}}{\ell\left(l_c,l_o\right)_{20,20} + \ell\left(l_c,l_o\right)_{20,00}\ell\left(l_c,l_o\right)_{00,20}} \tag{4}$$

which measures this correlation, would be non-zero. Figure 3 presents the solution of the listed multipoles for many choices of x. It is obvious that the quoted value of $\ell_{40}(l_c) \sim \left\langle \mathcal{I}_{40}^\dagger(l_c) \otimes \mathcal{I}_{20}^\dagger(l_o) \right\rangle / \left\langle \mathcal{I}_{00}^\dagger(l_c) \otimes \mathcal{I}_{20}^\dagger(l_o) \right\rangle$ is consistent with the graph up to x = 0.5. In combination with the fact that the RB hypothesis appears to hold for the orbital angular momentum, the results in Figure 3 indicate that our analysis can accurately determine the hexadecapole of the Ar^+ $3p^4(^1D)$ core.

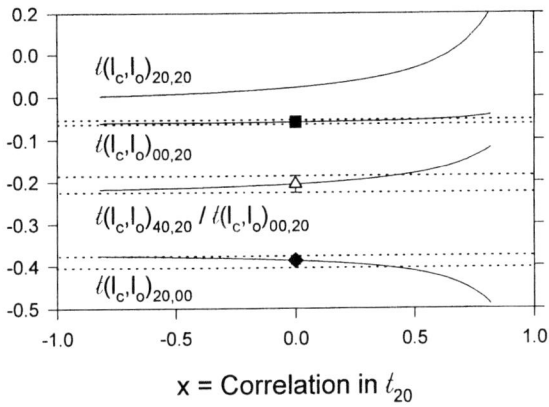

$$x = \text{Correlation in } \ell_{20}$$

Figure 3 The calculated values of the multipoles of the $3p^4(^1D)$ core and outer 4p electron, as a function of correlation (eq. (4)). The points correspond to the reported values assuming no correlation, and the dotted lines correspond to the 1σ error bars.

RESULTS AND CONCLUSIONS

Applying the method described in the previous section we were able to deduce values of the multipoles for the $3p^4(^1D)$ core and 4p outer electron subshells. These subshell multipoles are presented in Figure 1b. This method hinges on the applicability of the RB hypothesis. The measurements reported here would not be invalidated, even if the RB hypothesis did not hold rigidly true for the core/outer electron subsystems. The similarity of the ℓ_{20} $(L=2)$ multipoles for J = 5/2 and 7/2 in (Figure 1a) confirms this for fine-structure relaxation, as do our spin polarized measurements [11].

This work was supported by NSF Grant PHY-9704650.

REFERENCES

[1] I. Bray, Phys. Rev. A **49** (2), 1066 (1994).
[2] I. Bray, A.T. Steobivics, Phys. Rev. A **46** (11), 6995 (1992).
[3] N. Anderson and K. Bartschat, Adv. At. Mol. Phys **36**, 1 (1996).
[4] V. Karaganov, I. Bray, P.J.O. Teubner, and P. Farrell, Phys. Rev. A **54** (1), R9 (1996).
[5] P.A. Hayes, D.H. Yu, and J.F. Williams, J. Phys. B **31**, L193 (1998).
[6] V. Zeman et al., Phys Rev. A **58**, 1275 (1998).
[7] J.E. Chilton and C.C. Lin, Phys. Rev. A **60**, 3712 (1999).
[8] B.G. Birdsey et al., Phys. Rev. A **60**, 1046 (1999).
[9] X. Guo et al., J. Phys. B **32**, L155 (1999).
[9] H. Statz et al., J. App. Phys. **36**, 2278 (1965).
[10] K. Bartschat, K. Blum, and J. Kessler, J. Phys. B **14**, 3761 (1981).
[11] H.M. Al-Khateeb, B.G. Birdsey, T.J. Gay, Phys. Rev. Lett. **85** (19), 4040 (2000).
[12] K. Blum, *Density Matrix Theory and Applications* (Plenum, New York, 1996).
[13] J.M. Chambers, *Computational Methods for Data Analysis* (Wiley, New York, 1977).

Paul-trap promotion in He$^+$-He collisions

Gebhard von Oppen*, Marco Busch* and Ryszard Drozdowski[†]

* Techische Universität Berlin, Fakultät II, Institut für Atomare Physik
und Fachdidaktik, Hardenbergstraße 36, D-10623 Berlin, Germany

[†] University of Gdansk, Institute of Experimental Physics,
ul. Wita Stwosza 57, 80-952 Gdansk, Poland

Abstract. We investigated the charge distributions of excited He I states resulting from 50-200 keV He$^+$-He collisions. The pronounced asymmetry of the measured charge distributions confirms the assumption that Paul-trap promotion on the rotating saddle of a two-center field is a dominant excitation mechanism in ion-atom collisions.

INTRODUCTION

Single-electron excitation in ion-atom collisions is one of the elementary processes in atomic collision physics. Its basic features can be understood by considering a three-particle system as, for example, H$_2^+$, where an electron moves in the two-center field of the two protons. For ion-atom collisions the excitation cross sections are usually largest at intermediate energies, where the projectile velocity v_p is on the order of the Bohr velocity v_B of the bound electron. Therefore, this energy range is of particular interest.

Theoretically, however, collisions at intermediate energies are usually considered as complicate processes, which cannot easily be modelled theoretically. The basic theoretical approaches to collision processes, the molecular orbital (MO) approach and the Born approximation (BA), apply only to the extreme energy ranges, where either $v_p \ll v_B$ (MO) or $v_p \gg v_B$ (BA). But a well established and widely applicable theory for ion-atom collisions at intermediate energies is missing.

Nevertheless, there is the Paul-trap model [1], which gives at least some intuitive insight into the excitation process. According to this model one assumes that during a collision the electron gains energy by riding on the saddle of the two-center field. One imagines that the electron reaches the saddle at the moment of closest approach of the two collision partners. Afterwards the electron's motion is transiently stabilized on the saddle due to the rotation of the collision system through an

CP604, *Correlations, Polarization, and Ionization in Atomic Systems*
edited by D. H. Madison and M. Schulz
© 2002 American Institute of Physics 0-7354-0048-2/02/$19.00

angle $\pi/2$, and its energy is lifted together with the energy of the saddle between the separating ions. When the electron finally leaves the saddle to either side, it has gained enough energy to populate excited states of the target atom (direct excitation) or to form an excited projectile atom (transfer excitation).

The quantum dynamical evolution of the electron state on the saddle has been analyzed by Rost and Briggs [2]. As suggested by the classical model, where the electron transiently rides on the saddle between the ions, they find that the electron preferentially populates states with large electric dipole moments, where the charge center is shifted from the nucleus towards the receding collision partner. In particular, the parabolic Stark states $|n; n_1, n_2, m\rangle$ with either $n_1 = 0$ or $n_2 = 0$, are strongly populated.

Various experimental investigations on p-He collisions and highly-charged-ion (HCI)-He collisions [3,4,5] have shown that indeed this asymmetry of the charge distribution is typical for atomic electron states arising from ion-impact excitation at intermediate energies. We consider the experimental verification of this asymmetry as a strong support for the Paul-trap model. In particular, the experiments on p-He collisions, where collisional excitation of states with principal quantum numbers from $n = 3$ to $n = 8$ was investigated, confirm that this excitation mechanism is not an exception, but of general importance.

In the cases of p-He and HCI-He collisions, the projectile can be considered as a structureless, i.e. point-like particle affecting the state of the target atom only due to a time-dependent interaction $\mathcal{H}_{int}(t)$ resulting from the Coulomb field of the ion passing by on an approximately straight trajectory $\vec{r}(t)$. A somewhat more complicated collision system is He^+-He, where the binding energy of the electron bound to the ion is approximately equal to the binding energies of the target electrons. Therefore, the indistinguishability of the three electrons belonging to the collision system cannot be disregarded. Furthermore, also the two nuclei of the collision system are identical, if the most abundantly occurring isotope ^{4}He is used both as target and projectile.

In this work we study and discuss the influence of the electronic and nuclear symmetry of the He_2^+ system on the excitation process. Experiments were performed to measure the charge distributions of both excited-target and excited-projectile states. It turns out that both of them have the pronounced asymmetry predicted by the Paul-trap model. However, it also becomes obvious that the electronic symmetry affects the excitation process. Therefore, the He_2^+ system must not simply be considered as a three-particle system, where an electron is promoted on the Coulomb saddle of the two ionic He^+ cores. Rather, the presence of all three electrons has to be taken into account.

ELECTRONIC SYMMETRY IN He$^+$-He COLLISIONS

Collisional excitation of He atoms by point-like ions, as protons or HCI, is highly selective. According to Wigner's spin conservation rule, only singlet states are populated. However, both singlet and triplet states are populated by He$^+$-impact [6]. The cross sections σ_{exc} and σ_{trans} for direct and transfer excitation of the $1s4d$ singlet and triplet levels of He I are shown in Figure 1. In the intermediate energy range (20 keV $\lesssim E_{ion} \lesssim$ 150 keV) one finds: $\sigma_{exc}(4^1D) + \sigma_{trans}(4^1D) \approx \sigma_{exc}(4^3D) + \sigma_{trans}(4^3D)$. However, taking into account the statistical weights of singlet and triplet levels, one concludes that on the average excitation of a singlet state is three times more likely than excitation of a corresponding triplet state.

The fact that not only singlet states, but also triplet states are populated indicates the importance of electron exchange for He$^+$-He-collisions. In the presence of electron exchange, Wigner's spin conservation rule applies only to the total electron spin $S = \frac{1}{2}$ of the He$_2^+$ collision system. Therefore, the collision system can be left in states $|(s_1, s_2)S', s_3; S = \frac{1}{2}\rangle$ with $S' = 0$ as well as 1. That is the atomic electrons with spins s_1, s_2 can be in both singlet and triplet states, but these spins are entangled with the spin s_3 of the ionic electron in such a way that the total spin of the collision system is $S = \frac{1}{2}$.

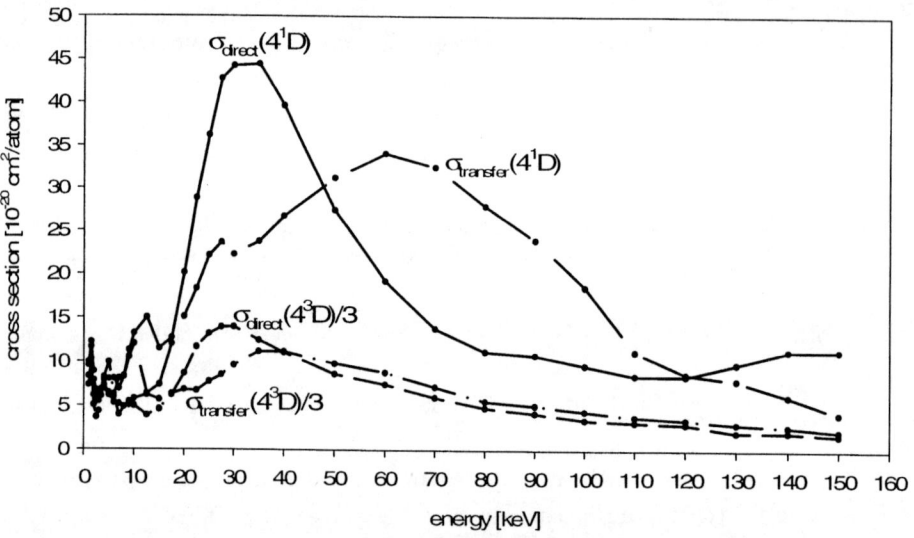

Figure 1: Cross sections of direct and transfer excitation of $1s4d$ ^1D and ^3D levels for He$^+$-He collisions [6].

Electron exchange is intimately related to the exchange symmetry of the three electrons. Assuming that Paul-trap promotion is the relevant excitation mechanism in He^+-He collisions, we have to analyze the promotion process for antisymmetrized three-electron states. Using the molecular orbital (MO) model, the $(He^+)_2$ state, on which the third electron is promoted, can be $|(1s\sigma)^2 \ ^1\Sigma_0\rangle$, $|(2p\sigma)^2 \ ^1\Sigma_0\rangle$, $|1s\sigma 2p\sigma \ ^1\Sigma_0\rangle$ or $|1s\sigma 2p\sigma \ ^3\Sigma_1, m_s\rangle$ with $m_s = \pm 1$ or 0.

A saddle suitable for electron promotion is formed only by the $(He^+)_2$ molecules in a $^3\Sigma_1$ state. Only in that case, the spatial state $|1s\sigma 2p\sigma\rangle$ is antisymmetric and, therefore, the two core electrons are spatially well separated. The antisymmetric MO state corresponds to the antisymmetric atomic orbital state, where each He nucleus binds one 1s electron and hence, a saddle sufficiently stable for electron promotion is formed.

Accordingly, we assume that Paul-trap promotion in He^+-He collisions proceeds as follows: Initially, the collision system is in a $|(s_1, s_2)0, s_3; \frac{1}{2}\rangle$ state with the atom in the $|1s^2 \ ^1S_0\rangle$ groundstate. During the collision a He_2^+ MO state is formed, but electron promotion is effective only in a $(He^+)_2$ two-center field, where the two electrons are in the $|1s\sigma 2p\sigma \ ^3\Sigma_1\rangle$ state. Therefore, the spin state of the excited collision system can be assumed to be $|s_1, (s_2, s_3)1; \frac{1}{2}\rangle$, where the spins s_2, s_3 of the core electrons couple to $S'' = 1$. By expanding this spin state into product states of the separated ion-atom system, one finds:

$$|s_1, (s_2, s_3)1; \tfrac{1}{2}\rangle = \frac{1}{2}\sqrt{3}\,|(s_1, s_2)0, \tfrac{1}{2}; \tfrac{1}{2}\rangle + \frac{1}{2}\,|(s_1, s_2)1, \tfrac{1}{2}; \tfrac{1}{2}\rangle$$

Accordingly one expects that singlet and triplet states are populated in the ratio 3:1. This result is in agreement with the experimental data.

These considerations suggest the proposition that Paul-trap promotion is the dominant excitation mechanism also in He^+-He collisions. For testing this proposition, experiments were performed to analyze the charge distributions of excited He I states resulting from direct and transfer excitation in He^+-He collisions.

MEASUREMENTS

Paul-trap promotion implies that the charge distribution of the collisionally excited state is asymmetric. The electron cloud is shifted towards the receding ionic collision partner. For analyzing the asymmetry of the charge distribution, we investigated collisional excitation in external electric fields -30 kV/cm$< F_z < +30$ kV/cm applied to the collision volume parallel ($F_z > 0$) and antiparallel ($F_z < 0$) to the ion beam. In particular, a beam of 50-200 keV He^+ ions was crossed with a thermal He-atomic beam. We measured the intensity $I_\lambda(F_z)$ of the He I spectral lines at $\lambda(1s4\ell - 1s2p \ ^3P) = 447$ nm emitted by excited target and projectile atoms as a function of the electric field F_z. These intensity functions are asymmetric with respect to the sign of F_z, if the charge cloud of the collisionally excited electron is not centered on the nucleus. If the charge cloud is centered downstream from the nucleus (i.e. if the electric dipole moment d_z of the atom is $d_z < 0$), typically an intensity function with $I_\lambda(F_z < 0) > I_\lambda(F_z > 0)$ is measured.

Three recordings of $I_{447}(F_z)$ are shown in Figure 2 for the electric-field ranges 15 kV/cm$\lesssim \pm F_z \lesssim 30$ kV/cm. This electric field range is of particular interest, because there is a series of singlet-triplet anticrossings of $1s4d$ and $1s4f$ Stark sublevels. At these anticrossings, singlet and triplet states are coupled and mixed due to the spin-orbit coupling of the excited electron. Therefore, a triplet-line component of the $1s4\ell - 1s2p\ ^3P$ transitions appears at the electric field of each anticrossing even if exclusively the singlet $1s4\ell$ states are populated. In the case of He$^+$-He collisions, where both singlet and triplet states are populated, the narrow intensity maxima at anticrossings indicate that the excitation rates for singlet states are larger than for triplet states.

For 200 keV He$^+$ impact the intensity function $I_{447}(F_z)$ is extremely asymmetric (Figure 2a). The anticrossing resonances measured for $F_z < 0$ are significantly larger than those measured for $F_z > 0$. This result indicates an asymmetric charge distribution of the collisionally excited $n=4$ states. At impact energies of 200 keV, direct excitation of target atoms is significantly more likely than transfer excitation of projectiles. Therefore, the measured asymmetry is essentially related to the charge distribution of excited target atoms. One deduces from the measured intensity function that the electric dipole moment $d_z(n=4)$ of the collisionally excited $n=4$ state is negative as expected according to the Paul-trap model.

A more symmetric intensity function was measured for 100 keV-He$^+$-He collisions (Figure 2b). However, this result does not indicate a symmetric charge distribution of the collisionally excited state. Rather, at an impact energy of 100 keV both target and projectile atoms contribute to the measured intensity function due to the fact that at this energy transfer excitation is much more likely than direct excitation.

This statement was proved by using different He isotopes for projectile and target beam. Figure 2c shows a recording of $I_{447}(F_z)$ for 75 keV-^{3}He$^+$-^{4}He collisions. Here, the relative velocity of projectile and target atom is the same as for 100 keV-^{4}He$^+$-^{4}He collisions. The asymmetry of the intensity function is now obvious. The anticrossing resonances of the target isotope ^{4}He are found almost exclusively for $F_z < 0$, whereas the anticrossing resonances of the projectile isotope ^{3}He are found for $F_z > 0$. Here we exploited the fact that due to the strong hyperfine coupling between nucleus and $1s$ electron different anticrossing resonances are observed for ^{4}He and ^{3}He $(n=4)$-states. The recording shown in Figure 2c indicates that both the excited target atoms and the excited projectile atoms have strongly asymmetric charge clouds, but with opposite electric dipole moments: $d_z < 0$ for the target and $d_z > 0$ for the projectile atoms. This result is an accord with the prediction of the Paul-trap model.

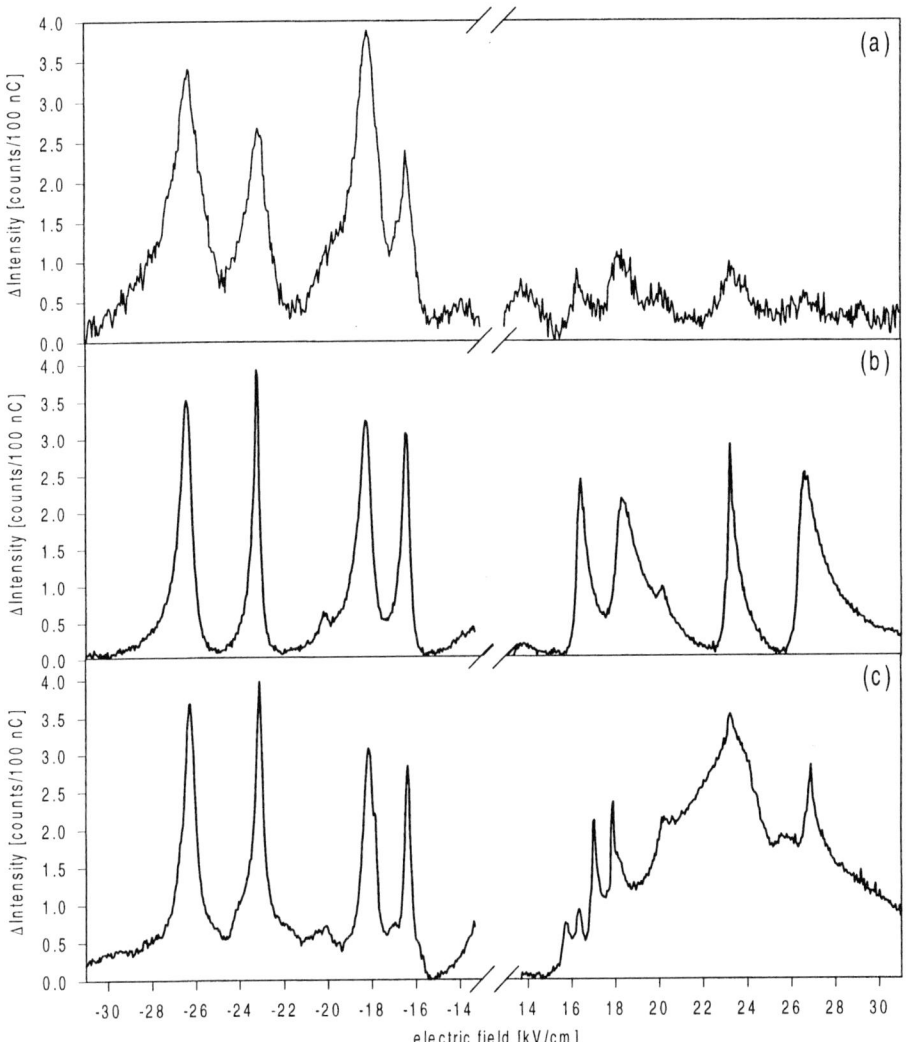

Figure 2: Intensity functions $I_{447}(F_z)$ measured for (a) 200 keV and (b) 100 keV $^4\text{He}^+$ impact and (c) 75 keV $^3\text{He}^+$ impact on thermal ^4He atoms.

NUCLEAR SYMMETRY OF $^4\mathrm{He}_2^+$

So far we assumed that the dynamical evolutions of the collision systems $^4\mathrm{He}^+ + ^4\mathrm{He}$ and $^3\mathrm{He}^+ + ^4\mathrm{He}$ are essentially equal provided the relative velocities of the collision partners are the same. However, this assumption is not fulfilled exactly. The collision systems differ in several respects: (i) The impact energies are different. In the center-of-mass (cm) system one finds $E_{cm}(^4\mathrm{He}^+ - ^4\mathrm{He})/E_{cm}(^3\mathrm{He}^+ - ^4\mathrm{He}) = 7/6$. (ii) The nuclei of $^3\mathrm{He}^+$ and $^4\mathrm{He}^+$ have different nuclear moments. (iii) The two nuclei of the collision system $^4\mathrm{He}_2^+$ are indistinguishable, whereas the nuclei of the system $^3\mathrm{He}^+$-$^4\mathrm{He}$ are distinguishable. There are good reasons to believe that (i) and (ii) can be disregarded in connection with our measurements. However, the nuclear symmetry of the $^4\mathrm{He}_2^+$ system deserves special attention.

To discuss Paul-trap promotion in the two-center field of two identical nuclei, let us - for the sake of simplicity - consider the one-electron system He^{++}-He^+. Without nuclear symmetry the initial state $|i\rangle$ of the collision system can be written approximately as:

$$|i\rangle = \exp(i\vec{K}\cdot\vec{R}_1)\cdot\exp(-i\vec{K}\cdot\vec{R}_2)\psi_{1s}(|\vec{r}-\vec{R}_2|) \tag{1}$$

It is a product of the plane-wave states of projectile nucleus 1 and target nucleus 2, to which a $1s$ electron is bound. (We neglect the translational motion of the electron.) For a collision system with nuclear symmetry, however, the initial state has to be symmetrized. For bosonic nuclei it reads:

$$|i\rangle_b = \frac{1}{\sqrt{2}}\left\{\exp(i\vec{K}\cdot\vec{R})\psi_{1s}(r_2) + \exp(-i\vec{K}\cdot\vec{R})\psi_{1s}(r_1)\right\}, \tag{2}$$

where we have put $\vec{R} = \vec{R}_1 - \vec{R}_2$ and $\vec{r}_i = \vec{r} - \vec{R}_i$. The state vectors of nucleus 1 and nucleus 2 are now entangled. In the first component of $|i\rangle_b$ particle 1 is the projectile and particle 2 the target nucleus, but in the second component particle 2 is the projectile and particle 1 is the target nucleus, to which the electron is bound. The question arises: to which extent does this symmetry-induced entanglement affect the collision process? With respect to Paul-trap promotion we rewrite the initial states $|i\rangle$ and $|i\rangle_b$ by introducing the asymptotic molecular orbitals:

$$|1s\sigma\rangle = \frac{1}{\sqrt{2}}\left\{\psi_{1s}(r_1) + \psi_{1s}(r_2)\right\} \tag{3}$$

$$|2p\sigma\rangle = \frac{1}{\sqrt{2}}\left\{\psi_{1s}(r_1) - \psi_{1s}(r_2)\right\} \tag{4}$$

One obtains:

$$|i\rangle = \frac{1}{\sqrt{2}}\exp(i\vec{K}\cdot\vec{R})\left\{|1s\sigma\rangle - |2p\sigma\rangle\right\} \tag{5}$$

$$|i\rangle_b = \cos(\vec{K}\cdot\vec{R})|1s\sigma\rangle - i\sin(\vec{K}\cdot\vec{R})|2p\sigma\rangle \tag{6}$$

Obviously, the state vectors $|i\rangle$ and $|i\rangle_b$ are remarkably different. For both the $|1s\sigma\rangle$ and the $|2p\sigma\rangle$ molecular orbital the relative nuclear motion is represented by a running wave $\exp(i\vec{K}\cdot\vec{R})$, if the nuclei are distinguishable. However, if the nuclei are identical, the relative nuclear motion is represented by standing waves. For $1s\sigma$ promotion, the initial motional state is a cosine (symmetric) standing wave, and for $2p\sigma$ promotion, it is a sine (antisymmetric) standing wave.

Experiments are in progress to investigate the influence of this symmetry-induced entanglement on the collision process. To this end, experiments on $^3\mathrm{He}^+$-$^4\mathrm{He}$ and $^4\mathrm{He}^+$-$^4\mathrm{He}$ collisions were performed, where care was taken that exclusively the decay of target atoms was observed. In that case, for both collision systems only the characteristic intensity functions of the thermal $^4\mathrm{He}$ atoms are measured. Any difference in the signal structure between the measurements on $^3\mathrm{He}^+$ and $^4\mathrm{He}^+$ impact would indicate that there is a difference between the two collisional excitation processes. So far, a significant difference has not been detected.

CONCLUSIONS

As for p-He and HCl-He collisions, we found that also for He^+-He collisions at intermediate energies the charge distribution of collisionally excited He I states is asymmetric. This result confirms the assumption that Paul-trap promotion is generally a dominant mechanism for single-electron excitation in ion-atom collisions in the intermediate energy range.

Presently the idea of Paul-trap promotion provides only an intuitive picture of the collision process allowing some qualitative conclusions. For evaluating excitation cross sections and coherence parameters also quantitatively, a sound quantum dynamical approach to this excitation mechanism is needed.

Acknowledgement. We thank The Polish Komitet Badan Naykowych for supporting this cooperation.

REFERENCES

1. von Oppen, G. et. al., *Aust. J. Phys.* **52**, 431 (1999).
2. Rost, J. M., and Briggs, J. S., *J. Phys. B* **24**, 4293 (1991).
3. Ashburn, J. R. et. al., *Phys. Rev. A* **41**, 2407 (1990).
4. Drozdowski, R. et. al., *J. Phys. B* **32**, 397 (1999).
5. Tschersich, M. et. al., *J. Phys. B* **32**, 5539 (1999).
6. Wolterbeek Muller, L. and de Heer, F. J., *Physica* **48**, 345 (1970).

Note added in proof: For explaining the relative (1:1) population of singlet and triplet levels in He^+ - He collisions, electron promotion on the singlet states of $(\mathrm{He}^+)_2$ cannot be disregarded. The Coulomb repulsion of the electrons supports that also for singlet states the core electrons separate with the nuclei during the final phase of the collision.

Progress with Optically Pumped Sources of Polarized Electrons

M.A. Rosenberry, H. Batelaan, J.P. Reyes, T.J. Gay

Behlen Laboratory of Physics, University of Nebraska, Lincoln, NE 68588-0111

Abstract. We report our work in developing new "turn-key" sources of polarized electrons. These sources operate by extracting the electrons from a discharge and polarizing them through optical pumping. Preliminary work demonstrates that beams of 4 µA with greater than 20% polarization are possible. Such devices could enormously simplify the running of many experiments in the fields of atomic structure, magnetic materials and biophysics.

INTRODUCTION

Polarized electrons are an extremely valuable tool in physics, and are used to study electron-atom scattering (e.g. in plasmas) [1], magnetism at surfaces (e.g. in nanostructures) [2], and the chiral structure of molecules [3]. At present the standard source of polarized electrons is a specially prepared GaAs crystal. A typical GaAs system produces currents of 20 µA with 30% polarization, though special crystals and extensive preparations can increase this to 100 µA at 70% polarization [4]. Such performance can also be obtained with a properly constructed flowing He* afterglow source [5].

Both of these sources have stringent vacuum requirements, so that maintenance of the system is time-consuming. More importantly, they are difficult to use, so that extensive training of their operating personnel is necessary. We have been developing sources of polarized electrons that operate at pressures of about a torr and require minimal operator sophistication. With the proper equipment, our source would be "turn-key" (i.e. trivial to operate) while providing an equivalent electron beam to present standard sources.

RUBIDIUM SPIN-EXCHANGE DEVICE

The body of our system is a Conflat™ 2¾" nipple that is 6" in length. To this we have attached inlets for rubidium, discharge gas, and the 400 V needed to run a cold cathode discharge. The system is heated to around 80 °C, giving a rubidium density of about 10^{12} per cc. The source chamber also contains between 0.4 and 2.0 torr of discharge gas (N_2, He, Ar or some combination). A weak electric field (1 V/cm) and a moderate magnetic field (400 G) act to steer the electrons towards the target region. En route, the electrons pass through the polarized rubidium vapor and undergo spin-

CP604, *Correlations, Polarization, and Ionization in Atomic Systems*
edited by D. H. Madison and M. Schulz
© 2002 American Institute of Physics 0-7354-0048-2/02/$19.00

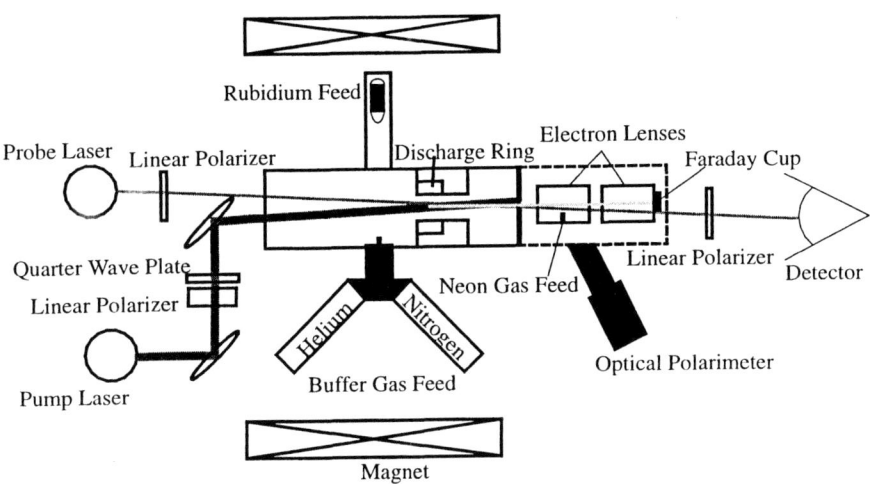

Figure 1: Rubidium Spin Exchange Apparatus

exchange collisions, causing them to acquire some of the rubidium polarization.

The source chamber is separated from the target region by an aperture of 0.9 mm, giving a pressure ratio of about 10^4. The target region also contains He or Ne at a pressure of 10^{-4} torr, which is excited by the extracted electron beam. The energy width of the electron beam (about 1eV) can be estimated from the target gas excitation function. The polarization of the light emitted by the target atoms is also used to determine the electron polarization through known formulae [6]. The electrons then continue along our lens train to a Faraday cup where the current is measured.

This system uses both a pump and a probe laser. The probe beam is at the Rb D2 transition (780 nm), and is used to determine the rubidium density and polarization through the optical Faraday rotation. At larger values of B (200 G), the rotation is dominated by the diamagnetic effect and determines the density, while at low values (20 G) the rotation is due to the paramagnetic effect and measures the polarization. The equations are given below, with δ being the detuning from resonance, $\Delta\varphi$ the rotation, L the cell length of 15 cm, and Γ is the natural linewidth of 5.4 MHz.

$$\eta_{Rb} = \frac{8\pi\delta^2(\Delta\varphi)}{L\Gamma\lambda^2\mu_B B} \quad ; \tag{1}$$

$$P_{Rb} = \frac{56\pi\delta(\Delta\varphi)}{3\eta_{Rb}L\Gamma\lambda^2} \quad . \tag{2}$$

The pump beam (at the D1 transition) optically pumps the rubidium [7]. For preliminary work we employed a 250 mW, 40 GHz wide dye laser for the pump beam. This device was highly temperamental, making "turn-key" operation impossible. We have since moved to a 10 W diode bar, which is temperature controlled with a homebuilt circuit. We have narrowed the frequency profile of this laser (see Figure 2)

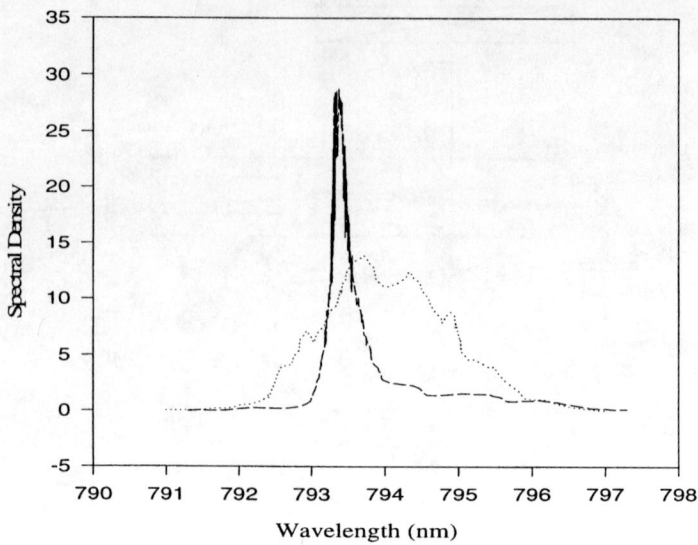

Figure 2: Narrowing with Grating Feed-back

according to the method of Walker et al. [8]. Once set up, this device provides the resonant light needed with only a few seconds of tweaking, as well as generating better rubidium polarizations than the dye laser.

We are also investigating the roles of the buffer gas. To begin with, the buffer gas determines polarization of the electrons. It does so by pressure broadening the rubidium absorption line, reducing the diffusion rates for both Rb and electrons, and moderating the energy of the electrons (which affects the spin exchange cross-section). The buffer gas also sets the discharge characteristics, which determine the current. Figure 3 shows how these effects compete: the rubidium polarization (which is proportional to the electron polarization) rises steadily with buffer gas pressure, while the current drops precipitously as the density increases.

Different buffer gases may affect the system in different ways. N_2 is the standard buffer gas for Rb optical pumping, since its vibrational levels allow it to effectively quench excited Rb to prevent radiation trapping [7]. Our buffer gas pressures (<1 torr) are sufficiently low that quenching effects are minimal, freeing us to consider other gases. For example, electrons diffuse through helium more readily than nitrogen, so that our currents improve when using a helium buffer gas. The rubidium polarization is somewhat reduced, but the overall figure of merit P^2I is still improved. Our best results to date (using the dye laser) were obtained in helium: 4uA with 23% polarization [7].

Argon, which has a Ramsauer minimum in its cross section, yields even larger currents and should give even better results. Alternately, a mix of gases could be used: nitrogen to moderate the electron energies and a heavy noble gas to improve optical pumping. Unfortunately we have been unable to test these hypotheses due to two

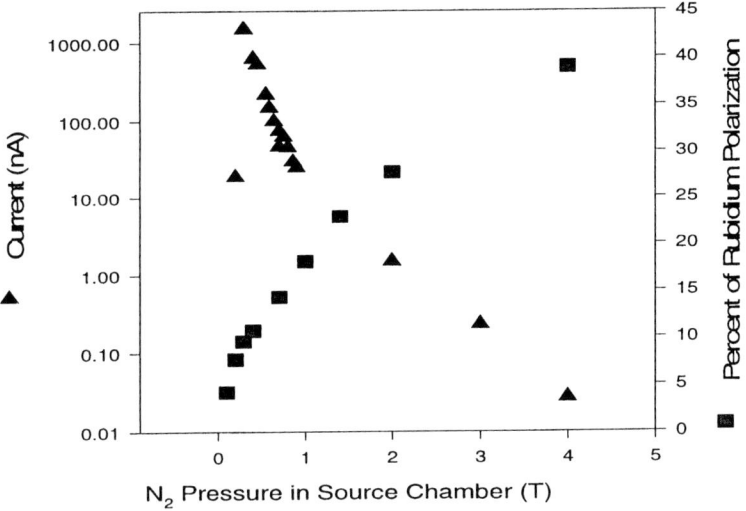

Figure 3: Current and Rubidium Polarization vs Nitrogen Pressure

difficulties. The first problem is the discharge, which can switch between different modes that have different characteristics. This has made systematic studies difficult. To overcome this, we are therefore building a new system that will generate the electrons with a filament, then polarize them through spin exchange. Work on this modification is in progress.

The other obstacle is that for some reason, we have been unable to produce electron polarizations of more than 2% using the diode laser. This has been very puzzling, but are hopeful that once the problem is resolved, the improved laser power and new gas mix will lead to a substantial improvement in the performance, making it nearly comparable with present standards.

METASTABLE NOBLE GAS SOURCE

We have also been developing a second possible source of polarized electrons. This system extracts polarized electrons directly from optically-pumped metastable atoms rather than through an intermediate species. It is therefore even easier to use than the one based on spin exchange with rubidium. Studies of this technique have been performed in the past using He, but resulted in only moderate performance; typically currents of tens of μA with less than 10% polarization [10].

It is not known exactly why such low values are obtained. However, one problem is certainly the difficulty in producing sufficiently large metastable densities while limiting the production of spurious (unpolarized) electrons. Our work leads us to believe that the key parameter is the ratio of metastable density to that of the ground state atoms. This parameter is crucial in setting the percentage of extracted electrons that have undergone a polarizing interaction. Therefore heavier noble gases, where these ratios are larger, should generate much better beams that He*.

To test this hypothesis we are developing a system that uses either argon or krypton instead of helium. The apparatus is extremely simple, being a glass tube connected to a small turbo pump via an aperture. Metal bands around the outside the cell create an RF discharge, while pump and probe lasers are used as described above. Typical discharge pressures are 30 mTorr of gas, and it is hoped that cleanliness requirements will not be as stringent as those for helium.

We have confirmed by direct measurement that the ratio (of metastable to ground state atoms) is better in Ar than that of He by an order of magnitude (see Figure 4). We are now in the process of determining how these modifications improve the electron polarization.

To conclude, our new source that relies on spin exchange with optically pumped rubidium to polarize our electron beam. This source shows great promise of being far easier to use and maintain than a GaAs crystal, making it highly desirable for many applications.

This work has been supported by NSF grant #PHY-9704650.

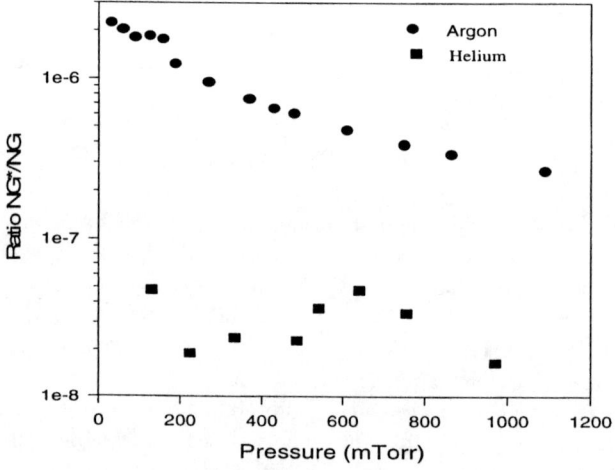

Figure 4: Metastable Fractions

REFERENCES

1. H.M. Al-Khateeb, B.G. Birdsey, and T.J. Gay, Phys. Rev. Lett. **85**, 4040 (2000)
2. See, e.g. M. Johnson, Science **260**, 320 (1993)
3. K.W. Trantham, M.E. Johnston, and T.J. Gay, J. Phys. B. **28**, L543 (1995)
4. *Polarized Gas Targets and Polarized Beams*, edited by R.J. Holt and M.A. Miller, AIP Conf. Proc. No. 421 (AIP, New York, 1998)
5. G.H Rutherford *et. al.*, Rev. Sci Instrum. **61**, 1460 (1990)
6. K. Trantham, M.S. thesis at University of Missouri-Rolla, 1993
7. W. Happer, Rev. Mod. Phys. **44** 169 (1972)
8. B. Chann, I. Nelson, T.G. Walker, "Frequency-Narrowed External Cavity Diode Laser Array Bar", preprint available at www-atoms.physics.wisc.edu
9. H. Batelaan, A.S. Green, B.A. Hitt, and T.J. Gay, Phys. Rev. Lett. **82**, 4216 (1999)
10. Ray Vandiver, Ph.D. Thesis, University of Missouri-Rolla, 1993
    M.V. McCusker, L.L Hadfield, and G.K. Walters , Phys. Rev. A **5** 177 (1972)

Electron Transfer to Individual Magnetic Substates of Multi-charged Ions

Hajime Tanuma

Department of Physics, Tokyo Metropolitan University, Tokyo 192-0397, Japan

Abstract. Polarization spectroscopy measurements of the $1s^2 3p$ states of lithium-like ions produced in collisions of helium-like ions with neutral gas target have been described for O^{6+} - He, N^{5+} - He, N^{5+} - H_2, and C^{4+} - H_2 systems. The degree of polarization P for the $3s\ ^2S_{1/2}$-$3p\ ^2P_{3/2}$ transition has been measured as a function of projectile velocities ranging between 0.34 and 0.55 au. The population ratio of magnetic substates in $^2P_{3/2}$ has been obtained from the polarization data. Collision system dependence observed in the degree of polarization is discussed.

INTRODUCTION

Charge transfer reaction in collisions of multi-charged ions with atoms and molecules is extensively investigated both theoretically and experimentally as one of the most active research subjects in the atomic and molecular physics. Identification of the initial and final states in the collision system is necessary in order to study the reaction mechanism in detail. For single electron transfer reaction, which can be represented by means of the following expression,

$$A^{q+} + B \to A^{(q-1)+}(nlm) + B^+,$$

the quantum numbers of the atomic orbital in which the electron is captured should be distinguished to specify the final states. The principal and azimuthal quantum numbers, n and l, can be analyzed by means of translational energy spectroscopy and optical spectroscopy, because the energy gaps between different nl states are larger than the energy resolution in each spectroscopy for many cases. On the other hand, magnetic substates with different magnetic quantum number, m, cannot be separated in energy by ordinary spectroscopic methods without a strong magnetic field, because all energies of different magnetic substates are completely degenerate in the absence of a magnetic field. For singly excited states, the measurement of the degree of polarization for photon emission is the only method for observing alignment which is determined by the population distribution of the magnetic substates of the excited states produced by electron transfer.

COLLISION SYSTEMS

In this paper, alignment of the $1s^2 3p\ ^2P$ states of lithium-like ions produced by the charge transfer in collisions of helium-like ions, these are O^{6+}, N^{5+}, and C^{4+}, with two-electron targets, namely He and H_2, is reported. In the following collision systems,

CP604, *Correlations, Polarization, and Ionization in Atomic Systems*
edited by D. H. Madison and M. Schulz

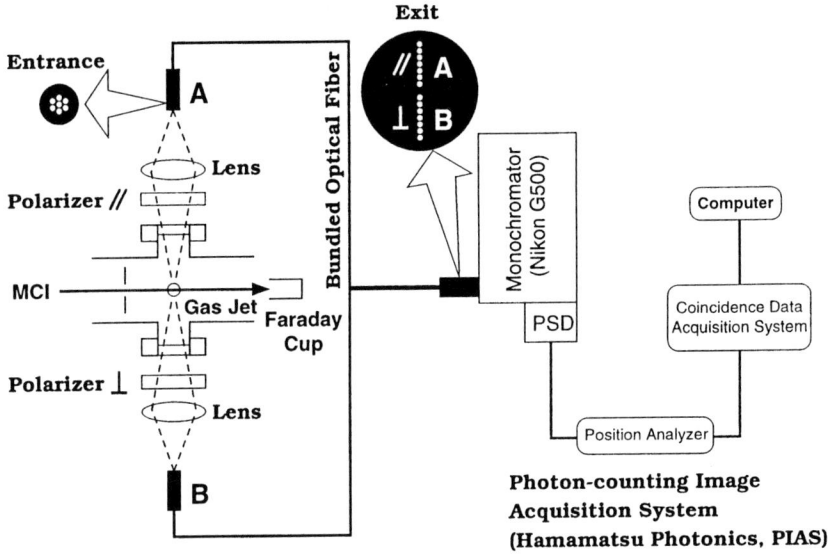

FIGURE 1. Overview of the experimental apparatus with the double-arm arrangement for measurements of the degree of polarization.

O^{6+} - He, N^{5+} - He, N^{5+} - H_2, and C^{4+} - H_2, the 3p state is directly populated by electron transfer and the cascades from the upper states can be neglected according to the cross section measurements for each nl states by means of the photon spectroscopy in the VUV region [1]. The alignment of the $C^{3+}(1s^2 3p\ ^2P)$ state produced in collisions of C^{4+} with H_2 had been investigated by Baptist *et al.* [2] and Hoekstra *et al.* [3] with VUV spectroscopy. Their results, however, are significantly different from each other.

EXPERIMENTAL PROCEDURE

The experimental arrangement is shown schematically in Fig. 1. Multi-charged ions O^{6+}, N^{5+}, and C^{4+} produced by a 14.25 GHz ECR (electron cyclotron resonance) ion source from pure O_2, N_2, and CH_4 gases, respectively. The ion beams extracted at high voltage of 7.5 - 20 kV from the ion source were analyzed by a magnet for the charge separation. The beam of the selected helium-like ions is directed into the collision chamber, where it intersects a gas jet of either He or H_2 at 90°. Typical ion beam currents were 2 - 15 μA as measured by a Faraday cup located beyond the collision region, and the target gas pressure in the collision chamber was held at about 1×10^{-4} Pa to maintain the single-collision conditions. optical radiation, emitted from the collision center and passing through synthetic quartz windows, is observed on both sides of the ion beam, at 90° to both the ion beam and the gas jet. Before the detection, the radiation passes through linear polarizers, oriented parallel to the ion beam on one side of the beam axis, and perpendicular to the ion beam on the other side. After passing through the polarizers, the radiation is forced by lenses into two branches of a single fiber optic bundle. The two branches are joined at the entrance slit of a monochromator (Nikon G500). In this way, both the parallel and perpendicular components of the emitted radiation are acquired simultaneously. The two components of the radiation are directed into the monochromator such that the radiation which is polarized parallel to the ion beam axis enters through the upper portion of the slit, while the radiation polarized perpendicular to the ion beam axis enters the spectrometer through the lower portion of the slit. After passing through the monochromator, the radiation is directed onto a position sensitive detector (PSD). The signal from the PSD is recorded with a photon-counting image acquisition system (PIAS, Hamamatsu Photonics).

Taking the ion beam direction as the quantization axis, the degree of polarization is defined as

$$P = \frac{I_{\parallel} i_{\perp} - I_{\perp} i_{\parallel}}{I_{\parallel} i_{\perp} + I_{\perp} i_{\parallel}}, \tag{1}$$

where $I_{\parallel}$ and $I_{\perp}$ are the intensities of the light passing through the respective polarizers in the parallel and perpendicular directions with respect to the ion beam direction. These intensities are normalized by $i_{\parallel}$ and $i_{\perp}$, which represent the intensities of light from an isotropic transition, passing through the same optical analyzers. The isotropic light used in this study is the line corresponding to the 3s $^2S_{1/2}$-3p $^2P_{1/2}$ transition. Some of experimental data with O^{6+} and N^{5+} ions have already published elsewhere [4, 5].

RESULTS AND DISCUSSION

Degree of polarization

Figure 2 shows a plot of the degree of polarization, P, for the line corresponding to the 3s $^2S_{1/2}$-3p $^2P_{3/2}$ transition versus projectile velocity in atomic units for four collision systems. Experimental errors, shown in the figure as bars, are estimated from only the statistical uncertainties in the photon counting. No significant collision velocity dependence is seen in this velocity range for all collision systems. The results, however,classify the collision systems into two categories: In the cases of O^{6+} - He and N^{5+} - H_2, P has a very small value and is close to 0 %. On the other hand, P has very large value around 20 % for the N^{5+} - He and C^{4+} - H_2 collisions. The result for $C^{3+}(1s^2 3p)$ in this study shows fairly good agreement with that of Hoekstra et al., but the detailed discussion about the comparison of data is omitted for want of space.

Population distribution in magnetic substates

The relation between the degree of polarization and the population distribution in the magnetic substates has been described in the literature [6]. In this experiment, we observe the photon emission from the 3p $^2P_{3/2}$ state which has four magnetic substates having different M_J values, namely -3/2, -1/2, +1/2, and +3/2. Because of the cylindrical symmetry around the ion beam axis, the population of M_J=-1/2 and -3/2 states should be

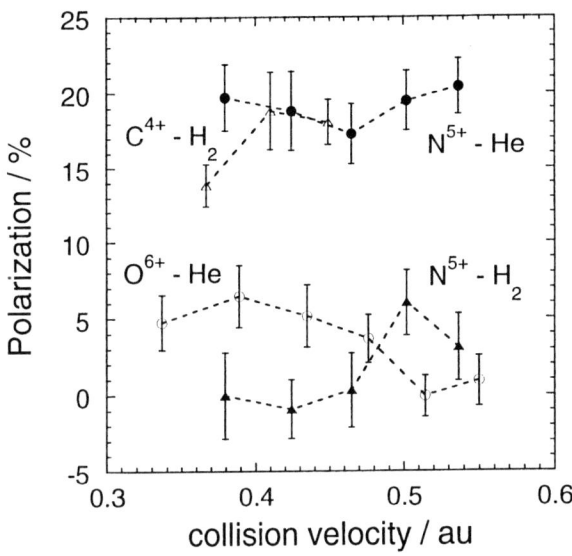

FIGURE 2. The degree of polarization for the line corresponding to the 3s $^2S_{1/2}$-3p $^2P_{3/2}$ transition as a function of the projectile velocity.

equal to that of $M_J=+1/2$ and $+3/2$, respectively. Therefore, using some formulas in the paper by Fano and Macek [6], we can deduce that the population ratio $\sigma_{1/2}$ to $\sigma_{3/2}$ in the 3p $^2P_{3/2}$ state is given as a function of the degree of polarization P for the transition from this state to $^2S_{1/2}$ by

$$R_{M_J} = \frac{\sigma_{1/2}}{\sigma_{3/2}} = \frac{3+3P}{3-5P}. \tag{2}$$

Figure 3 shows the population ratio R_{M_J} as a function of the collision velocity. Assuming the capture cross section to 3p $^2P_{3/2}$ is given as 2/3 of that to the 3p state, which had been measured by VUV spectroscopy, we can estimate the absolute value of the capture cross section to the individual magnetic substate using this population ratio.

Collision system dependence

The large positive values of P in the N^{5+} - He and C^{4+} - H_2 collisions can be understood as a general characteristic of the electron transfer reaction in collisions of multi-charged ions with neutral targets, which had been reported for various collision systems [7, 8, 9, 10, 11, 12, 13, 14, 15]. On the other hand, the very small P values for O^{6+} - He and N^{5+} - H_2 systems is difficult to explain with a simple model. However, as can be seen in Table 1, the crossing distances between the initial and final channels in the diabatic potential curves of the collision systems show the correlation with the P values, where the crossing distance is roughly estimated from the entrance channel

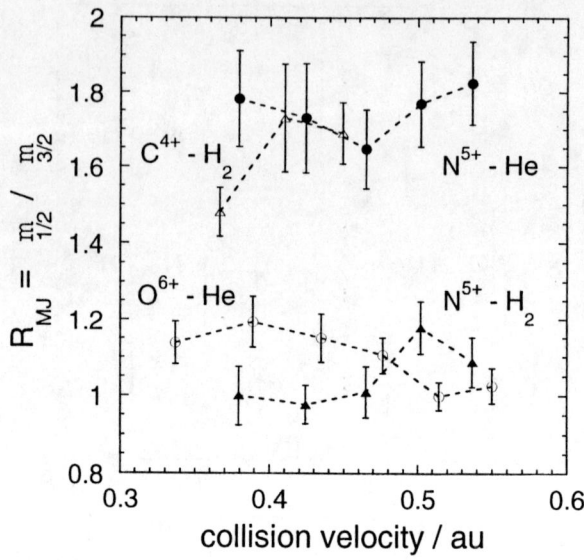

FIGURE 3. Population ratio $\sigma_{1/2}/\sigma_{3/2}$ of the magnetic substates in the 3p $^2P_{3/2}$ state as a function of the projectile velocity.

TABLE 1. Crossing distance in diabatic potential curves, electric field at the crossing, and P values for each collision system.

Collision system	Crossing distance / Å	Electric Field / GVm^{-1}	P
O^{6+}-He	2.3	27	small
N^{5+}-H$_2$	2,6	22	small
N^{5+}-He	4.1	8.6	large
C^{4+}-H$_2$	5.0	5.9	large

potential without any interactions between the projectile and the target and the exit channel potential taking only the Coulombic repulsion between ions into account. In this table, we also show the strengths of the electric field at the potential crossing point by the residual target ion produced by the single electron transfer to the projectile ion, which the excited projectile ion feels just after the electron capture. As a highly probable cause of the small P values, we are considering the mixing of different nl states through the linear Stark effect in electric fields by residual target ions. The importance of this effect has been shown in theoretical calculation by Salin [16]. It is quite reasonable to think that stronger electric field with smaller distance should make the initial alignment smaller, because the mixing of different nl states should creates a near-uniform distribution in magnetic substates.

ACKNOWLEDGMENTS

This work became possible due to the contribution of graduate students, T. Hayakawa, H. Kano and C. Verzani (Kansas State University), and due to the collaboration with H. Watanabe (Japan Science and Technology Corporation), B. D. DePaola (Kansas State University) and N. Kobayashi. The author and coworkers would like to thank K. Sato (National Institute for Fusion Science) for allowing us to use the photon counting system. This work was financially supported in part by the Japanese Ministry of Education, Science, Sports and Culture, Grant-in-Aid for International Science Research (University-to-University Cooperative Research).

REFERENCES

1. Dijkkamp, D., Ćirić, D., Vlieg, E., de Boer, A., and de Heer, F. J., *J. Phys. B: At. Mol. Phys.* **18**, 4763-4793 (1985).
2. Baptist, R., Bonnet, J. J., Chauvet, G., Desclaux, J. P., Dousson, S., and Hitz, D., *J. Phys. B: At. Mol. Phys.* **17**, L417-L421 (1984).
3. Hoekstra, R., Suraud, M. G., de Heer, F. J., and Morgenstern, R., *J. Physique* **50**, C1/387-392 (1989).
4. Tanuma, H., Hayakawa, T., Verzani, C., Kano, H., Watanabe, H., DePaola, B. D., and Kobayashi, N., *J. Phys. B: At. Mol. Opt. Phys.* **33**, 5091-5098 (2000).
5. Hayakawa, T. Lomsadze, R. A., Verzani, C., Watanabe, H., Tanuma, H., DePaola, B. D., and Kobayashi, N., *Physica Scripta* **T92**, (2001) *in press*.
6. Fano, U., and Macek, J. H., *Rev. Mod. Phys.* **45**, 553-573 (1973).
7. Vernhet, D., Chetioui, A., Wohrer, K., Rozet, J. P., Piquemal, P., Hitz, D., Dousson, S., Salin, A., and Stephan, C., *Phys. Rev. A* **32**, 1256-1259 (1985).
8. Lembo, L. J., Danzmann, K., Stoller, Ch., Meyerhof, W. E., and Hänshe, *Phys. Rev. Lett.* **55**, 1874-1876 (1985).
9. Lembo, L. J., Danzmann, K., Stoller, Ch., Meyerhof, W. E., and Hänshe, *Phys. Rev. A* **37**, 1141-1151 (1988).
10. Roncin, P., Adjouri, C., Gaboriaud, M. N., Guillemot, L., Barat, M., and Andersen, N, *Phys. Rev. Lett.* **65**, 3261-3264 (1990).
11. Liu, C. J., Dunford, R. W., Berry, H. G., and Church, D. A., *Phys. Rev. A* **43**, 572-574 (1991).
12. Schippers, S., Hoekstra, R., Morgenstern, R., and Olsen, R. E., *J. Phys. B: At. Mol. Opt. Phys.* **29**, 2819-2836 (1996).
13. Laulhé, C., Jacquet, E., Boduch, P., Chantepie, M., Cremer, G., Gherardi, N., Husson, X., Lecler. D., and Pascale, J., *J. Phys. B: At. Mol. Opt. Phys.* **30**, 1517-1530 (1997).
14. Jacquet, E., Kucal, H., Bazin, V., Boduch, P., Chantepie, M., Cremer, G., Laulhé, C., Lecler. D., and Pascale, J., *Phys. Rev. A* **62**, 022712 (2000).
15. Bazin, V., Boduch, P., Chantepie, M., Cremer, G., Jacquet, E., Kucal, H., Lecler. D., and Pascale, J., *Phys. Rev. A* **62**, 052706 (2000).
16. Salin, A., *J. Physique* **45**, 671-680 (1984).

Theoretical and Experimental Evidence of the Heisenberg Angular Quantization in Helium ?

Yu.V.Popov[1] and L.U.Ancarani[2]

[1]*Institute of Nuclear Physics, Moscow State University, Moscow 119899 Russia*
[2]*L.P.M.C., Institut de Physique, Technopôle 2000, 57078 Metz, France*

Abstract. The results of several recent experiments which probe the asymptotitc part of the helium wavefunction are compared. They all seem to indicate that the mutual angle between the two electrons is asymptotically quantized, and takes a value in the range $120^0 - 150^0$, thus supporting a recently proposed idea.

In our recent paper [1] we have studied the structure of the formal solution of the helium S-bound states. In particular, the analysis of the asymptotic behaviour showed that when the two correlated electrons "move" far from the nucleus they do so with a constant mutual angle θ_{12}. Here we illustrate how this property is supported by several recent experiments on the helium atom in its ground state.

First of all, it seems to us important to remind that there is no formal solution to the three-body problem represented by the two electrons and the helium nucleus. Moreover, no probing wave function used up to now in practical calculations has the correct behaviour both at small and large hyperradius ρ. The quality of the helium wavefunction can be judged with several criteria like having the energy close to the experimental value, having a small energy variance, fulfilling the cusp conditions, having the correct asymptotic behaviour, or having a "local" energy as constant as possible. Depending on the process one looks at, the wavefunction may be particularly probed in a specific region of the six-dimensional space $(\vec{r}_1, \vec{r}_2)$. From a theoretical point of view, the wave function describing the initial state of helium appears in the calculation of matrix elements for a given physical process, so that the information on the whole space is included, and somehow "averaged out" by the integration. By this we mean that, a probing wave function which is not a formal solution of the Hamiltonian may provide a reasonable description in a large portion of the $(\vec{r}_1, \vec{r}_2)$ space, being poor in others (for example not satisfying the cusp conditions). In many situations, however, the deficiencies do not show up in the final calculation of physical quantities, for example because they are in a region of the space relatively less relevant for the studied process, or because they somehow cancel out in the calculations.

CP604, *Correlations, Polarization, and Ionization in Atomic Systems*
edited by D. H. Madison and M. Schulz

Simple bound state wavefunctions, used with success for calculations of elastic and inelastic scattering, are not necessarily satisfactory to describe the details of many-electrons ionization processes. In many experimental configurations for (e,2e), (γ,2e) or (e,3e) processes, conventional wave functions (Hylleraas type or multi-configuration Hartree-Fock type) are also adequate, giving theoretical results in reasonable agreement with the measurements. However, as noted in [2-3], this is not always the case. In this contribution, we comment on a few experiments which particularly probe the asymptotic behaviour of the initial helium wave function and put in evidence an interesting property.

Because of the long range nature of the Coulomb interaction, the motion of the two electrons is correlated out to infinity. To observe and probe the behaviour of the wave function at large distances from the nucleus, one needs experimental situations where the transferred momentum is small. In [1] we have already mentioned the (e,2e) experiments in the high energy dipolar regime [4] with small ejected energies and momentum transfers. The persistent discrepancy in the vicinity of the recoil peak may be attributed [2] to the asymptotic behaviour of the wave functions. Let us now turn to some other recent set of experiments.

First, the (e,3e) measurements at high impact energies E_0 (3-8 keV) and small momentum transfers (dipolar kinematics) of the Freiburg [5] and the Paris [6-8] groups. The measurements of both these groups show that the two slow ejected electrons (labeled 1 and 2, and can be safely considered as originating from the bound target) tend to escape at a specific mutual angle θ_{12}, in particular when they escape with the same energy. More precisely, in the Freiburg experiments (based on the COLTRIMS technique [9]), for $E_0 \sim 3$ keV, the two-fold differential ionization cross section as a function of θ_{12} and of the energy partition $E_1/(E_1 + E_2)$ illustrates that the peak appears at $E_1 = E_2$ and θ_{12} in the range $130 - 150^0$ (Fig. 2 in [5]). For a momentum transfer smaller than 0.5 a.u., the most probable mutual angle is about 135^0 (Fig. 3 in [5]). In one of the Paris measurements, for $E_0 \sim 5.5$ keV and $E_1 = E_2 = 10$ eV, the 5DCS plotted as a function of the mutual angle θ_{12} - irrespective of the direction of each electron - peaks again at about the same angle (Fig. 4 in [6]). Furthermore, still in dipolar kinematics ($E_0 \sim 5.5$ keV, momentum transfer 0.22 and 0.24 a.u.), when one looks at the 5DCS for a given angle of one the ejected electrons, the other will predominantly be observed at this angle $\pm$ a quantity of the order of $120^0 - 140^0$ [9]. In a further recent experiment at $E_0 \sim 1$ keV [8] the two electrons are detected in the symmetric geometry where they emerge with equal energy (10 eV) and at opposite angles with respect to the incident direction. The cross section seem to indicate a trend qualitatively consistent with the previous ones.

Secondly, there has been many (γ,2e) coincidence measurements with relatively small energies above threshold (for a recent review we refer to [10]. In this process, the two ejected electrons definitely originate from the helium atom; moreover the photo-double ionization takes place only if the initial state is correlated, and therefore (γ,2e) measurements provide direct information on the electron-electron correlation. As it is well known, this process is similar to (e,3e) in the limit of high incident energy dipolar kinematics i.e. at small momentum transfers, and we would then expect similar results. Without going into any details we indeed observe, again, in many experimental data for equal energy sharing and low ejected energies a mutual angle θ_{12} which is rather constant, and in the range $120^0 - 140^0$ (see [10] and references therein).

In the theories for the double ionization processes considered above, the principal matrix element describes the transition between the initial and the final three-body states. For small momentum transfers, the fast projectile does not disturb substantially the target, but leaves enough energy to make the two bound electrons escape into the continuum. The long-range Coulomb interaction that influences their propagation from the atomic volume to the detectors can be scrutinized within a semiclassical approach [11]. For slow escaping electrons, the distortion depends on the initial mutual angle: the larger the angle the smaller the deflection from straight paths. More important deflections may appear for electrons with small mutual angle, but this situation is less favoured in the first place because of the cusp property of the exact bound state wavefunction. We may therefore consider that large mutual angles (*e.g.* in the range $120^0 - 150^0$) measured by the detectors correspond to practically the same angle of departure from the atomic volume.

In a further and quite different experiment [12], the Frankfurt group has measured the angular distribution of fragments of the reaction $p + He \rightarrow H + e + He^{++}$ using again the COLTRIMS technique. Since the incident proton (which becomes an hydrogen atom) gets only very slightly deflected (scattering angles below 5.5×10^{-4} rad), it means it passes far from the helium nucleus ($\sim 2-3$ a.u.) and therefore provides us with yet another probe of the asymptotic behaviour on the helium wavefunction. The experimental conditions suggest that one of helium electrons is resonantly captured by the proton ($v_e \approx v_p$). For all proton energies (range 150-1400 keV) it was found that the angle θ_{12} between the ejected electron and the hydrogen is approximately constant, about $135^0 \pm 15^0$ [13].

All the experimental evidence presented above seem to indicate that, at large distances from the nucleus, the angle separating the two electrons is somehow fixed at a given value in the range $120 - 150^0$ although always with a certain spread. One has also to take into account the uncertainty of each measurement and the influence of the final state. Since in the independent electrons model such property does not appear it can then be identified as a signature of their correlation. No "traditional" wave function proposed or used so far predicts such a behaviour. In the paper [1] we have put forward the idea that "the triangle in the energy space formed by two vectors

$$\vec{p_1} = \sqrt{2\varepsilon}\,\cos\theta\frac{\vec{r_1}}{r_1}\,, \qquad \vec{p_2} = \sqrt{2\varepsilon}\,\sin\theta\frac{\vec{r_2}}{r_2}\,, \qquad (1)$$

with fixed mutual angle θ_{12} fully determines the asymptotic dynamics of both electrons. Since the sides of the triangle are quantum numbers, the angle θ_{12} becomes perhaps the asymptotical quantum number of the system" (ε is the energy, θ fixes the energy sharing, and p_1 and p_2 are the "momenta" of electrons 1 and 2).

In paper [1], this constant asymptotic angle is a mathematical integral of motion of a partial differential equation; however, there is no straightforward way to attribute a numerical value to the physical angle θ_{12}. Only the full mathematical resolution of the Schrödinger equation, with inclusion of boundary conditions, could provide the quantization of the energy and of this mutual angle. As the energy is concerned, this is similar to the text-book case of the hydrogen atom: the energy is a constant of motion for the electron-nucleus system; by imposing boundary conditions on the possible solutions one obtains the well known quantization of the energy. Formally, a similar procedure can be undertaken for helium (for example following the algorithm suggested in [1]); however, this is no easy task, and no rigorous solution of the three-body helium problem is yet available. In the meantime, one may take the experimental value of the energy $E = 2.904$ a.u. and think of some additional quantization rule from which a consistent picture may be derived, in particular with respect to the observed asymptotic property.

In analogy to the quantization of the energy levels of the hydrogen atom, attempts have been made to quantize the two-electron periodic orbits for the helium ground state, in particular Heisenberg's proposition to Sommerfeld in 1922 (see the detailed and interesting discussion of the historical developments in Sect. II.A of [14]). This approach failed (and was then abandoned) as it was based on a series of unproven beliefs and heuristic rules. We would like to point out that our suggested quantization of the asymptotic mutual angle is of a different nature since it originates from a rigorous study of the asymptotic solutions of the Schrödinger equation [1].

Further information was gained from the multipolar analysis proposed in [2] and [3]. Indeed, it indicates that high multipoles may provide a considerable contribution to the FBA matrix element. Since these terms are important at large distances, one may say that both the correlated electrons in the helium ground state "sit" mainly far from nucleus. Such a picture reminds us of the structure of halo–states in nuclear physics [15] and is quite different from the Hartree–Fock description. Although surprising, this novel view of the helium atom may be able to explain the experiments of Lahmam-Bennani's experimental team at small momentum transfer and small energy of ejected electron, in particular the (e,2e) experiment (ratio of binary to recoil peaks) [4] and (e,3e) experiment ("mutual angle mode") [7].

We have presented a series of qualitative observations on experimental results which seem to support the picture of a constant mutual angle predicted in [1]. Further investigations to fully understand this property is under way. For the moment, we may say that coincidence measurements such as those described above illustrate that it may be necessary to reconsider the simple atomic models used up to now, and possibly to accept the idea of an "asymptotic angular" quantization although not necessarily with Heisenberg's approach.

ACKNOWLEDGEMENTS

We are grateful to A. Lahmam-Bennani, H. Schmidt–Böcking and collaborators for providing their experimental results prior to publication, and for fruitful discussions.

REFERENCES

1. Yu.V. Popov and L.U. Ancarani, Phys.Rev. A **62**, 42702 (2000).
2. Yu.V. Popov *et al* , Few Body Systems suppl. **10**, 235 (1999).
3. Yu.V. Popov, C. Dal Cappello and L.U. Ancarani, in *Many-particle spectroscopy of atoms, molecules, clusters and surfaces*, Kluwer Academics, London, 2001, pp. 291-306.
4. C. Dupré *et al* , J. Phys. B **25**, 259 (1992).
5. A. Dorn et al. Phys. Rev. Lett. **82**, 2496 (1999).
6. I. Taouil et al. Phys. Rev. Lett. **81**, 4600 (1998).
7. A.S. Kheifets *et al* , J. Phys. B **32**, 5047 (1999); A. Lahmam-Bennani *et al* , Phys. Rev. A **59**, 3548 (1999).
8. A. Lahmam-Bennani, private communication.
9. R. Dörner *et al* , Phys. Rep. **330**, 95 (2000).
10. J.S. Briggs and V. Schmidt, J. Phys. B **33**, R1 (2000).
11. Yu.V. Popov and J.J. Benayoun, J. Phys. B **14**, 3513 (1981); Yu.V. Popov and V.F. Erokhin, J. Phys. Lett. A **97**, 280 (1983).
12. V. Mergel *et al* , *Many-particle spectroscopy of atoms, molecules, clusters and surfaces*, Kluwer Academics, London, 2001.
13. H. Schmidt-Böcking *et al* , private communication.
14. G. Tanner, K. Richter and J-M. Rost, Rev. Mod. Phys. **72**, 497 (2000).
15. M.V. Zhukov *et al* , Phys. Rep. **231**, 151 (1993).

On the Tricomi Expansion in Rydberg Collisions

D. P. Dewangan

Physical Research Laboratory, Navrangpura, Ahmedabad 380 009, India

Abstract. Expressions of the first Born cross sections for Rydberg transitions contain hypergeometric functions of large arguments. Their numerical computation becomes difficult due to severe cancellation errors. The use of their asymptotic expression overcomes this problem and helps analytical studies. We examine two such asymptotic approaches, the Tricomi expansion and Jacobi polynomial representation methods. We show that the asymptotic expression of the Jacobi polynomial representation gives remarkably accurate results over the entire range of the physical momentum transfer whereas that of the Tricomi expansion method gives a good estimate only in the region near the minimum momentum transfer.

INTRODUCTION

It is well known that by using conventional methods the first Born amplitude for transition between arbitrary states may be expressed as a series containing terminating hypergeometric functions, i.e. a terminating series of algebraic terms. For high Rydberg states the series contains many terms and its numerical computation poses difficulty because of severe cancellation errors. As a result, the conventional expression becomes unsuitable for numerical evaluation. In addition, the conventional expression does not usually lead to an asymptotic formula of the first Born amplitude. These difficulties can be surmounted if compact expressions can be derived. It turns out that finding compact expressions of the first Born amplitude in a form suited to the study of transitions between high Rydberg states is a challenging problem and so far such expressions have been obtained for only a few sets of transitions. It is found that these expressions contain the terminating hypergeometric functions whose 2 or 3 arguments are large, and in general, the hypergeometric functions are of such a form that even their asymptotic expression is not available in the literature. Therefore, considerable attention has been paid to deriving their asymptotic form [1]. Of particular interest is an asymptotic expansion method [2] based on the Tricomi expansion, that has been employed in the calculation of the first Born cross section [2-3]. Recently, another approach [4] is developed in which the terminating hypergeometric function is expressed as a Jacobi polynomial [5]

CP604, *Correlations, Polarization, and Ionization in Atomic Systems*
edited by D. H. Madison and M. Schulz
© 2002 American Institute of Physics 0-7354-0048-2/02/$19.00

and then the known properties of the latter are exploited to obtain an asymptotic expression of the terminating hypergeometric function of large arguments.

Evidently, the accuracy of the first Born cross section depends on these asymptotic expressions. Since similar hypergeometric functions also appear in quantum mechanical matrix elements of other physical quantities of Rydberg atoms, their asymptotic expression is of much wider interest. The main aim of this work to study the region of validity of the asymptotic expressions of the Jacobi representation and Tricomi expansion methods. Through this article atomic units are used.

FIRST BORN THEORY FOR RYDBERG TRANSITIONS

The first Born cross section for excitation of hydrogen by electron impact from the initial state of the principal quantum number n to the final state of the principal quantum number n' is written as [3]

$$\sigma_{nn'} = \frac{8}{n^2 k^2} \int_{k-k'}^{k+k'} \frac{G(q)}{q^3} dq, \tag{1}$$

where $\mathbf{k}$ is the initial wave vector, $\mathbf{k}'$ is the final wave vector, $\mathbf{q} = \mathbf{k} - \mathbf{k}'$ is the momentum transfer so that the incident energy is $k^2/2$. The quantity $G(q)$ is the sum of the squared form factor,

$$G(q) = \sum_{lml'm'} |\langle n'l'm'| \exp(i\mathbf{q} \cdot \mathbf{r})|nlm\rangle|^2, \tag{2}$$

where l and m are the angular momentum and magnetic quantum numbers of the initial state of hydrogen and l' and m' are the corresponding quantum numbers of the final state. An expression of $G(q)$ is obtained [2-3] as

$$G(q) = \frac{1}{q} \int_0^q A(q')dq', \tag{3}$$

where $A(q)$ is expressed in terms of the hypergeometric functions of the form $F(-n; -n'; 1; z)$ and their derivatives with respect to q. Here

$$z = -\frac{4}{(nn')\left[\left(\frac{\Delta}{nn'}\right)^2 + q^2\right]} = -\frac{4nn'}{\Delta^2 + p^2} = -y, \quad \Delta = n' - n, \quad p = qnn'. \tag{4}$$

When Δ/n and p/n are small, z is large (negative number) and out of 4 arguments of $F(-n; -n'; 1; z)$, 3 arguments, namely $-n$, $-n'$ and z, are large. However, when Δ/n is small and q is finite, $z \to -4/(qnn')^2$, which is a small negative number.

To simplify the calculation of $A(q)$, the Tricomi asymptotic expansion method is used to write (correcting the printing errors of [2-3])

$$F(-n; -n'; 1; z) \longrightarrow C(-1)^n y^{(n+n')/2} e^{\frac{n}{2y}} J_\Delta(x), \tag{5}$$

where

$$x^2 = \frac{2n(n+n'+1)}{y},$$

$$C = 2^{\Delta/2}[n(n+n'+1)]^{-\Delta/2}\frac{\Gamma(n'+1)}{2\Gamma(n+1)}\left[1+\frac{\Gamma^2(n+1)}{\Gamma(n+\Delta+1)\Gamma(n-\Delta+1)}\right].\quad(6)$$

Recently, the first Born amplitude for transition between arbitrary circular states is derived in a compact form containing only one terminating hypergeometric function [4]. Using the relationship between the hypergeometric function and the Jacobi polynomial, an expression of the first Born amplitude containing only one Jacobi polynomial is obtained. The known properties of the Jacobi polynomials [5] are then exploited to deduce the asymptotic form and other properties. This method has also been used to study the behaviour of the dipole matrix element as well as the first Born amplitude for transition between arbitrary parabolic states. In particular, the Jacobi polynomial representation method has been used to deduce the formula of the dipole matrix element of the correspondence principle method from the corresponding quantum mechanical formula [3]. Therefore, it is of interest to examine the accuracy of the asymptotic expression of the asymptotic expression of $F(-n; -n'; 1; z)$ given by the method based on the Jacobi polynomial representation.

To represnt $F(-n, -n'; 1; z)$ as a Jacobi polynomial , it is convenient to introduce the variable θ,

$$\cot^2\frac{\theta}{2} = -z = y, \quad \sin^2\frac{\theta}{2} = \frac{\Delta^2+p^2}{\lambda^2+p^2}, \quad \cos^2\frac{\theta}{2} = \frac{4nn'}{\lambda^2+p^2}, \quad \lambda = n+n'. \quad (7)$$

Using a standard formula [5], one obtains

$$F(-n, -n'; 1; z) = \left(-\sin^2\frac{\theta}{2}\right)^{-n}P_n^{(\Delta,0)}(\cos\theta). \quad (8)$$

For $\alpha > -1$ and β an arbitrary *real* number, the asymptotic expression of $P_n^{(\alpha,\beta)}(\cos\theta)$ for large n is given by [5]

$$P_n^{(\alpha,\beta)}(\cos\theta) = \frac{\Gamma(n+\alpha+1)}{n!\left[n+\frac{(\alpha+\beta+1)}{2}\right]^\alpha}\sqrt{\frac{\theta}{2}}\frac{J_\alpha\left(\left[n+\frac{\alpha+\beta+1}{2}\right]\theta\right)}{\left(\sin\frac{\theta}{2}\right)^{\alpha+\frac{1}{2}}\left(\cos\frac{\theta}{2}\right)^{\beta+\frac{1}{2}}} + \theta^{\frac{1}{2}}O\left(\frac{1}{n^{\frac{1}{2}}}\right). \quad (9)$$

We see from Eq. (7) that for $(\Delta, p) \ll \lambda$, $\theta/ \sim O(1/\lambda)$ and the remainder $\theta^{\frac{1}{2}}O(n^{-\frac{1}{2}})$ is of order n^{-1}. But when θ is not too small, the remainder is of order $n^{-\frac{1}{2}}$.

Application of (9) to (8) leads to

$$F(-n, -n'; 1; z) = (-1)^n\frac{2^\Delta}{(n+n'+1)^\Delta}\frac{\Gamma(n'+1)}{\Gamma(n+1)}\sqrt{\frac{\theta}{2}}\frac{J_\Delta\left(\left[\frac{n+n'+1}{2}\right]\theta\right)}{\left(\sin\frac{\theta}{2}\right)^{n+n'+\frac{1}{2}}\left(\cos\frac{\theta}{2}\right)^{\frac{1}{2}}}. \quad (10)$$

Next, we show that in the limit as $\Delta/n \to 0$ and $p/n \to 0$, the asymptotic expressions (5) and (10) become equivalent. We have already seen that in this limiting case y i.e. $\cot^2(\theta/2)$ is very large, meaning that $\theta/2$ is very small. Hence we may put $\tan(\theta/2) \approx \sin(\theta/2) \approx \theta/2 \approx 1/\sqrt{y}$. Consequently,

$$\sqrt{\frac{\theta}{2}} \frac{1}{\left(\sin\frac{\theta}{2}\right)^{n+n'+\frac{1}{2}} \left(\cos\frac{\theta}{2}\right)^{\frac{1}{2}}} \approx \frac{1}{\left(\sin\frac{\theta}{2}\right)^{n+n'}} \to y^{\frac{(n+n')}{2}}.$$

We also have $e^{n/(2y)} \to 1$. We may note that $(n+n'+1) \equiv \sqrt{(n + n' + 1)^2} \approx \sqrt{2n(n + n' + 1)}$. Hence $(n+n'+1)\theta/2 \approx \sqrt{2n(n + n' + 1)/y} = x$. Therefore, $J_\Delta([n + n' + 1]\theta/2) \to J_\Delta(x)$. We also see from (6) that in this limit, $C \to 1$. In Eq. (10) in the same limit, the factor $\frac{2^\Delta}{(n+n'+1)^\Delta} \frac{\Gamma(n'+1)}{\Gamma(n+1)} \to 1$. Consequently, the Jacobi and Tricomi asymptotic expressions assume the same limiting form, $F(-n; -n'; 1; z) \longrightarrow (-1)^n y^{(n+n')/2} J_\Delta(x)$.

However, an important case of practical importance is that when n and n' are not very large and q is not very small. Then the two expressions, (5) and (10), become different.

RESULTS AND CONCLUSION

To test the accuracy of the expressions (5) and (10), we have done some numerically 'exact' calculations of $F(-n; -n'; 1; z)$ using extended (quadruple) precision. To keep a check on the accuracy of the computation, we have transformed $F(-n; -n'; 1; z)$ into 4 other forms, namely $F(-n; -n; 1 + \Delta; 1/z)$, $F(-n; -n; -n - n'; 1 - z)$, $F(\cdots; 1/[1 - z])$ and $F(\cdots; z/[1 - z])$. We have numerically calculated all these 5 forms of the hypergeometric functions together with the expression (8), its Jacobi polynomial representation. It is found that, in general, the direct numerical computation of $F(-n; -n'; 1; z)$ gives a very large value and many times overflow occurs. Therefore we have computed the reduced quantity $F(-n; -n'; 1; z)/y^{(n+n')/2}$. In these calculations, the expression (8) and the form containing $F(-n; -n; 1 + \Delta; 1/z)$ are found to be the most stable. The exact numerical value was taken when at least two out of these six expressions yielded the numbers agreeing in not less than ten (usually many more) significant figures. Table 1 gives the results of our calculations when the incident electron energy E is 2 times the threshold energy (i.e. $E = 1/n^2 - 1/n'^2$ au). We see that the expression (5) is remarkably accurate over the entire range of the physical momentum transfer q $[q_{min} = k - k', q_{max} = k + k']$. In contrast, the expression (5) based on the Tricomi expansion gives unrealistically large values for large momentum transfer. The main source of the large number is the factor $e^{\frac{n}{2y}}$. It is only for small momentum transfer in the region $q \sim q_{min} = k - k'$ that the expression (5) provides a good estimate.

In conclusion, the asymptotic expression of the Jacobi polynomial representation method is striking accurate.

TABLE 1. Comparison of the exact scaled $_2F_1(-n, -n', 1, z)/|z|^{(n+n')/2}$ for $n \to n'$ transition for the incident electron energy $E/E_{th} = 2$ when the momentum transfer q is $k - k', k$ and $k + k'$ with the corresponding results of the Jacobi and Tricomi methods.

Transition $n \to n'$	q	Exact value of $\frac{_2F_1(-n,-n',1,z)}{y^{(n+n')/2}}$	Asymptotic Jacobi expression (10)	Asymptotic Tricomi expression (5)
$10 \to 12$	$k - k'$	0.59433	0.59262	0.74352
$10 \to 12$	k	1.6471	1.6516	9.5466
$10 \to 12$	$k + k'$	-5.8144	-5.4215	8422.9^a
$15 \to 17$	$k - k'$	-0.46889	-0.46790	-0.61516
$15 \to 17$	k	-0.38965	-0.38047	-3.3009
$15 \to 17$	$k + k'$	1.8627	2.1230	10698.0
$50 \to 52$	$k - k'$	-0.35119	-0.35120	-0.40216
$50 \to 52$	k	-1.0049	-1.0040	-7.32889
$50 \to 52$	$k + k'$	-2.96662	-2.90904	7891.0^a
$100 \to 102$	$k - k'$	0.058547	0.058578	0.047577
$100 \to 102$	k	0.94403	0.94420	6.7288
$100 \to 102$	$k + k'$	34.2847	38.2831	10050.8
$200 \to 202$	$k - k'$	-0.086238	-0.086247	-0.089841
$200 \to 202$	k	0.10056	0.10067	0.62085
$200 \to 202$	$k + k'$	9.0745	9.0682	-

^{a} $J_\Delta(\sqrt{2n(n + n' + 1)/y})$ and $J_\Delta([n + n' + 1]\theta/2)$ are of opposite sign.

REFERENCES

1. Goreslavskii S.P., and Krainov V.P., Laser Phys **9**, 1266-1269 (1999); Bersons I., and Kulsh A., Phys Rev A **55**, 1674-1682 (1997).
2. Lebedev, V.S., and Beigman I.L., *Physics of Highly Excited Atoms and Ions*, Springer, 1998; Beigman I.L., and Lebedev, V.S., Phys. Rep. **250**, 95 (1995).
3. Sobel'man I.I., Vainshtein L.A., and Yukov E.A., *Excitation of Atoms and Broadening of Spectral Lines*, Springer, 1995.
4. Dewangan D.P., J. Phys. **B 31**,L379 (1998); Phys. Lett. A**254**, 197 (1999); Phys. Lett. A **258**, 285 (1999).
5. Szegö G., *Orthogonal Polynomials*, American Mathematical Society, New York, 1959.

The role of final state potentials in the distorted-wave Born approximation calculation of inner-shell ionization processes

Marco Kampp*, N. C. Pyper† and Colm T. Whelan**

*Department of Applied Mathematics and Theoretical Physics, University of Cambridge, Silver Street, Cambridge CB3 9EW, United Kingdom
†University Chemical Laboratory, Lensfield Road, Cambridge, CB2 1EW, United Kingdom
**Department of Physics, Old Dominion University, Norfolk, Virginia 23529-0116, USA

Abstract. We study the significance of the final state potential in the distorted-wave Born approximation (DWBA) calculation of inner-shell ionization of Mg. The final state potential is generated using the Hartree-Fock method. We found that the triple differential cross section (TDCS) is sensitive to the model used to describe the ion seen by the slow scattered electron.

MOTIVATION

The inner-shell ionization process at intermediate energies has previously been studied theoretically by Zhang et al. [1] in order to describe experiments performed on the 2p ionization of argon in various coplanar geometries. In recent times experimental techniques have become available which allow the study of non coplanar geometries. In particular, the experimental group at the University of Maryland are considering the 2p ionization of magnesium in a geometry where the slow outgoing electrons are detected on the circumference of a cone whose symmetry axis lies perpendicular to the direction of the incoming electron. The fast outgoing electron is detected in coincidence at a fixed angle.

The advantage of non coplanar geometries is that they allow the theorist to focus on particular experimental arrangements where delicate effects can be seen which would be swamped in a more conventional arrangement. For example, if one choses to work in a geometry which is symmetric about the direction of momentum transfer, the prediction from the first Born approximation is constant and higher order effects should therefore manifest themselves in the shape of the cross section (see for example [2, 3]).

An abundance of previous work has shown that inner-shell experiments of this type are well described by using the distorted-wave Born approximation (DWBA). The use of this approach for a magnesium target inevitably raises the question of the state of the remaining magnesium ion to be used to construct the direct and static exchange potentials experienced by the slow outgoing electron. A priori there are three primary candidates for this potential namely those produced by

1. the *fully relaxed* ion.
   This ion has the outermost configuration $2p^6 3s^1$.

CP604, *Correlations, Polarization, and Ionization in Atomic Systems*
edited by D. H. Madison and M. Schulz
© 2002 American Institute of Physics 0-7354-0048-2/02/$19.00

2. the *frozen configuration* ion.

   This species is the lowest energy state of the ion having the configuration $2p^5 3s^2$. This potential would be appropriate if the individual shells had time to relax to their optimal form for the ion before the slow ejected electron travelled a significant distance from the ion.

3. the *fully frozen* ion.

   This is the species having the configuration $2p^5 3s^2$ where the orbitals have exactly the same form as those in the neutral atom. This potential would be appropriate if the time taken for any relaxation process was much larger than that for the slow outgoing electron to leave the atom.

Here we first present and compare the triple differential cross sections (TDCS) predicted for magnesium using these three different potentials. We then present arguments indicating the potential providing the physically most realistic approximation.

METHODS

The numerical Mg^+ wavefunctions corresponding to the three configurations described above were computed using the Oxford Dirac-Fock programme [4] with the velocity of light artificially increased to produce large components which are essentially the same as those resulting from a standard non-relativistic Hartree-Fock calculation. These large components were renormalized to unity. For each of these three wavefunctions the spherically symmetric part of the direct potential generated by the many-electron bound state was then calculated numerically using the standard techniques originally introduced by Hartree [5]. This potential was then used to construct the Furness-McCarthy [6] local approximation to the exchange potential as used previously by Zhang et al. [1].

It is also possible to generate these potentials analytically if one has an analytic expansion of the Roothaan [7] type in which the Hartree-Fock orbitals are expanded in a basis set of Slater functions. Wavefunctions of this type have been presented by Clementi and Roetti [8] for the ground states of both the neutral magnesium atom and the ion having configuration $2p^6 3s^1$.

The potential and TDCS calculated using these analytic methods agreed perfectly with those computed by generating the numerical form of the orbitals from the Clementi and Roetti wavefunction. This perfect agreement provides strong evidence that our computer programs function correctly.

RESULTS

We applied the three different potentials to generate the three corresponding descriptions of the final state for a scattering process with an incoming electron with energy $E_i = 750$ eV, a fast scattered electron with energy $E_f = 472$ eV and a slow ejected electron with energy $E_s = 221$ eV. The fast scattered electron is detected at a fixed angle of $\theta_f = -18°$ in the plane defined by the incoming and fast outgoing electrons.

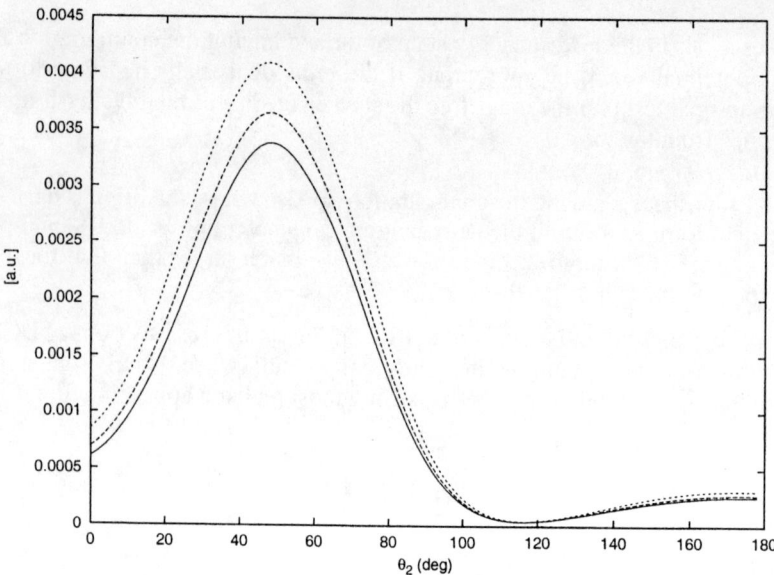

FIGURE 1. The TDCSs computed for the inner-shell ionization of magnesium. The incoming electron has an energy of $E_i = 750$ eV and the slow ejected electron is detected with an energy of $E_s = 221$ eV in the plane defined by the incident and fast outgoing electrons. The fast scattered electron is detected at a fixed angle of $\theta_f = -18°$. The solid line is the TDCS derived considering the fully frozen ion, the dashed line that considering the frozen configuration ion and the dotted line shows the prediction derived considering the completely relaxed ion.

For the incoming electron it is obvious to use the potential generated by the neutral atom. For the fast outgoing electron we used the potential generated by the same neutral atom on the grounds that, for those very short times after the inital collision in which the fast scattered electron is moving through the atom, the wavefunction for the slow ejected electron will still be very similar to that of a 2p electron in the neutral target.

The TDCS calculated using the fully frozen ion potential generated from the neutral magnesium Clementi and Roetti wavefunction was essentially the same as that predicted when the frozen ion potential was calculated from the fully numerical Hartree-Fock wavefunction generated using the Oxford Dirac-Fock program as described above. This shows that the Clementi and Roetti wavefunction is of essentially Hartree-Fock quality even in the context of the present scattering problem.

We used the three different ionic species described above to represent the potential for the slow outgoing electron and calculated the TDCS both in coplanar asymmetric geometry and the geometry specific to the experiments performed at the University of Maryland.

The results show that although the TDCS predicted using the three potentials are similar in overall shape their magnitude at small angles can differ by as much as a third. The TDCS decreases on passing from the potentials generated using the fully relaxed ion to that from the frozen configuration ion to that derived from the fully frozen ion.

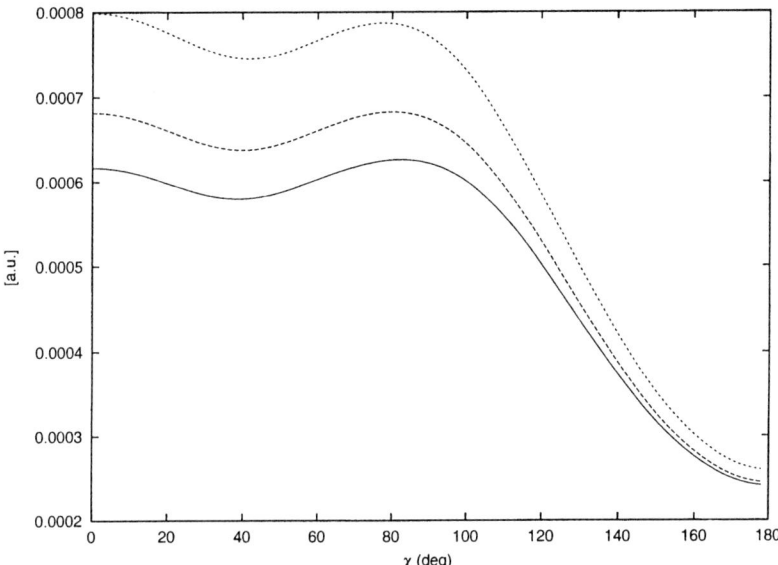

FIGURE 2. The TDCSs computed for the inner-shell ionization of magnesium. Same kinematics as in Figure 1, but the slow electron detected on the circumference of a cone as in the Maryland experiment.

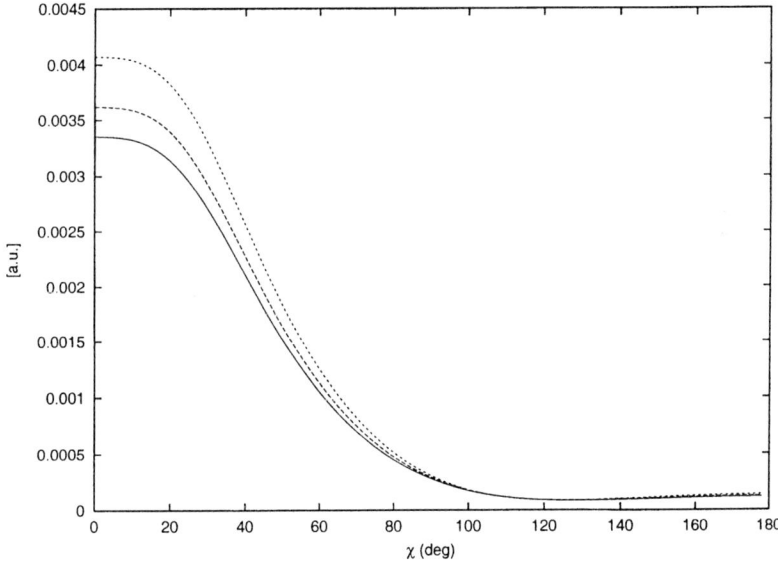

FIGURE 3. The TDCSs computed for the inner-shell ionization of magnesium. Same process as in Figure 2, but the symmetry axis of the cone pointing in the opposite direction as in the Maryland experiment.

DISCUSSION

For distances where the electron is outside the ion the potential is purely Coulombic and is therefore independent of the ion wavefunction used in its construction. Inside the atom, however, the more relaxed the potential the greater the screening and therefore the potential is less attractive. The probability of ionization will be lower when the electron sees a more attractive potential. This explains the decrease in the TDCS on progressing from the fully relaxed to the frozen configuration to the fully frozen potential. Since the difference in the potentials are only significant when the slow ejected electron is inside the atom, one might expect that the frozen potential provides the closest description of reality. This argument would appear to be entirely independent of any considerations of timescale for the ejection and relaxation processes.

ACKNOWLEDGMENTS

We are grateful to our colleagues at the University of Maryland for many useful discussions.

M. K. gratefully acknowledges financial support from the Gottlieb-Daimler and Karl-Benz Stiftung, Germany.

REFERENCES

1. X. Zhang, Colm T. Whelan and H.R.J. Walters, J Phys B **25** (1992) L457-L462
2. P. Marchalant et al., Talk given at this conference
3. M. Kampp et al., to be published
4. I.P.Grant, B.J.McKenzie, P.H.Norrington, D.F.Mayers and N.C.Pyper, Comput Phys Commun **21** (1980) 207
5. D.R. Hartree, The calculation of atomic structures, Ch5 p 77-79, 1957, J. Wiley, New York
6. J.B. Furness and I.E. McCarthy, J. Phys. B. **6** (1973) 2280
7. C.C.J. Roothaan, Rev. Mod. Phys. **23** 1951 p 69
8. E. Clementi and C. Roetti, Atomic Data and Nuclear Data Tables **14** (1974) 177

Momentum Distribution of Fragment Ions Produced in Collisions of Ar^{q+} with CH_4 and CF_4

Kenji Motohashi and Seiji Tsurubuchi

Department of Applied Physics, Tokyo University of Agriculture and Technology,
2-24-16 Koganei, Tokyo 184-8588, Japan

Abstract. Three-dimensional momentum-distributions of fragment ions produced in collisions of Ar^{q+} with CH_4 and CF_4 were measured with the energy-gain of the projectile ions. The yield of multiply charged atomic ions increased with increasing incident charge state of the projectile. The energy-gain spectra of CH_4 and CF_4 consist of two or three components corresponding to different ionic states of the parent molecule. This result suggests that the multiply charged fragment ions are produced from the multiply charged parent molecule via the Auger ionization. The momentum-distribution of the H^+ produced from single charge transfer of $Ar^{8+}+CH_4$ with the energy-gain of 17 eV was measured.

INTRODUCTION

Fragmentation of a molecule by charge transfer with a highly charged ion is one of the fundamental processes in the fusion and in the high temperature plasma. The kinetic energy release distributions (KERD) of the fragment ions provide an important information about the potential surface and the dissociation dynamics [1], but in general intermediate states of the parent molecule produced after the charge transfer collision can not be determined by only measuring the KERD. They provide no information about the projectile ion. The translational energy spectroscopy provides the information of internal energy change of the charge transfer collisions rather than that of fragmentation [2]. It is possible to make clear the relationship between an intermediate state of a molecule and the following dissociation path ways by measuring the translational energy and the KERD in coincidence. On the other hand, the momentum imaging spectroscopy is a useful technique for many studies in atomic physics [3]. We have developed a new momentum-imaging analyzer of fragment ions combined with a translational energy analyzer of the projectile ions. In this paper, we presented the momentum distributions of fragment ions specified by the energy-gain of the charge changed Ar^{q+} (q=8-12).

EXPERIMENTAL

The momentum imaging analyzer is schematically illustrated in Fig. 1. This apparatus consists of a time of flight spectrometer with double extraction field and a $45°$ parallel plate energy analyzer. The projectile ion which passed through the collision region is energy analyzed by the parallel plate energy analyzer. The energy

CP604, *Correlations, Polarization, and Ionization in Atomic Systems*
edited by D. H. Madison and M. Schulz
© 2002 American Institute of Physics 0-7354-0048-2/02/$19.00

resolving power is about 1×10^3. This signal provides a trigger pulse for a digital storage oscilloscope (DSO). The TOF spectrometer (20 cm) has no meshes in order to form a focal plane of the momentum image at the MCP surface. The effective diameter of the MCP is 40 mm. Target gases are injected from a capillary tube with 0.2 mm inner diameter. There are two deflectors before and after the TOF spectrometer to adjust the trajectory of incident ions, since the projectile ions are bent by the strong electric field (210 V/cm) in the TOF extraction region. The position sensitive detector includes a conventional wedge and strip anode. The waveforms of the position signals in each event are recorded by the DSO. Highly charged Ar ions are extracted by a compact EBIS developed in our laboratory [4]. This EBIS is able to produce bare Ne, hydrogen-like Ar, and nitrogen-like Kr ions. An Ar^{12+} beam with an intensity of 5×10^4 (cps) collimated by an orifice with 2 mm diameter was used in this experiment.

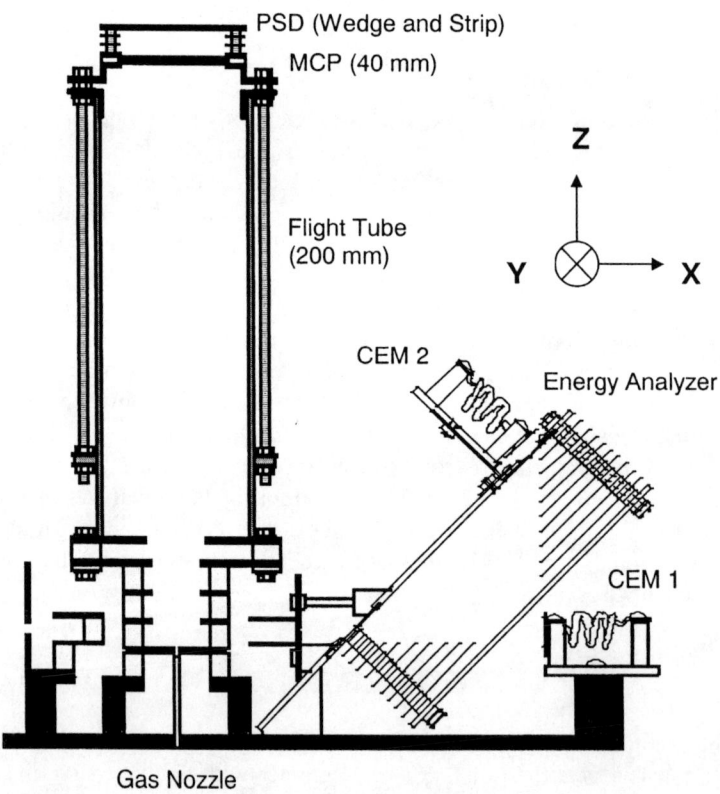

FIGURE 1. The schematic illustration of the 3D momentum imaging analyzer.

RESULTS AND DISCUSSIONS

The time of flight spectra of the fragment ions in single electron transfer of $Ar^{q+}+CF_4/CH_4$ (q=8, 9, and 11) are shown in Fig. 2. Relative intensities of the C^{2+} and F^{2+} ions increase with increasing projectile charge state in the case of $Ar^{q+}+CF_4$ collision. These results may be explained by the following considerations. When increasing the incident charge state, the projectile ion could capture inner electrons of the target molecule. The multiply charged atomic fragments were dissociated from the multiply charged parent molecules after the Auger ionization. The energy-gain spectra of single charge transfer in $Ar^{q+}+CF_4 / CH_4$ (q=8-12) at the collision energy of 8 keV are shown in Fig. 3. Several lines in this figure represent the states of $Ar^{(q-1)+*}$ evaluated under the assumption that the projectile ion captures a $1t_1$ electron of CF_4 or a $1t_2$ electron of CH_4. These electrons have the lowest binding energies (16.2 eV [5] and 12.6 eV [6] respectively). There is no striking difference between the spectrum of q=8 and that of q=9 in the case of CF_4. The maximum peaks and the second one (which are labeled as r, and s in Fig. 3(a)) have 10-20 eV and 30-40 eV of energy-gain respectively. The spectra of $Ar^{11+}+CF_4$ and $Ar^{12+}+CF_4$ were remarkably changed from those of Ar^{8+} and Ar^{9+}. The third components (which are labeled as u and w) were clearly observed in both spectra. The shifts of the peaks to the high-energy side were also observed. In the case of the CH_4 target, all spectra consist of two components. The second components were labeled as x, y, and z in Fig. 3(b). There are remarkable differences among the incident charge states. These results suggest that the incident ions with different charge state capture the target electrons into the excited states with different principal quantum number. It is expected that each component separated by 10-20 eV observed in both targets corresponds to the ionic state of the CH_4^{n+*} including the inner-hole states or the doubly excited states. These multiply-charged parent molecules dissociates into the fragments.

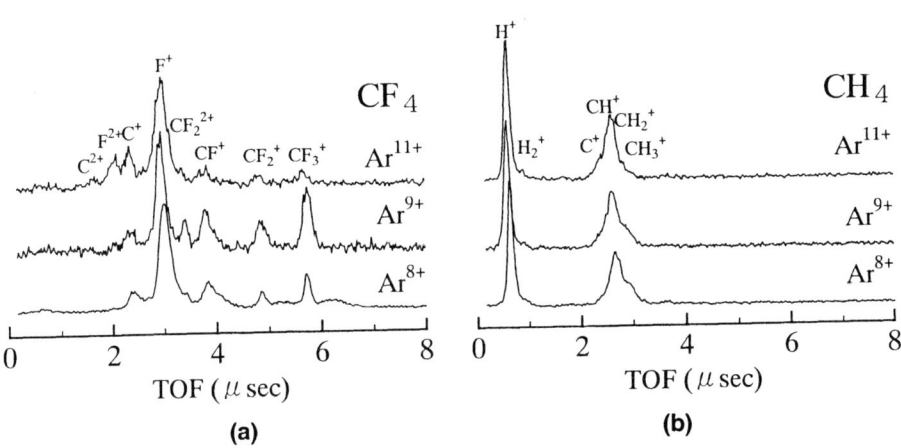

FIGURE 2. The TOF spectra of the fragment ions produced in single electron transfer of $Ar^{q+}+CF_4$ (a) / CH_4 (b) (q=8, 9, and 11).

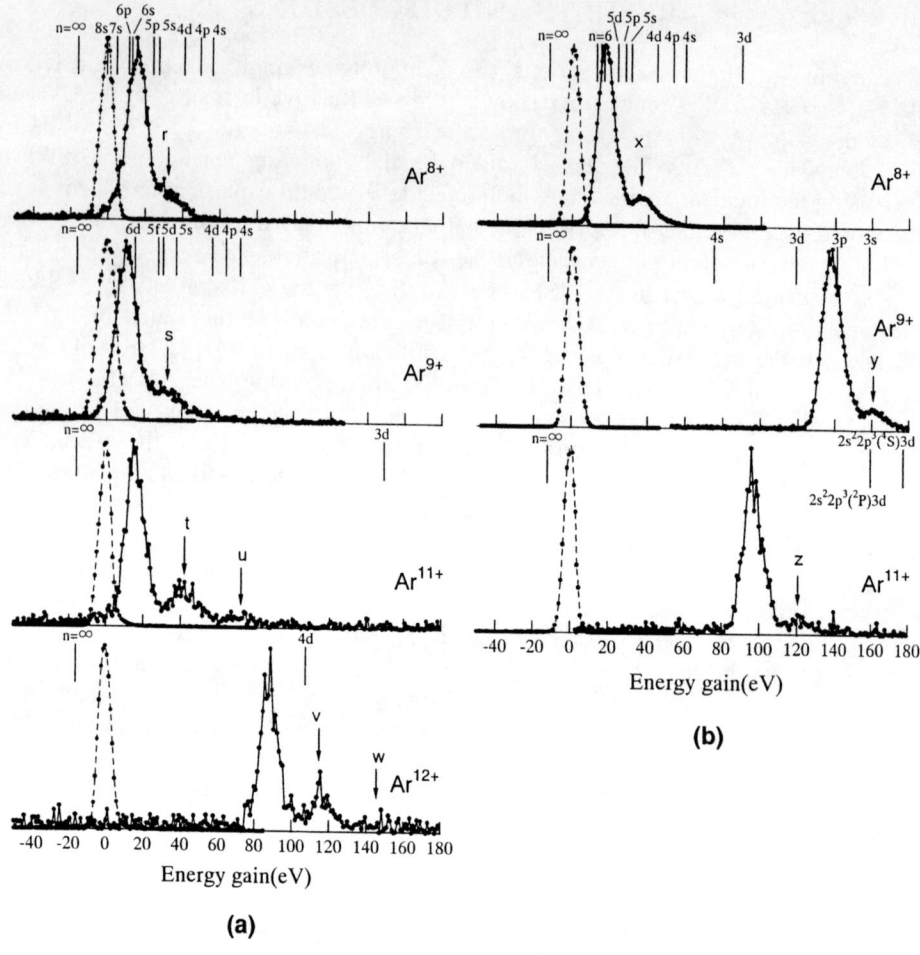

FIGURE 3. The energy-gain spectra of single charge transfer in $Ar^{q+}+CF_4$ (a) (q=8-12) / CH_4 (b) (q=8-11) at the collision energy of 8 keV. The broken lines represent the primary ions.

The momentum image of the H^+ ions produced in collision of Ar^{8+} with CH_4 is shown in FIGURE 4. This momentum image was measured in coincidence with the Ar^{7+} having an energy-gain of 17 eV. The momentum $\sqrt{p_x^2 + p_y^2}$ exceeding 25 a.u. can not be detected by the MCP with 40 mm diameter. There is a faint structure in the xy momentum distribution. Two components of the momentum lower than 10 a.u. and higher than 15 a.u. are observed. The fast component of the momentum distributes circularly in the x-y plane, although the slow component slightly distributes along with the x-axis or the forward direction of the projectile's incident velocity.

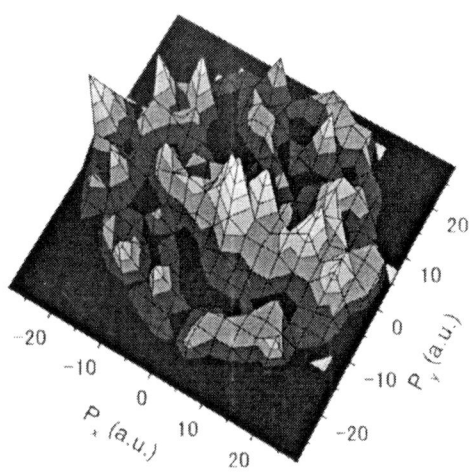

FIGURE 4. Momentum images of H^+ dissociated from CH_4 in collision with Ar^{8+}. The positive x-direction is the primary beam axis. The positive z-direction is along the extraction field of TOF axis, and the y-axis is perpendicular to both the beam and the TOF axes.

CONCLUSIONS

We have developed a 3D-momentum imaging analyzer combined with an energy-gain analyzer. In this paper, first results measured with this apparatus were presented. Intensities of the multiply charged atomic fragments increase with increasing the incident charge states of the projectile. There are two or three components in the energy-gain spectra of CH_4 and CF_4. It is expected that they correspond to different ionic states of the target. Two components are observed in the H^+ momentum distribution produced by $Ar^{8+}+CH_4$ with the energy-gain of 17 eV.

ACKNOWLEDGMENTS

Financial support of the Grant-in-Aid for Scientific Research (C) from the Japan Society for the Promotion of Science is gratefully acknowledged.

REFERENCES

1. Folkerts, H. O., Bliek, F. W., de Jong, M. C., Hoekstra, R., and Morgenstern, R., *J. Phys. B: At. Mol. Opt. Phys.* **30**, 5833-5847 (1997).
2. Leputsch, P., Dumitriu, D., Aumayr, F., and Winter, H. P., *J. Phys. B: At. Mol. Opt. Phys.* **30**, 5009-5024 (1997).
3. Parker, D. H., and Eppink, A. T. J. B., *J. Chem. Phys.* **107**, 2357-2362 (1997).
4. Motohashi, K., Moriya, A., Yamada, H., and Tsurubuchi, S., *Rev. Sci. Instrum.* **71**, 890-892 (2000).
5. Brundle, C. E., Robin, M. B., and Basch, H., *J. Chem. Phys.* **53**, 2196-2213 (1970).
6. Catham, H., Hils, D., Robertson, R., and Gallagher, A., *J. Chem. Phys.* **81**, 1770-1777 (1984).

(e, 2e) Triple Differential Cross Section of Argon at 64.6 eV

Y. Khajuria, L. Q. Chen, X. J. Chen and K. Z. Xu

Open Laboratory of Bond Selective Chemistry
Department of Modern Physics; University of Science and Technology of China
Hefei; Anhui- 230027; People's Republic of China

Abstract. The distorted wave Born Approximation with a modified semiclassical exchange potential has been used to calculate the triple differential cross section of Ar ($3p^6$) in the coplanar to perpendicular plane geometry. Like helium a strong interference minima in the cross-section has been observed for argon atom for all the incident electron angles in contrast to the early predictions. The triple differential cross section for the case of argon at low and high scattering angles is in very good agreement with the present experiment. Inclusion of post collision interaction gives a dip at 45° (characteristic feature of p-orbital targets), which has not been observed exactly at 45° by the experiment and calculations without post collision interaction. The agreement with experiment is quite good at high scattering angles, but some discrepancies are still there.

INTRODUCTION

Electron impact atom ionization provides a very interesting diversity of phenomena because of the wide range of kinematical situations available to the three body final state. The incident electron knocks out the target electron with the remainder of the target atom acting as an inert core. An (e, 2e) reaction is the measurement of the electron impact ionization process where both the existing electrons are detected in coincidence. It is a measurement almost at the limit of what is quantum mechanically predictable and its description presents a substantial challenge to theory. The (e, 2e) technique has been applied to a wide range of targets and kinematical arrangements since the first experimental studies of this type Ehrhardt et al [1] and Amaldi et al [2]. Such experiments in which the kinamatics of the collision are fully determined, provide the most complete test of theories of (e, 2e) reactions to study in detail the collision dynamics. The testing and development of current theories has been further challenged by including the wide range of kinematical options employed for example by Pochat et al [3], Hawely-Jones et al [4], Röder et al [5], Murray [6] and Murray and Read [7,8,9,10,11]. The standard Distorted Wave Born Approximation (DWBA) is now well known and has been shown to be capable of predicting accurate (e, 2e) cross sections for a wide variety of atomic targets above about 50eV (Whelan et al

CP604, *Correlations, Polarization, and Ionization in Atomic Systems*
edited by D. H. Madison and M. Schulz
© 2002 American Institute of Physics 0-7354-0048-2/02/$19.00

[12], Khajuria and Tripathi [13]). Allan et al [14] noticed that the inclusion of post collision interaction (PCI) and polarization of atomic targets play important role in determining the structure.

In the present work, the DWBA a with modified semiclassical exchange potential in which the polarization has been taken implicitly, has been applied to study the behavior of the cross sections for Argon atom. Previous measurements and calculations for helium have shown a strong interference minima in the triple differential cross section (TDCS) curve at an incident electron angle of 67.5°. Hence it is worthwhile to see whether the same interference minima appears for argon ionization. The effect of post collision interaction has also been studied.

THEORY

The theory and the formulation used here are the same as given in our earlier works. Both the spin averaged static exchange potential (SASEP) by Furness and McCarthy [15], and the modified semiclassical exchange potential (MSCEP) of Gianturco and Scialla [16] have been used to calculate the TDCS for ionization helium and the MSCEP has been used to calculate the triple differential cross section for argon. The exchange potential of Furness and McCarthy [15] is local in character and contains a physically unreasonable repulsive component in it. Also, their approximation of disregarding the local momentum of the bound electron is not tenable at low impact energies. Gianturco and Scialla [16] obtained an expression for the exchange potential that is local in character after taking these aspects into account. The MSCEP is given by

$$V_{vex} = \frac{1}{2}[E - V_D(r) + \frac{3}{10}(3\pi^2\rho(r))^{2/3} - \{[E - V_D(r) + \frac{3}{10}(3\pi^2\rho(r))^{2/3}]^2 + 4\pi\rho(r)\}^{1/2}]$$

(1)

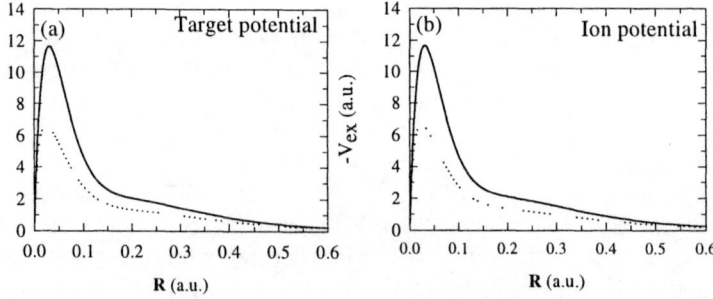

FIGURE 1. Exchange potentials for electron-impact ionization of argon. The theoretical curves are: (——) SASEP and (········) MSCEP

where $\rho(r)$ is the electron charge density. The direct distorting potential $V_D(r)$ is the usual static potential obtained from the target orbitals (McCarthy [17]). The inclusion of the local momenta of the bound electron by Gianturco and Scialla [16] reduces the local velocity of the incoming electron with the result that the corresponding correlation term weakens the attractive part of the interaction. The exchange part of both potentials has been calculated at a collision energy of 64.6 eV for both the incident electron-target potentials and ejected electron-ion potentials for argon. The results are shown in figure 1. It is clearly seen from the figure that the additional term reduces the attractive interaction between the colliding partners.

RESULTS AND DISCUSSION

The TDCS of helium at 64.6 eV incident energy has already been discussed in detail in our earlier work [18] and need not to be discussed in again. The main findings of the previous results are summarized as: (i) the existence of a deep interference minima in the TDCS; (ii) good agreement between the calculated and measured TDCS at the common point for all the incident electron angles; and (iii) the MSCEP approach yields good agreement with experiment. Consequently, it can be said that the addition of the term which decreases the attractive interaction between the colliding partners represents a more realistic interaction potential.

Figure 2 shows the theoretical results normalized to the experimental data for electron impact ionization of argon. The six incident electron directions were chosen in accordance with the experiment. These were $\Psi = 0°, 7.5°, 15°, 30°, 45°, 60°$. The present results have been normalized at $\xi = 90°$ since the cross section is same at this angle for all the incident electron directions. This common point lies before the minima as compared to helium, where it lies after the minima for all the incident electron angles. Murray and Read have found a dip in the cross section for $\Psi = 0°$ at $\xi = 50°$, which is a characteristic feature of the p-state bound electron. The present calculations also show a dip near $\xi = 50°$. This dip can be explained using the impulse approximation (McCarthy and Wiegold [19]), which assumes that at the point of impact the ion recoil momentum is equal but opposite to the bound electron momentum $\mathbf{q}$. For the symmetric calculations ($\Psi = 0°$) the magnitude of $\mathbf{q}$ is given by

$$q = \left[\left(2k_s \cos\xi - k_s \sin\Psi\right)^2 + \left(k_0 \sin\Psi\right)^2 \right]^{1/2} \qquad (2)$$

and from this relation for an incident electron energy of 64.6 eV and ejected electron energy of 24.4 eV, the value of ξ is 45°. Murray and Read explained this shift as due to post collision interactions, but the present calculations with post collision interactions clearly show a dip at $\xi = 45°$ in figure 2a(dotted line). For other incident electron angle directions, this dip has not been observed since the component ($k_0 \sin \Psi$) of the incident electron momentum always lies out of the detection plane. An interesting feature in the present calculations is the presence of another dip at $\xi = 75°$, for $\Psi = 0°$, 7.5°, 15°, 30°. This dip shifts towards smaller scattering angles and lies at 65° and 55° for $\Psi = 45°$ and 60° respectively. This dip has not been observed by the present experiment for $\Psi = 0°$, 7.5°, 15°, 30°, 45°. However for $\Psi = 60°$, there is a small dip in the cross section, but the magnitude of the calculated dip is not in accord with the experiment. There are no other theoretical results available for comparison.

There exists another dip in the cross section near $\xi = 100°$. As compared to helium, (where the dip position changes with changing incident electron angle) the dip position remains same for incident electron angles of 0°, 7.5°, 15°, 30°, 45° and 60°. The different nature of the triple differential cross section might be explained from figure 1. In the case of helium [18], the magnitude of the peak for the electron-ion potential is smaller than the peak for the electron-target potential. For argon, on the

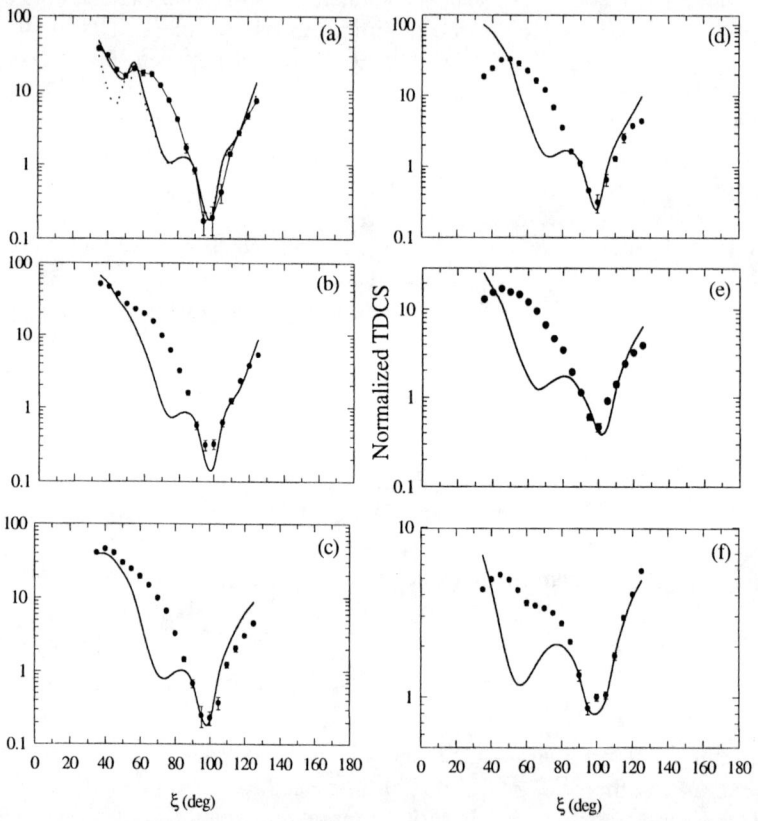

FIGURE 2. TDCS for electron-impact excitation of argon atom at 64.6 eV with the detection energies of (24.4, 24.4) eV for six different incident electron directions: $\Psi = $ (a) 0°, (b) 7.5°, (c) 15°, (d) 30° (e) 45°, and (f) 60°. The solid line represents the present calculations normalized to the experimental data and dots denote the experimental results of Murray and Read. The dotted line in figure (a) represents the calculations with post collision interaction.

other hand, the magnitude is almost the same for the electron-target and electron-ion potentials (figure 1) i.e. the exchange part of the potentials in the initial and final channel is the same and hence yields a dip always at the same position. It has been seen for helium that these dips in the TDCS appear due to the interference between the different scattering amplitudes. The TDCS has been calculated by switching off the electron-nucleus, the electron-passive electron interaction and both of these interactions from the interaction potential. Figure 3 shows the resulting cross section for $\Psi = 0°$ in the angular region of 90°-110°. From figure 3 it is clear that the TDCS without the electron-nucleus interaction, without the electron-passive electron interaction and without both of these is smooth, whereas the TDCS with all of these shows a dip in that region. Similar effects have been found for the incident electron angle of 60° for both the dips at 55° and 100°. Rasch et al [20], using the DWBA, have predicted that (for Ar (2p)) such interference effects should be masked by the different magnetic sublevels and they have shown that the summation over all three sublevels (m= 0, ±1) causes the dip to disappear. The present calculations clearly show that the dips appear in the TDCS of Ar (3p) are due to the strong interference between the different scattering amplitudes, contrary to the predictions of Rash et al [20]. In the perpendicular plane geometry ($\Psi = 90°$), the cross section has to be symmetric about the incident direction, which is quite obvious from the present results. Figure 4 shows the TDCS of Ar (3p) in the perpendicular plane geometry. As required by symmetry, the forward and backward scattering angle peaks are of equal magnitude and are exactly separated by 180°. The magnitude of these peaks is in accordance with the experiment, but the position is shifted by 5°. In conclusion, the agreement between experiment and the present calculations is good except in the

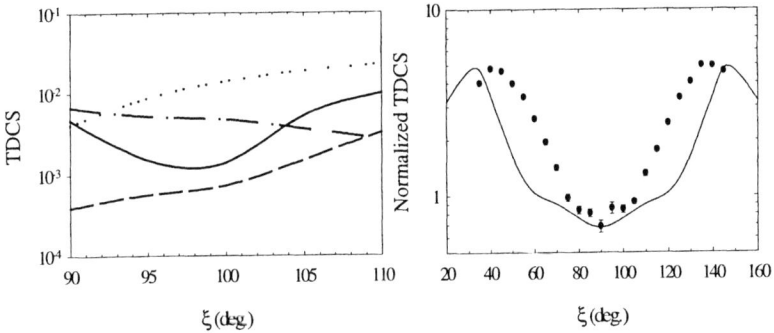

FIGURE 3. Dotted line represents the TDCS calculated with the distorting potential minus the electron-nucleus interaction, dashed line the distorting potential minus the electron-passive electron interaction and the chained line the distorting potential minus both of these. The solid line corresponds to total distorting potential

FIGURE 4. TDCS in the perpendicular plane geometry. Solid line is the present calculation and dots denote the experimental results due to Murray and Read.

301

angular region of 65°-80° for coplanar geometry. This angular range increases with increasing incident electron angle. At higher scattering angles, the agreement is quite good for all the incident electron angles. There are no other theoretical or experimental results available at this incident electron energy.

ACKNOWLEDGEMENT

Financial supports for this research was provided by the national Science Foundation of China. One of the authors, Y. Khajuria is grateful to the Chinese Science Academy (CSA) for providing the necessary grants.

REFERENCES

1. Ehrhardt H, Schulz M, Tekaat T, and Willmann K, *Phys. Rev. Lett.* **22** (1969) 89.
2. Amaldi Jr. U, Egidi A, Marconero R, and Pizzella G, *Rev. Sci. Instr.* **40** (1969) 1001.
3. Pochat A, Tweed R J, Peresse J, Joachain C J, Piraux B and Byion F W, *J. Phy .B: At. Mol.Opt. Phys* **16**(1983) 775.
4. Hawley –Jones T J, Read F H, Cvejanovec S, Hammond P and King G C, *J. Phy .B: At. Mol.Opt. Phys.* **25** (1992) 2393.
5. Roders,Ehrhardt H, Bray I, Farsa DV, and McCarthy I E, *J. Phy .B: At. Mol.Opt. Phys.* **29**(1996) 2103.
6. Murray A J, *in (e,2e) and Related process*, edited by Whelan C T, Walters H R J, Lahman-Bennan A and Ehrhardt H, 372.
7. Murray A J and Read F H, *Phys.Rev.Letts.* **69** (1992) 2912.
8. Murray A J and Read F H, *J. Phy .B: At. Mol.Opt. Phys.* **26**(1993) 369.
9. Murray A J and Read F H, *Physical .Rev. A.* **47** (1993) 3724.
10. Murray A J and Read F H, *J. Phy .B: At. Mol.Opt. Phys* **33** (2000) 2859.
11. Murray A J and Read F H, *J. Phy .B: At. Mol.Opt. Phys* **33** (2000) L297.
12. Whelan C T, Walters H R J, Allan R J, and Zhang X , *in (e,2e) and Related process*, edited by Whelan C T, Walters H R J, Lahman-Bennan A and Ehrhardt H,
13. Khajuria Y and Tripathi D N, *J. Phy .B: At. Mol.Opt. Phys* **31** (1998) 2359.
14. Allan R J,Whelan C T, and Walters H R J, *int.symp. on (e,2e) collisions, double photoionization and related processes, J. Physique IV* **3** (1993) 39 .
15. Furness J B, and McCarthy I E, *J. Phy .B: At. Mol.Opt. Phys.* **6** (1973) 2280.
16. Gianturco F A, Scialla S, *J. Phy .B: At. Mol.Opt. Phys* **20** (1987) 3171.
17. McCarthy I E, *Aust. J. Physics* **48** (1995) 1.
18. Khajuria Y and Tripathi D N, *Phys. Letter A* **260** (1999) 360.
19. McCarthy I E, and Weigold E, *Rep. Prog. Physics* **54** (1991) 789.
20. Rash J, Whelan C T, Allan R J, Lucey S P, and Walters H R J, *Phys. Rev. A.* **56** (1997) 2280.

302

CONFERENCE PROGRAMS

(e,2e) Ionization Program

THURSDAY, JULY 26, 2001

08:30 - 09:00   **JOINT OPENING SESSION**

09:00 - 10:30   **STRONG FIELDS** (Chair: Erich Weigold)

09:00 - 09:30   Chris Greene (USA)
Multiphoton processes in a two-dimensional model of helium

09:30 - 10:00   Joachim Ullrich (Germany)
Multiple Ionization and (e,2e)-like Mechanisms in Intense Laser Fields

10:00 - 10:30   Andreas Becker (Canada)
Interplay of electron correlation and intense field dynamics in laser induced double ionization

10:30 - 11:00   **Coffee Break**

11:00 - 12:30   **DOUBLE PHOTO IONIZATION** (Chair: Lorenzo Avaldi)

11:00 - 11:30   Uwe Becker (Germany)
Interference structure in double photoionization

11:30 - 12:00   Laurence Malegat (France)
Absolute differential cross sections for double photoionization of two-electron systems from the HRM-SOW method

12:00 - 12:30   Bertold Kraessig (USA)
A complete experiment on photo-double-ionization of helium

12:30 - 14:00   **Lunch**

14:00 - 15:30   **(e,2e) SESSION 1** (Chair: Gunter Baum)

14:00 - 14:30   Colm Whelan (United Kingdom/USA)
Fragmentation Processes

14:30 - 15:00   Birgit Lohmann (Australia)
Low energy inner-valence ionization of the rare gases

| 15:00 - 15:30 | Boghos B. Joulakian (France) |
| | *Simple ionisation of atomic and diatomic lithium by electron impact* |

| 15:30 - 16:00 | **Coffee Break** |

| 16:00 - 17:30 | **(e,2e) Session 2** (Chair: Andrew Murray) |

| 16:00 - 16:30 | Julian Lower (Australia) |
| | *(e,2e) Collisions with Polarized Electrons and Excited, Oriented and Spin-Polarized Targets* |

| 16:30 - 17:00 | Stephane Mazevet (USA) |
| | *Initial state resolved (e,2e) experiments* |

| 17:00 - 17:30 | Steve Jones (Australia) |
| | *Fast Algorithm for Solving Three-Body Quantum Collision Problems* |

FRIDAY - JULY 27, 2001

| 09:00 - 10:30 | **IONIZATION OF MOLECULES** (Chair: Yenyou Zheng) |

| 09:00 - 09:30 | Chris Brion (Canada) |
| | *Chemical Applications of Binary (e,2e) Spectroscopy* |

| 09:30 - 10:00 | John Briggs (Germany) |
| | *Complete fragmentation of few-body Coulomb systems* |

| 10:00 - 10:30 | Danielle Dowek (France) |
| | *Vector correlations in dissociative photoionization of diatomic molecules induced by linearly polarized light in the VUV energy range* |

| 10:30 - 11:00 | **Coffee Break** |

| 11:00 - 12:30 | **(e,2e) FROM SOLIDS** (Chair: Alexander Dorn) |

| 11:00 - 11:30 | Gianni Stefani (Italy) |
| | *Electron-electron coincidence experiments: from atoms to solids* |

| 11:30 - 12:00 | Jamal Berakdar (Germany) |
| | *On the electronic correlation in finite and extended systems* |

304

12:00 - 12:30	Maarten Vos (Australia)
	Electron correlation effects in materials as observed by electron momentum spectroscopy

12:30 - 14:00	**Lunch**

14:00 - 15:30	**(e,2e) SESSION 3** (Chair: Anatoli Kheifets)

14:00 - 14:30	Mark Baertschy (USA)
	Fully Correlated Wavefunctions Describing Two Outgoing Electrons at Low Energies

14:30 - 15:00	Penny Scott (United Kingdom)
	Electron impact ionisation using an R-matrix approach

15:00 - 15:30	Igor Bray (Australia)
	Close-coupling approach to ionization processes

15:30 - 16:00	**Coffee Break**

16:00 - 17:30	**(e,3e)** (Chair: Jack Moore)

16:00 - 16:30	Azzedine Lahmam-Bennani (France)
	New results in (e,3e) and (e,3-1e) double ionisation of He and H_2

16:30 - 17:00	Mike Coplan (USA)
	Distinguishing between Target Structure and Ionization Mechanisms in (e,3e) Experiments

17:00 - 17:30	Bernard Piraux (Belgium)
	Double ionization of atoms by electron and photon impact: a progress report

---

SATURDAY - JULY 28, 2001

09:00 - 10:30	**NEW DEVELOPMENTS - 1** (Chair: Jerry Peacher)

09:00 - 09:30	Jim Feagin (USA)
	Reaction Imaging with Interferometry: Circular dichroism in photo single ionization angular distributions

09:30 - 10:00	Holger Kollmus (Germany) *Electron-Electron Interaction in Projectile Ionization: A New Way to Explore (e,2e) on Ions*
10:00 - 10:30	Ugo Ancarani (France) *Role of the He bound state wavefunction in the interpretation of coincidence experiments*
10:30 - 11:00	**Coffee Break**
11:00 - 12:30	**NEW DEVELOPMENTS - 2** (Chair: Patricia Selles)
11:00 - 11:30	Horst Schmidt-Bocking (Germany) *Visualization of the He ground state momentum wave function*
11:30 - 12:00	Michael Schulz (USA) *Coplanar and Non-Coplanar Triple Differential Cross Sections for Fast Ion-Atom Collisions*
12:00 - 12:30	Bob DuBois (USA) *Differential information about multiple ionization of atoms by positrons and electrons*

Electron-Photon Polarization and Correlation Program

08:30 - 09:00   **Joint Opening Session**

09:00 - 10:30   **ELECTRON-ATOM COLLISIONS 1** (Chair: Max Standage)

09:00 - 09:30   Albert Crowe (United Kingdom)
Complete experiments for electron excitation of pseudo-two-electron atoms

09:30 - 10:00   Dmitry Fursa (Australia)
Electron-photon correlations in electron-impact excitations of alkaline-earth atoms

10:00 - 10:30   Rajesh Srivastava (India)
Electron excitation of the D states of Mg: Calculation of the Stokes Parameters

10:30 - 11:00   **Coffee Break**

11:00 - 12:30   **ELECTRON-ATOM COLLISIONS 2** (Chair: G. Friedrich Hanne)

11:00 - 11:30   Danica Cvejanovic (United Kingdom)
Excitation of the 3D states of helium by electron impact

11:30 - 12:00   Morty Khakoo (USA)
Differential cross-section ratio parameters for the electron impact excitation of the rare gases, as a test of target wavefunctions used in scattering models

12:00 - 12:30   S. Tsurubuchi (Japan)
Excitation of the $2p_9$ State of Rare Gases by Electron Impact

12:30 - 14:00   **LUNCH**

14:00 - 15:30   **CORRELATION FOR PHOTON IMPACT** (Chair: Alain Huetz)

14:00 - 14:30   Bernd Zimmermann (Germany)
Alignment of photo-ions for the autoionization analysis of doubly excited states

14:30 - 15:00	Orhan Yenen (USA) *Quantifying Relativistic Effects from Alignment and Orientation Studies in Photoionization*
15:00 - 15:30	Kouichi Soejima (Japan) *Chiral Electron Pairs in Photo-double Ionization of Closed Shell Atoms*
15:30 - 16:00	**Coffee Break**
16:30 - 17:30	**AUGER PROCESSES** (Chair: Volker Schmidt)
16:00 - 16:30	Markus Drescher (Germany) *On the feasibility of a complete Auger decay experiment*
16:30 - 17:00	Marc Simon (France) *Auger angular distribution and non-dipole effects from core excited fixed in space molecules*
17:00 - 17:30	Giorgio Turri (USA) *Experimental measurement of the angular correlations in Auger cascade following the 1s -> 3p resonant excitation in neon*

FRIDAY - JULY 27, 2001

09:00 - 10:35	**ELECTRON-ATOM COLLISIONS 3** (Chair: Peter Teubner)
09:00 - 09:30	Bill McConkey (Canada) *Balmer-Alpha Polarization Studies*
09:30 - 10:00	Michael Went (Australia) *Elastic scattering of spin polarised electrons from krypton*
10:00 - 10:30	Victor Karaganov (Australia) *Superelastic electron scattering from alkali atoms*
10:30 - 11:00	**Coffee Break**

11:00 - 12:30	**EXCITATION/IONIZATION** (Chair: Timothy Reddish)

11:00 - 11:30	Jim Williams (Australia)
	Excitation-with-ionization by spin-polarised electrons

11:30 - 12:00	Klaus Bartschat (USA)
	Simultaneous Ionization–Excitation: A Challenge for Theory and Experiment

12:00 - 12:30	Pascale Marchalant (United Kingdom)
	The ionization of atomic helium-complementary studies using different charged particles

12:30 - 14:00	**Lunch**

14:00 - 15:30	**SURFACES AND SOLIDS** (Chair: Allan Stauffer)

14:00 - 14:30	Sergey Samarin (Australia)
	Scattering of Spin-Polarized Electrons from Ferromagnetic Surface: (e,2e) Experiment

14:30 - 15:00	Natasha Fominykh (Germany)
	Correlation studies at surfaces and interfaces

15:00 - 15:30	Roberto Gotter (Italy)
	X-Ray Photoelectron Diffraction as Measured by APECS (Auger Photoelectron Coincidence Spectroscopy) form the Cu(111) Surface

15:30 - 16:00	**Coffee Break**

16:00 - 17:30	**NEW DEVELOPMENTS** (Chair: Durga Dewangan)

16:00 - 16:30	Henk Heideman (Netherlands)
	Polarization in Collisions between Ultra-cold Sodium Atoms

16:30 - 17:00	Nikolay Kabachnick (Russia)
	New developments in the problem of complete experiment for photoinduced Auger processes

17:00 - 17:30	Bernd Lohmann (Germany)
	Recent developments in the theory of electron emission

09:00 - 10:30   **POLARIZED ELECTRON IMPACT** (Chair: Timothy Gay)

09:00 - 09:30   Carsten Herting (Germany)
Orientation of heavy atoms by polarized-electron-impact excitation

09:30 - 10:00   Dehong Yu (Australia)
Spin dependent interactions in the excitation of atoms and molecules

10:00 - 10:30   B.G. Birdsey (USA)
Determination of hexadecapole moments for the $3p^4$ (^1D) core of argon II
excited in polarized e^--Ar collisions[*]

10:30 - 11:00   **Coffee Break**

11:00-12:30   **RELATED TOPICS** (Chair: Peter Zetner)

11:00 - 11:30   Gebhard Von Oppen (Germany)
Symmetry-induced entanglement in homonuclear ion-atom collisions

11:30 - 12:00   Mark Rosenberry (USA)
Progress with Optically Pumped Sources of Polarized Electrons

12:00 - 12:30   Hajime Tanuma (Japan)
Electron transfer to individual magnetic substates of multicharged ions

POSTER SESSION

p+He Transfer Ionization Process: Theory and Experiment
L.U. Ancarani (France)

Direct Photo-Double Ionization of He and Unequal Energy Sharing Conditions at 40eV Excess Energy
L. Avaldi (Italy)

Measurement of Spin-Asymmetries in Elastic and Inelastic Electron-Cesium Scattering at Intermediate Energies
Günter Baum (Germany)

Excitation of Metallic Clusters: Size Dependent, Shape Dependence and Correlation Effects
J. Berakdar (Germany)

Comparison of Hartree-Fock and Furness-McCarthy Continuum Waves
D.A. Biava (USA)

Theoretical Study of Double Photoionization of Alkaline Earths: The Case of Calcium
F. Citrini (France)

On the Tricomi Expansion in Rydberg Collisions
D.P. Dewangan (India)

(e,3e) and (e,γ2e) Experiments on Helium
A. Dorn (Germany)

Correlated Electron Dynamics of Artificial Atoms in Laser Fields
N. Fominykh (Germany)

The Relativistic Cross Sections of Multiphoton Bremsstrahlung in the Strong Laser Fields
T.R. Hovhannissian (Armenia)

All Electron Calculations of the Multiply Differential Cross Section of the Simple Ionisation of Lithium by Electron Impact
B. Joulakian (France)

Multiply Differential Cross Section of the Simple Ionisation of Diatomic Lithium Li_2 by Fast Electron Impact
B. Joulakian (France)

On the Inner Shell Ionization by Electron Impact
Marco Kampp (United Kingdom)

Excitation of Metallic Clusters: Size Dependence, Shape Dependence and Correlation Effects
Oleg Kidun (Germany)

311

Triple Escape from Simple Atoms
Hubert Klar (Germany)

Spin-Polarized (e,2e) Spectroscopy of a Ferromagnetic Fe(110) Surface: Resolution in Momentum Space
A. Morozov (Germany)

Spin-Polarized (e,2e) Spectroscopy of W(001) Surface: Spin-Orbit Coupling Induced Asymmetry
A. Morozov (Germany)

Momentum Distribution of Fragment Ions Produced in Collisions of Ar^{q+} with CH_4 and CF_4
K. Motohashi (Japan)

Single Ionization of Helium by Slow Electron Impact
B. Najjari (Germany)

Distorted Wave BBK Calculation for Heavy Inert Gases
A. Prideaux (USA)

An entire set of absolute cross sections for double photoionization of helium from the HRM-SOW method
P. Selles (France)

Construction of a Binary (e,2e) Apparatus with Position-Sensitive Detectors and its Application to Molecules
Masahiko Takahaski (Japan)

An Investigation of Basic Symetries in (e,2e) Measurements on Magnesium
R.W. van Boeyen (USA)

Controlled Correlated Collisions and Interaction of a Moving Neutral Atoms with a Periodical Modulated Surface at Condition of Intratomic Electromagnetic Resonance
Vladimir I. Vysotskii (Ukraine)

Energy-Sharing Following Low-Energy Ionisation of Argon
D.K. Waterhouse (Australia)

Structure in the (e,2e) TDCS of Ar for Low Incident Energy?
D.K. Waterhouse (Australia)

Electron Exchange in Inner-Shell Ionisation-with-Excitation of Zinc Atoms by Polarised Electrons
D.H. Yu (Australia)

Spin Up-Down Asymmetry in the Excitation of Kr $5p'[3/2]_2$ State By Polarized Electrons
D.H. Yu (Australia)

Calculation for (e,2e) Ionization of the 2p Orbital of Argon
Y. Zhou (China)

LIST OF PARTICIPANTS

ANCARANI, Lorenzo Ugo
Institut de Physique, Labo LPMC
Universite de Metz
1 Bd Arago, Technopole 2000
57078 Metz Cedex 3
France

AVALDI, Lorenzo
Instito di Metolodologie Avanzate Inorganiche
del Consiglio Nazionale delle Ricerche
Area della Ricerca di Roma
CP10, 00016 Monterotondo Scalo
Italy

BAERTSCHY, Mark
JILA
University of Colorado
440 UCB
Boulder, CO   80309-0440

BARTSCHAT, Klaus
Drake University
Department of Physics & Astronomy
Des Moines, IA   50311

BAUM, Günter
Fakultät für Physik
Universität Bielefeld
Postfach 100 131
Universitätsstr. 25
Bielefeld D-33615
Germany

BECKER, Andreas
Département de Physique
Université Laval
Québec, PQ
G1K 7P4 Canada

BECKER, Uwe
Fritz-Haber-Institut der MPG
Abteilung Oberflächenphysik
Faradayweg 4-6
D-14195 Berlin
Germany

BERAKDAR, Jamal
Max-Planck-Institut für Mikrostruktur
Physik
Weinberg 2
06120 Halle
Germany

BIAVA, Dominic
University of Missouri-Rolla
Physics Department
1870 Miner Circle
Rolla, MO   65409-0640

BIRDSEY, Benjamin
University of Nebraska-Lincoln
116 Brace Lab
Lincoln, NE   68588

BRAY, Igor
Centre for Atomic, Molecular and Surface Physics
School of Mathematical & Physical Sciences
Murdoch University
Perth 6150
Australia

BRIGGS, John S.
Universität Freiburg
Fakultät für Physik
Universität Freiburg
D-79104 Freiburg
Germany

BRION, Chris
University of British Columbia
Department of Chemistry
2036 Main Mall
Vancouver, BC V6T 1Z1
Canada

CHEN, Zhangjin
University of Missouri-Rolla
Physics Department
1870 Miner Circle
Rolla, MO 65409-0640

CITRINI, France
Laboratoire de Spectroscopie Atomique et
Ionique
Université Paris-Sud
Bâtiment 350, 91405 Orsay Cedex
France

COPLAN, Michael
Institute for Physical Science & Technology
University of Maryland
CSS Building
College Park, MD   20742

CROWE, Albert
University of Newcastle
Department of Physics
Newcastle upon Tyne
NE1 7RU
United Kingdom

CVEJANOVIC, Danica
Atomic, Molecular and Laser Manipulation Group,
Dept. of Physics & Astronomy
The University of Manchester
Manchester, Lancashire M13 9PL
United Kingdom

DEWANGAN, Durga P.
Physical Research Laboratory
Theoretical Physics Division
Navrangpura
Ahmedabad 380 009
India

DORN, Alexander
University of Freiburg
Faculty of Physics
Hermann-Herder-Str. 3
D-79104 Freiburg
Germany

DOWEK, Danielle
LCAM - UMR Université Paris-Sud et
Bât. 351
Université Paris-Sud
F-91405 Orsay Cedex
France

DRESCHER, Markus
Universität Bielefeld
Fakultaet für Physik
Universitätsstr. 25
D-33615 Bielefeld
Germany

DuBOIS, Robert
University of Missouri-Rolla
1870 Miner Circle
Rolla, MO   65409-0640

EDAH, Gaston
FYAM
Department de Physique
Chamin due Cyclotron 2
Louvain-la-Neuve B1348
Belgium

FEAGIN, James M.
California State University-Fullerton
Department of Physics
Fullerton, CA   92834

FOMINYKH, Natasha
Max-Planck-Institut für Mikrostruktur Physik
Weinberg 2,
06120 Halle
Germany

FOSTER, Matt
University of Missouri-Rolla
Physics Department
1870 Miner Circle
Rolla, MO 65409-0640

FURSA, Dmitry
The Flinders University
School of Chemistry, Physics & Engineering
Electronic Structure of Materials Centre
GPO Box 2100
Adelaide 5001, Australia

GAO, Junfang
University of Missouri-Rolla
Physics Department
1870 Miner Circle
Rolla, MO   65409-0640

314

GAY, Timothy J.
University of Nebraska
Behlen Laboratory of Physics
Lincoln, NE   68588-0111

GOTTER, Roberto
INFM-Instituto Nazionale per Ia Fisica della Mat.
  Laboratorio TASC-INFM
SS 14KM 163.5, I-34012 Basovizza
Trieste
Italy

GRADZIEL, M.
National University of Ireland-Maynooth
Department of Experimental Physics
Maynooth CO Kildare
Ireland

GREEN, Adam
University of Nebraska-Lincoln
116 Brace Lab
Lincoln, NE   68588

GREENE, Chris
University of Colorado
Department of Physics and JILA
UCB 440
Boulder, CO   80309-0440

HANNE, G. Friedrich
Universität of Münster
Physikalisches Institut
Wilhelm-Klemm Str. 10
D-48149 Münster
Germany

HEIDEMAN, Henk
Utrecht University
Debye Institute, Buys Ballotlaboratorium
Department of Physics
P.O. Box 80000
3548 CC Utrecht
The Netherlands

HERTING, Carsten
University of Münster
Physikalisches Institut
Wilhelm-Klemm-Str. 10
D-48149 Münster
Germany

HOVHANNISYAN, Tigran
Yerevan State University
Plasma Physics Laboratory
1, A. Manukian, 375049 Yerevan
Armenia

HUETZ, Alain
Université Paris Sud
Laboratoire de Spectroscopie Antomique et
Ionique, Bat 305
Centre dΠOrsay 91405, Cedex
France

HUSSEY, Martyn
University of Manchester
AMLaM Group, Schuster Lab
Manchester M13 9PL
Great Britain

JAMES, Kenneth
California State University-Fullerton
Physics Department
Fullerton, CA   92831

JONES, Stephen
Murdoch University
Centre for Atomic, Molecular & Surface Phys.
Division of Science & Engineering
Murdoch University
Perth 6150, Australia

JORDAN-THADEN, Brandon
University of Nebraska-Lincoln
116 Brace Lab
Lincoln, NE   68588

315

JOULAKIAN, Boghos
University of Metz
Institut de Physique
1 Bld Arago
Technopôle 2000
57078 Metz Cedex 3, Lorraine
France

KABACHNIK, Nikolay
Institute of Nuclear Physics
Moscow State University
Moscow 119899
Russia

KAMPP, Marco
University of Cambridge
Dept. of Applied Mathematics & Theoretical
Physics (DAMTP)
Silver Street, Cambridge CB3 9EW
United Kingdom

KARAGANOV, Victor
The Flinders University
Department of Physics
GPO Box 2100
Adelaide, SA 5001
Australia

KHAKOO, Murtadha
California State University-Fullerton
Department of Physics
800 N. State College
Fullerton, CA   92834

KHAYYAT, Khaldoun
University of Missouri-Rolla
Physics Department
1870 Miner Circle
Rolla, MO   65409-0640

KHEIFERTS, Anatoli
RSPhysSE
Australian National University
Canberra ACT 0200
Australia

KIDUN, Oleg
Max-Planck Institut fuer Mikrostrukturphysik
Weinberg 2
06120 Halle/Saale
Germany

KLAR, Hubert
Universität Freiburg
Fakultät für Physik
Hermann-Herder-Straße 3
79104 Freiburg
Germany

KOLLMUS, Holger
Max-Planck-Institut für Kernphysik
Saupfercheckweg 1
69117 Heidelberg, Germany

KONO, Hirohiko
Tohoku University
Katahira 2-1-1
Department of Chemistry
Graduate School of Science
Sendai 980-8578 Miyagi
Japan

KRÄSSIG, Bertold
Argonne National Laboratory
Bldg. 203, 9700 S. Cass Avenue
Argonne, IL   60439

LAHMAM-BENNANI, Azzeddine
Laboratoire des Collisions
Atomiques et Moléculaires
(LCAM, UMR Université Paris-Sud et CNRS N°
8625)
Batiment 351,Université de Paris
F-91405 Orsay Cedex
France

LOHMANN, Bernd
Westfälische Wilhelms-Universität Münster
Institut für Theoretische Physik
Wilhelm-Klemm-Str. 9
D-48149 Münster
Germany

316

LOHMANN, Birgit
Griffith University
School of Science
Kessels Road
Nathan, Queensland
Australia 4111

LOWER, Julian
The Australian National University
AMPL, Res. School of Phys. Sciences & Engr.
IAS, Canberra ACT 0200
Australia

MACKENZIE-ROSS, Heather
Flinders University
SoCPES GPO Box 2100
Adelaide SA-5001
Australia

MADISON, Don
University of Missouri-Rolla
107 Physics Department
1870 Miner Circle
Rolla, MO   65409-0640

MALEGAT, Laurence
Université Paris Sud
Laboratoire de Spectroscopie Atomique et
Ionique
Bat. 350, Center d'Orsay
91405, Orsay Cedex, France

MARCHALANT, Pascale
Dept. of Applied Mathematics and Theoretical
Physics
University of Cambridge
Silver Street
Cambridge CB3 9EW
United Kingdom

MAZEVET, Stephane
Los Alamos National Laboratory
Los Alamos, NM 87544

McCONKEY, J.W.
University of Windsor
Physics Department
2965 Orion Cres
Windsor, Ontario
N9B 3P4 Canada

MOORE, John F. (Jack)
University of Maryland
Department of Chemistry
College Park, MD   20742

MOROZOV, Andrey
Max-Planck-Institut für Mikrostrukturphysik
Weinberg 2
D-06120 Halle
Germany

MOTOHASHI, Kenji
Tokyo University of Agriculture and Technology
Dept of Applied Physics
Naka-cho 2-24-16, Koganei
Tokyo 184-8588
Japan

MURRAY, Andrew
University of Manchester
Schuster Laboratory
Manchester M13 9PL
United Kingdom

NAJJARI, Bennaceur
Max-Planck-Institut für Kernphysik
Saupfercheckweg 1
D-69117 Heidelberg,
Germany

O'NEILL, Raymond
National University of Ireland-Maynooth
Dept. of Experimental Physics
Maynooth CO. Kildare
Ireland

PEACHER, Jerry
University of Missouri-Rolla
Physics Department
1870 Miner Circle
Rolla, MO   65409-0640

PIRAUX, Bernard
Université Catholique de Louvain
Laboratoire de Physique Atomique
 et Moléculaire
2 Chemin du Cyclotron
B-1348 Louvain-la-Neuve
Belgium

PRAVICA, Luka
University of Western Australia
Physics Department
35 Stirling Highway
WA 6009
Australia

PRIDEAUX, Andy
University of Missouri-Rolla
Physics Department
1870 Miner Circle
Rolla, MO   65409-0640

REDDISH, Timothy
Newcastle University
Physics Department
Newcastle upon Tyne and Wear
NE1 7RU
United Kingdom

ROCHE, Peter
DAMPT
Silver Street
Cambridge CB3 9EW
United Kingdom

ROSENBERRY, Mark
Behlen Laboratory of Physics
University of Nebraska-Lincoln
116 Brace Lab
Lincoln, NE   68588-0111

SAMARIN, Sergey
University of Western Australia
Physics Department
Nedlands, Perth WA 6907
Australia

SCHMIDT, Volker
Universität Freiburg
Fakultät für Physik
Hermann-Herder-Str. 3
Freiburg D-79104
Germany

SCHMIDT-BÖCKING, Horst
Universität Frankfurt
Institut für Kernphysik
August-Euler-Str. 6
60486 Frankfurt
Germany

SCOTT, Penny
Queens University-Belfast
Dept. of Applied Maths & Theor. Physics
Belfast BT7 1NN
Ireland

SCHULZ, Michael
University of Missouri-Rolla
Physics Department
1870 Miner Circle
Rolla, MO   65409-0640

SELLES, Patricia
Laboratoire de Spectroscopie Atomique et Ionique
Université Paris-Sud
Batiment 350
91450 Orsay
France

SIMON, Marc
Centre Universitaire
Beliment 2090 B.P. 14
Orsay Cedex 91898
France

SOEJIMA, Kouichi
Graduate School of Science & Technology
Niigata University
Igarashi-ninocho 8050
Niigata-shi, Niigata 950-2181
Japan and
Bat 350
Orsay 91406 Essonne
France

318

SRIVASTAVA, Rajesh
University of Roorkee
Department of Physics
Roorkee-247667
India

STANDAGE, Maxwell
Griffith University
PMB 50 Gold Coast Mail Centre CE9726
Gold Coast, Queensland 9726
Australia

STAUFFER, Allan
York University
Department of Physics & Astronomy
Toronto, Ontario M3J 1P3
Canada

STEFANI, Gianni (Giovanni)
Universita di Roma Tre
Dipartimento di Fisica E. Amaldi
Via della Vasca Navale 84
00146 Roma
Italy

TABANLI, Muzaffer
University of Missouri-Rolla
Physics Department
1870 Miner Circle
Rolla, MO 65409-0640

TAKAHASHI, Masahiko
Tohoku University
Institute of Multidisciplinary
Research for Advanced Materials
2-1-1 Katahira, Sendai 980-857
Japan

TANUMA, Hajime
Tokyo Metropolitan University
Department of Physics
1-1 Minami-Ohsawa
Hachioji, Tokyo 192-0397
Japan

TEUBNER, Peter
Flinders University
GPO Box 2100
Adelaide, SA 5001
South Australia

TSURUBUCHI, Seiji
Tokyo University of Agriculture & Technology
Dept.of Applied Physics,Faculty of Technology
Naka-cho 2-24-16, Koganei
Tokyo 184-8588, Japan

TURRI, Giorgio
Western Michigan University & Lawrence
Berkeley National Laboratory
Advanced Light Source Division
1 Cyclotron Road
MS-7-222
Berkeley, CA   94720

UDAGAWA, Yasuo
Katahira 2-1-1
Aoba-ku
Sendai 980-877
Japan

ULLRICH, Joachim
Universität Freiburg
Fakultät für Physik
Max-Planck-Institut für Kernphysik
Saupfercheckweg 1
D-69117 Heidelberg
Germany

VAID, Deepak
University of Missouri-Rolla
Physics Department
1870 Miner Circle
Rolla, MO   65409-0640

VAN BOEYEN, Roger William
University of Maryland
Department of Chemistry
College Park, MD   20742

VON OPPEN, Gebhard
Institut für Atomare Physik und Fachdidaktik
PN3-1
Technische Universität
Berlin 10623
Germany

VOS, Maarten
Atomic & Molecular Physics Laboratories
Research School of Physical Sciences & Engr.
The Australian National University
Canberra ACT 0200
Australia

VYSOTSKII, Vladimir I.
Kiev Shevchenko University
Radiophysical Faculty
Vladimirskaya Ul., 64
252033, Kiev
Ukraine

WARRINGTON, Nicholas
University of Western Australia
Physics Department
Nedlands WA6009
Western Australia

WATANABE, Noboru
Tohoku University
Katahira 2-1-1
Institute of Multidisciplinary Research for
Advanced Materials
Sendai 980-857 Miyagi
Japan

WATERHOUSE, David K.
University of Western Australia
Centre for Atomic, Molecular and Surface Physics
Stirling Highway
Crawley, WA 6009
Nedlands, Perth
Australia

WEIGOLD, Erich
Australian National University
RSPHYSSE, Bldg. 60
Acton-200
Australia

WENT, Michael
Griffith University
School of Science
Laser Atomic Physics Laboratory
Nathan, Queensland 4111
Australia

WHELAN, Colm
Old Dominion University
Department of Physics
Norfolk, VA   23529-0116

WILLIAMS, James
The University of Western Australia
35 Stirling Highway
Nedlands
Perth, WA 6009
Australia

WILLIAMS, Tige
University of Missouri-Rolla
Physics Department
1870 Miner Circle
Rolla, MO   65409-0640

YENEN, Orhan
University of Nebraska-Lincoln
Department of Physics & Astronomy
Lincoln, NE   68588-0111

YU, Dehong H.
The University of Western Australia
Centre for Atomic, Molecular & Surface Physics
Physics Department
Nedlands 6907, Crawley
Western Australia

ZETNER, Peter
University of Manitoba
Physics Department
Room 301 Allen Building
Winnipeg, Manitoba R3T 2N2
Canada

ZHENG, Yenyou
2036 Main Mall
Vancouver BC
V6T 1Z1
Canada

ZHOU, Yajun
Institute of Atomic & Molecular Physics
National Laboratory of Theoretical &
 Computational Chemistry
Jilin University
Changchun 130023
China

ZIMMERMANN, Bernd
I. Physikalisches Institut
Justus-Liebig-Universität Giessen
Heinrich-Buff-Ring 16
35792 Giessen,
Germany

AUTHOR INDEX

A

Al-khateeb, H. M.. 250
Amelink, A., 217
Ancarani, L. U., 115, 276

B

Baertschy, M. D., 1, 76
Bartschat, K., 202
Batelaan, H., 264
Bauer, J., 108
Becker, A., 6
Berakdar, J., 32, 64, 120, 210
Biava, D. A., 18
Blum, K., 229
Boudali, F., 24
Bray, I., 90, 145
Brenot, J. C., 45
Brion, C. E., 38
Brown, D., 139
Brunger, M. J., 196
Bruno, P., 210
Burke, P. G., 82
Busch, M., 256

C

Cederquist, H., 120
Chen, L. Q., 297
Chen, X. J., 297
Cocke, C. L., 120
Cooper, G., 38
Cooper, J. W., 102
Coplan, M. A., 102
Crowe, A., 139, 156
Cvejanović, D., 139, 156

D

Dewangan, D. P., 281
Doering, J. P., 102
Dörner, R., 120
Doudna, C., 133

D (continued)

Dowek, D., 45
Drozdowski, R., 256
DuBois, R. D., 133
Duguet, A., 96

E

Ehresmann, A., 172
Ernst, A., 64

F

Faisal, F. H. M., 6
Fang, Y., 202
Feldmann, S., 238
Feng, R., 38
Fominykh, N., 210
Foster, M., 127
Fursa, D. V., 90, 145

G

Gay, T. J., 250, 264
Geers, S., 238
Gotter, R., 52
Granger, B. E., 1
Greene, C. H., 1
Guyon, P. M., 45

H

Hanne, G. F., 238
Haynes, M. A., 18
Heideman, H. G. M., 217
Henk, J., 210
Herting, C., 238
Houver, J. C., 45

I

Iacobucci, S., 52

323

J

Jagutzki, O., 120
Jones, S., 127
Joulakian, B., 24

K

Kabachnik, N. M., 223
Kamer, S., 172
Kampp, M., 286
Karaganov, V., 196
Khajuria, Y., 297
Khakoo, M. A., 162
Khayyat, Kh., 133
Kheifets, A. S., 70
Kidun, O., 64
Kleiman, U., 229
Kobayashi, H., 166
Krässig, B., 12

L

Lafosse, A., 45
Lagmago-Kamta, G., 108
Lahmam-Bennani, A., 96
Larsen, M., 162
Lebech, M., 45
Liebel, H., 172
Lin, P., 45
Lloyd, C., 133
Lohmann, B., 18, 190, 229
Lower, J., 32
Lucchese, R., 45
Lüdde, H. J., 120

M

MacGillivray, W. R., 190
Madison, D. H., 18, 127
Mazevet, S., 32
McCarthy, I. E., 38
McConkey, J. W., 184
McEachran, R. P., 18, 151, 190
Mergel, V., 120
Mickat, S., 172
Moore, J. H., 102

Moshammer, R., 127
Motohashi, K., 291

N

Najjari, B., 24

O

Olson, R. E., 127

P

Perumal, A. N., 127
Piraux, B., 108
Popov, Yu. V., 115, 120, 276
Pyper, N. C., 286

R

Reyes, J. P., 264
Rosenberry, M. A., 264
Ruocco, A., 52

S

Schartner, K.-H., 172
Schmidt, H. T., 120
Schmidt, L., 120
Schmidt-Böcking, H., 120
Schmoranzer, H., 172
Schuch, R., 120
Schulz, M., 127
Scott, M. P., 82
Scott, N. S., 82
Shi, Z., 38
Soejima, K., 178
Srivastava, R., 151, 229
Stauffer, A. D., 151
Stefani, G., 52
Stelbovics, A. T., 90
Still, T., 82